Disinfection By-Products

Disinfection By-Products

Relevance to Human Health

Steve E Hrudey and Jeffrey WA Charrois

Publishing
London • New York

australian water association

Published by **IWA Publishing**
Alliance House
12 Caxton Street
London SW1H 0QS, UK
Telephone: +44 (0)20 7654 5500
Fax: +44 (0)20 7654 5555
Email: publications@iwap.co.uk
Web: www.iwapublishing.com

First published 2012

British Library Cataloguing in Publication Data
A CIP catalogue record for this book is available from the British Library

ISBN 9781843395195 (Paperback)
ISBN 9781780400624 (eBook)

Co-Published by
Australian Water Association (AWA)
Level 6, 655 Pacific Highway, PO Box 222,
St Leonards NSW 1590,
Australia.
AWA ISBN 9781921335181

Contents

About the Editors

Dr. Steve E. Hrudey is Professor Emeritus of Analytical and Environmental Toxicology in the Faculty of Medicine and Dentistry at the University of Alberta and he has been engaged in research and environmental risk management practice for 40 years. He has served on a number of expert panels including the Research Advisory Panel to the Walkerton Inquiry into a fatal drinking water disease outbreak in Canada, an expert panel on safe drinking water for Canada's First Nation communities and an international expert panel advising the U.S. Army Corps of Engineers regarding water quality risks for the Washington, D.C. drinking water supply. In 2002, Professor Hrudey chaired a Health Canada international expert panel reviewing evidence on health effects of trihalomethanes and in 2010 he chaired the Royal Society of Canada Expert Panel on environmental and public health impacts of Canada's oil sands industry. He has consulted for governments, industry, First Nations and environmental NGOs.

Dr. Hrudey was elected an International Water Association Fellow in 2010, a Fellow of the Society for Risk Analysis in 2007 and a Fellow of the Academy of Sciences, Royal Society of Canada in 2006. Professor Hrudey has been selected for the 2012 A.P. Black Award, the top research award of the American Water Works Association. He has authored or co-authored over 160 refereed journal articles, 8 books, 26 book chapters, 19 expert panel reports, and over 100 other publications in the environmental health sciences. In 2004 with Elizabeth J. Hrudey, he co-authored IWA Publishing best-seller, Safe Drinking Water – Lessons from Recent Outbreaks in Affluent Nations.

Dr. Jeffrey Charrois is an Associate Professor and the Director of the Curtin Water Quality Research Centre (CWQRC) within the Department of Chemistry at Curtin University. Dr. Charrois completed his PhD at the University of Alberta in Public Health Sciences, after which he was awarded an NSERC Industrial

Research and Development Fellowship (2006–2008), which he undertook at the Alberta Research Council (Canada). A/Prof Charrois' main areas of research relate to improving water quality, through the development of ultra-trace analytical methods for detecting emerging contaminants, such as *N*-nitrosamines. Dr. Charrois was invited by the US Environmental Protection Agency (US EPA) to be a reviewer during the development of US EPA Analytical Method 521. In 2006 he was awarded the inaugural "Donald R. Stanley Best Environmental Paper Award" by the Canadian Society of Civil Engineering (in honour of a leader in the transfer of research and technology into the practice of environmental engineering).

A/Prof. Charrois is a member of the Swan River Trust's Technical Advisory Panel (Western Australia). He is currently Vice Chair of the International Water Association's Australia branch as well as a member of the American Water Works Association.

CORRESPONDING AUTHOR AFFILIATIONS

Ina Kristiana, Research Fellow, Curtin Water Quality Research Centre, Perth, Australia

Richard Walker, Manager, Drinking Water Quality, Water Corporation of Western Australia, Perth, Australia

Stuart Krasner, Principal Environmental Specialist, Metropolitan Water District of Southern California, La Verne, CA, U.S.A

William A. Mitch, Associate Professor, Yale University, Department of Chemical and Environmental Engineering, New Haven, CT, U.S.A

Kathryn L. Linge, Senior Research Fellow, Curtin Water Quality Research Centre, Perth, Australia

Jessica M. Boyd, PhD Candidate, Analytical & Environmental Toxicology, Faculty of Medicine & Dentistry, University of Alberta, Edmonton, Canada

Xing-Fang Li, Professor, Analytical & Environmental Toxicology, Faculty of Medicine & Dentistry, University of Alberta, Edmonton, Canada

Andrew R. Humpage, Senior Biochemist, Australian Water Quality Centre, SA Water, Adelaide, Australia

Richard J. Bull, Professor Emeritus, College of Pharmacy, Washington State University, Pullman, WA, U.S.A

Acknowledgements

We acknowledge the financial support of the Australian Water Association (AWA) for the workshop held at OzWater'11 in Adelaide, May 2011, which formed the basis for developing this book. Additionally, we acknowledge the financial support of Curtin University, Water Corporation (Western Australia), GHD Group Pty Limited (Perth office), Water Quality Research Australia (WQRA), and the Western Australia Branch of the AWA, for their financial support of the sister DBP workshop hosted by the Curtin Water Quality Research Centre (CWQRC) at Curtin University in Perth, May 2011.

We also acknowledge the support of the Canadian Water Network, the Natural Sciences and Engineering Research Council of Canada, Alberta Health and Wellness, the Water Research Foundation, and the Australian Research Council for supporting our research, which has formed the foundations to pursue this topic. Likewise, we acknowledge the support of the National Collaborating Centre on Environmental Health for its financial support of a comprehensive review of the public health significance of disinfection by-products in drinking water, which provided the foundations for Chapter 11 (cited therein as Hrudey 2008).

Finally, but most gratefully, we wish to acknowledge the priceless logistical support of our research and dissemination in this area provided by Dianne Sergy as well as the careful proof reading and input provided by Elizabeth J. Hrudey, Jessica Boyd, Katerina Carastathis and Ina Kristiana.

List of Abbreviations

AOX	Adsorbable organic halogen
AOP	Advanced oxidation process
AOM	Algal organic matter
ADWG	Australian Drinking Water Guidelines
ASIR	Age-standardized incidence rate
ASMR	Age-standardized mortality rate
BIF	Bromine incorporation factor
BCAA	Bromochloroacetic acid
BCAN	Bromochloroacetonitrile
BDCM	Bromodichloromethane
$BMDL_{10}$	Lower 95% confidence estimate of the 10% benchmark dose
BMX	Brominated MX analogues
CSF	Cancer slope factor
CCL	Candidate Contaminant List
C-DBP	Chlorination disinfection by-product
CI	Confidence interval
Ct	Disinfectant residual concentration – contact time
CNX	Cyanogen halide
DALY	Disability adjusted life-year
DBAA	Dibromoacetic acid
DBAN	Dibromoacetonitrile
DBCM	Dibromochloromethane
DCAN	Dichloroacetonitrile
DBP	Disinfection by-product
DHAA	Dihaloacetic acids
DHAN	Dihaloacetonitriles

DMA	Dimethylamine
DNA	Deoxyribonucleic acid
DOC	Dissolved organic carbon
EfOM	Effluent organic matter
ECD	Electron capture detector
GAC	Granular activated carbon
GC	Gas chromatography
GSH	Glutathione
HAA5	Five regulated haloacetic acids
HAL	Haloacetaldehyde
HAAms	Haloacetamides
HAA	Haloacetic acid
HAN	Haloacetonitrile
HBQ	Halobenzoquinone
HiWATE	Health Impacts of Long-Term Exposure to Disinfection By-Products in Drinking Water
HK	Haloketone
HNM	Halonitromethane
IARC	International Agency for Research on Cancer
ICR	Information Collection Rule
I-DBP	Iodinated disinfection byproduct
IRIS	Integrated Risk Information System
I-THM	Iodinated trihalomethane
IUGR	Intrauterine growth retardation
LC	Liquid chromatography
LLE	Liquid-liquid extraction
LOAEL	Lowest Observed Adverse Effect Level
LRAA	Locational running annual average
LTD_{10}	Lower 95% confidence limit on TD_{10} the dose estimated to cause tumors in 10% of experimental animals
MCL	Maximum Contaminant Level
MCLG	Maximum Contaminant Level Goal
MOA	Mode of action
MRL	Method reporting limit
MS	Mass spectrometer
MX	Mutagen X, 3-chloro-4-(dichloromethyl)-5-hydroxy-2(5H)-furanone
NCI	National Cancer Institute
N-DBP	Nitrogenous disinfection by-product
NDMA	*N*-Nitrosodimethylamine
NHMRC	National Health and Medical Research Council (Australia)
NORS	National Organics Reconnaissance Survey

NOM	Natural organic matter
NOAEL	No Observed Adverse Effect Level
NRMMC	Natural Resource Management Ministerial Council (Australia)
NSRL	No Significant Risk Level
OR	Odds ratio
PAR	Population attributable risk
PAC	Powdered activated carbon
POE	Point of entry
POU	Point of use
PolyDADMAC	Polydiallyldimethylammonium chloride
QSTR	Quantitative Structure Toxicity Relationship
QSAR	Quantitative Structure-Activity Relationship
RfD	Reference dose
RR	Rate ratio, Risk ratio, Relative risk
RO	Reverse osmosis
RAA	Running annual average
SGA	Small for gestational age
SPE	Solid-phase extraction
SPME	Solid-phase microextraction
SUVA	Specific UV absorbance
TBM	Bromoform (tribromomethane)
TCM	Chloroform (trichloromethane)
TDI	Tolerable daily intake
THM	Trihalomethane
THM4	Four regulated trihalomethanes (chlorinated & brominated)
THMFP	Trihalomethane formation potential
TONO	Total *N*-nitrosamine
TOC	Total organic carbon
TOX	Total organic halogen
UCMR	Unregulated Contaminant Monitoring Regulation
$UV_{abs\ 254}$	Ultraviolet absorbance at 254 nm
US EPA	US Environmental Protection Agency
US SDWA	US Safe Drinking Water Act
WTP	Water treatment plant
WHO	World Health Organisation

Preface

The genesis of this book was a series of discussions with Professor Don Bursill in November 2010 that ultimately led to two disinfection by-product (DBP) workshops that took place in Australia in May 2011. In Adelaide at OzWater'11—the Australian national water conference and exhibition, which was presented and organised by the Australian Water Association—a workshop titled, *Disinfection By-Products (DBPs) – Relevance to Human Health* was held. The workshop featured international speakers from Canada, the United States and Australia.

In Perth, the Curtin Water Quality Research Centre (CWQRC) based at Curtin University hosted a sister workshop to the Adelaide event titled, *Disinfection By-Product (DBP) Workshop: Emerging Issues in DBP Research and the Way Forward*. The Perth-based workshop featured several of the same speakers as in Adelaide in addition to some who provided some local perspectives.

Disinfection by-products (DBPs) were detected in drinking water over 38 years ago. Since then identification of DBP species has closely paralleled advances in analytical chemistry. In addition to a number of regulated DBPs (such as trihalomethanes and haloacetic acids), notable DBP classes of current interest include: halonitromethanes, haloamides, halogenated furanones, haloaldehydes, haloquinones, as well as *N*-nitrosamines and iodo-DBPs. Improvements in extraction, separation, and detection technologies have improved our ability to identify DBP species that were once difficult, if not impossible, to detect by gas chromatography methods. Moreover, liquid chromatography/mass spectrometry applications are providing new insights into the monitoring of nonvolatile, high-molecular-weight, highly polar, hydrophilic, and thermally labile target compounds in drinking water. However, simply finding over 600 individual DBP species in drinking water to date, does not necessarily constitute a health risk in and of itself. As with any chemical exposure, it is the dose delivered relative to an adverse effect level that matters.

Public concerns regarding exposure to DBPs and adverse health outcomes stem from toxicology and epidemiology studies, which have suggested elevated risks of developing certain cancers as well as ongoing speculation about potentially adverse reproductive outcomes with increased exposure to DBPs in drinking water. Minimizing DBPs capable of causing the adverse human outcomes identified in epidemiology studies, such as bladder cancer, remains a priority for public health and a focus for DBP researchers. This book offers an overview of nearly four decades of research, regulation, and water treatment advances in the area of DBPs. A frank assessment of the epidemiology and toxicology evidence is very informative when put in quantitative terms. It is now prudent to reflect on where the DBP agenda has been and more importantly consider where it should be going in the future.

While there are a variety of written works targeted at research scientists related to DBPs, there is still a lack of literature aimed specifically at drinking water professionals (engineers, chemists and public health professionals) who work on the front lines dealing with drinking water quality issues on a day-to-day basis. These front-line workers must deal with risk-risk tradeoff issues everyday as well as questions concerning DBP risk management in addition to all the other regulatory and water quality issues they must manage. Although a topic as complex as DBPs is certainly not amenable to simplistic explanations, this book aims to provide drinking water professionals with a pragmatic assessment of the current research evidence and emerging issues concerning DBPs and public health.

Steve E. Hrudey
Canmore, AB, Canada
Jeffrey W. A. Charrois
Perth, WA, Australia
May 2011

Foreword

Don Bursill

With more than 40 years professional experience in water quality management, I have seen firsthand the impact that the developments in analytical chemistry have had on public water supply management and the changes in community perceptions relating to its safety that have been an outcome of the resulting increased information. In my early days in the field, life was pretty simple with "comprehensive" water quality assessments of drinking water samples generally only involving a relatively small list of physical (pH, conductivity, temperature), aesthetic (colour and turbidity, and sometimes odour), inorganic (calcium, magnesium, sodium, chloride, sulphate, etc.) and microbiological indicator organisms (coliforms, *E. coli*, plate counts). At that time organic analyses were mainly undertaken on wastewater samples and were mostly confined to measurements such as total organic carbon and biochemical oxygen demand. One of the earliest organic measurements to find broader use in water monitoring was the carbon chloroform extract[1] – again a very crude and insensitive measure of organic contamination.

The development of atomic absorption spectrophotometry soon led to increasing attention on heavy metals and the possibility of pollution of water sources, particularly by mining operations and other heavy industries. The increasing availability of data on low levels of environmental exposure to heavy metals though various mechanisms, including water pollution, placed pressure on the existing knowledge of the health risks involved, particularly in the long term.

[1]Adsorption of organics from a large water sample onto activated carbon and desorption into a small volume of chloroform for gas chromatography (with or without mass spectrometry) analysis.

Mercury contamination in the aqueous environment and the widespread reporting of the problems in Minimata, Japan provided a good example where complex chemistry, bioaccumulation and other factors needed to be sorted out before a reasonable understanding of the health risks could be determined and the severe health outcomes evident in Minimata could be adequately explained.

During that period in the 1960s and early 1970s gas chromatography was becoming more commonly used in water laboratories. The availability of highly sensitive detectors such as electron capture detectors (ECD) for gas chromatography (GC) led to a rapid increase in interest in compounds such as pesticides. The organo-chlorine pesticides and other chlorinated hydrocarbons generally have good volatility for GC analysis, are easily extracted and concentrated into non-polar solvents and have high electron capture properties. These factors enabled these compounds to be detected at extremely low concentrations in environmental samples, including water. Inevitably consumer concerns about poisons being present in water supplies led to much greater attention to the regulations relating to pesticide use by health, environmental and other relevant regulatory agencies.

All this came at a time of increased interest in the environment and so when a chlorinated "solvent" such as chloroform and related compounds were found in chlorinated drinking water in 1974, this news was rapidly and widely taken up by the mass media and broadcast worldwide. Here, in South Australia, we saw community representations to government to sort the issue out with high priority. It added considerable pressure on the government of the day to go ahead with a large investment in water treatment facilities that had been under consideration for several years. The DBP issue seemed to put this investment decision beyond doubt.

The contributors to this book have comprehensively documented the developments outlined very briefly above and this is an important aspect of this publication as it puts the whole issue of DBPs in proper perspective – better than has been done to date. Notwithstanding the extensive interest in the issue of DBPs in public water supplies and the level of research undertaken, this issue remains on the priority list for most water supply authorities and continues to be an issue of contention among consumers. The reasons for this include the great complexity of the chemistry and the toxicology, the lack of any magic bullet solution and the understandable fear in the community about the possibility of developing cancer from the consumption of drinking water – no matter how small that risk might be.

Much has been learnt on this issue over the past more than 30 years. We have a greater understanding of the toxicology, but there are relatively few definitive answers and new DBPs are reported regularly where toxicological information is unavailable or lacking. We have a much better understanding of water treatment processes and disinfection. The options available for treatment and disinfection have greatly expanded and new technologies are regularly being developed. However, there are no practical, economic technology options available that can offer, with complete assurance, effective disinfection and distribution system protection against microbiological contamination whilst providing no risk at all with respect to disinfection by-products. This is the essence of what has kept this issue charging along for more than three decades.

We now have a broad acceptance of the risk-based approach to the management of public water supplies that is at the core of water quality guidelines such as the World Health Organisation's Guidelines for Drinking-water Quality and the Australian Drinking Water Guidelines. This guidance provides a means of working through all the potential hazards that may be of concern in a water supply system from the source of the water to the customer and to assess what we know about the risks involved with these hazards in each particular system. Success with this approach places the emphasis squarely with knowing the entire water system very thoroughly and relies on cooperation among the various players involved in water system operation and its regulation. Few things are known with absolute certainty, but a considered risk management approach provides the mechanism to achieve a safe reliable water supply that is associated with very low public health risk – without incurring unreasonably high costs. Where the knowledge is considered inadequate to make reasonable risk management decisions, the risk assessment process provides an excellent basis for the identification and prioritisation of further research. This logical approach applies equally to the DBP issue.

This book seeks to provide a comprehensive discussion and evaluation of the most significant knowledge we have to date on DBP formation and management in public water systems. In my view, it achieves this objective extremely well and in so doing, it is a valuable resource for those involved in water quality management and the associated regulatory functions. It is also an excellent reference for those involved in communicating with consumers about this issue. This book helps enormously in dealing with the complexities and placing this wealth of knowledge in appropriate perspective.

It is absolutely true that microbiological safety of drinking water is paramount and should not be compromised. In comparison, the risks posed by DBPs are very small and in some cases may ultimately be found to not cause adverse health effects. In my view it is also true that the effort put in to understanding DBP formation, the related public health implications, the improvement in understanding of conventional water treatment processes and the development of new treatment technologies has contributed greatly to improved water quality and safety over this past some thirty years. For example, enhanced coagulation for supplies with moderate to high DOC levels seeks to use conventional treatment chemistry to achieve optimal removal of DBP precursors using the better knowledge we have of the chemistry and kinetics of coagulation. Enhanced coagulation and other organics removal technologies are successful in reducing many DBPs, although not all. Whether that changes the public health risk for DBP exposure substantially is still an open question. However, there is no doubt that disinfection with chlorine can be made more effective at the treatment plant and that the management of distribution system chlorine residuals can now be controlled within much tighter operational levels. Consequently, the microbiological quality delivered to the customers is enhanced substantially whilst minimising exposure to DBPs.

The improvements in water treatment process development and operations, as outlined above, have been a substantial advance, but I must also pose a note of caution with respect to the "innovative" disinfection alternatives that are regularly put forward as a potential solution to DBP formation issues. These options need to be evaluated with great skill and caution. There is no good purpose in reducing or avoiding known DBP formation only to create increased microbiological risks or new chemical risks from the formation of other, as yet unknown by-products. When faced with claims of a wonderful new process option it is sobering to reflect on the lists of advantages and disadvantages that are attached to all currently accepted water disinfection processes. None are perfect and special care needs to be applied when faced with newer, less understood process options. The best option needs to be selected after a detailed and well informed assessment based on a comprehensive knowledge of the system under consideration.

There are many more examples where the pursuit of the DBP issues has had wider benefits for water quality management. Perhaps one day our knowledge of DBPs in water systems will be perfect. In the meantime, the information in this book can help achieve the optimal outcomes both in water quality and public health terms from our water systems.

Chapter 1

Disinfection By-Products (DBPs) as a public health issue

S. E. Hrudey and J. W. A. Charrois

1.1 DBPs AND THE CHLOROFORM SAGA

In 1970, when one of us (SEH) was taking a masters degree in public health engineering in London, a lecturer on drinking water treatment told the class that there was not much need for further research in drinking water treatment because we essentially knew all we needed to know to produce safe drinking water. As outrageous as this statement may sound to the water research community today, and we should all subscribe to the caution to "*never say never*", that view reflected some common perspectives 40 years ago. After all, the developed nations had pretty much eliminated major waterborne disease (mainly cholera and typhoid) epidemics that had plagued major urban areas since the industrial revolution. This was largely achieved through a combination of vastly improved sanitation, conventional treatment (coagulation and filtration) of surface waters and disinfection, generally by chlorination. Waterborne outbreaks still occurred in developed countries, albeit less frequently, often with the pathogen remaining unidentified (Hrudey and Hrudey, 2004).

The most common cause of bacterial drinking water outbreaks in many countries, *Campylobacter*, was not recognized as a common cause of human gastroenteritis until 1977, the protozoan pathogen *Giardia* which proved to be a common cause of drinking water outbreaks during the 1970s and 1980s was largely unknown, *Cryptosporidium* would not be recognized as a waterborne pathogen until 1984 and the serious and often fatal *Escherichia coli* O157:H7 would not be recognized as a waterborne pathogen until 1989 and would not be widely recognized as a drinking threat until it killed seven in Walkerton, Ontario in May 2000 (Hrudey and Hrudey, 2004). Clearly, hindsight shows that drinking water research over the past 40 years has been valuable for addressing microbial risk.

The reader will have to judge comparatively how well we have done in improving our understanding of DBP health risks over the same period.

Given the apparent complacency about safe drinking water reflected by that drinking water lecturer in 1970, imagine what a revelation it was when Dutch chemist, Johannes Rook (1974) first published his meticulous evidence showing that chlorination of drinking water produced chloroform and three other trihalomethanes (bromodichloromethane – BDCM, dibromochloromethane – DBCM and bromoform – TBM) in treated drinking water at readily detectable concentrations. This landmark discovery was followed closely and independently by Bellar *et al.* (1974) working for the U.S. Environmental Protection Agency (US EPA). Jim Symons, who was Chief of the US EPA Drinking Water Research Division at the time, has nicely documented much of that early history from a research and regulatory perspective (Symons, 2001a,b). The context of that time was an emergence of public concerns over environmental contamination, reflected by the creation of the US EPA in December 1970, and similar environmental regulatory agencies elsewhere throughout the early 1970s, combined with a prevalent and growing popular belief that cancer could be conquered simply by eliminating trace contaminants from food, air and water (Hrudey, 2009). The latter is interesting given the common denial of the role of tobacco as a cause of cancer, which was apparent at that time. A feature story in the journal *Science* (Marx, 1974) referring to research revealing a long list of trace organic compounds newly discovered in New Orleans drinking water was titled: "*Drinking Water: Another Source of Carcinogens?*"

The chloroform story continued to unfold with confirmation of widespread THM occurrence in a national survey of halogenated organics mandated by the U.S. Safe Drinking Water Act (SDWA) (Symons *et al.* 1975) and publication by the U.S. National Cancer Institute (NCI) of a rodent cancer bioassay on chloroform (NCI, 1976). The latter was performed with extremely high doses as a daily bolus (slug) dose of chloroform dissolved in corn oil directly delivered to the gut of experimental animals. The required high concentrations of chloroform administered could not be dissolved in water because of its limited water solubility. Using the corn oil dosing technique an overwhelming tumour response was achieved (liver tumours in mice, kidney tumours in rats). Within four months of these findings being published, the U.S. Food and Drug Administration banned the use of chloroform in cosmetics and its widespread use in consumer products ranging from mouthwash to toothpaste was ended. This is somewhat ironic because chloroform had also been widely used as an anaesthetic from the mid 1800s into the early 1900s. Dr. John Snow, the public health icon who established in the 1850s with epidemiologic evidence that human waste contaminating drinking water was responsible for cholera epidemics in London, actually earned his livelihood as an anaesthetist using chloroform (Vinten-Johansen *et al.* 2003). Snow campaigned to overcome chlorophobia (fear of chloroform) because London was experiencing negative publicity that chloroform was being used for mugging, rape and murder.

As outlined in Chapter 2, following Rook's discovery about THMs, guidelines and regulations limiting THMs in drinking water first began to appear in 1978 and 1979, while epidemiology studies seeking evidence of chlorinated drinking water causing cancer were initiated and have continued over the past 35 years (Chapter 11).

The chloroform saga has come full circle during this period. The initial extremely high dose cancer bioassay results on chloroform (NCI, 1976) provided a premise that was widely cited throughout the late 1970s and early 1980s that chloroform and by extension THMs in drinking water posed an authentic cancer risk. The NCI (1976) corn oil dosing results were later found to be much more carcinogenic than continuous dosing of chloroform in drinking water (Bull *et al.* 1986). A study providing high concentrations of chloroform (up to 1,800,000 μg/L) dissolved in drinking water (Jorgenson *et al.* 1985) produced no significant carcinogenic response. High doses of chloroform in corn oil were found by direct experimentation to produce cytotoxicity (causing cell death) and resulting cell proliferation (Larson *et al.* 1994, 1995). These findings demonstrating a plausible threshold mechanism for cancer were supported by extensive evidence showing virtually no mutagenic activity for chloroform (Golden *et al.* 1997).

According to the U.S. SDWA, the Maximum Contaminant Level Goal (MCLG) should be set at the maximum level of a contaminant in drinking water at which no known or anticipated adverse health effects would occur, and which allows an adequate margin of safety. US EPA policy for carcinogens in drinking water had required a MCLG to be zero (Pontius, 2000). A US EPA expert review panel recommended abandoning a MCLG of zero for chloroform and replacing it with a limit based on an estimated threshold (ILSI, 1997). If implemented, no cancer slope factor (CSF) would be used for estimating cancer risk below the chloroform threshold. In response, the US EPA (1998a) proposed to raise the MCLG to 300 μg/L in accordance with this expert advice. Many interveners protested this precedent-setting measure and the 1998 US EPA Final Stage 1 DBP Rule withdrew the proposal to change the MCLG for chloroform from zero to 300 μg/L (Pontius, 2000).

Because the SDWA requires the US EPA to use the best available science in setting standards and regulations, the chlorine industry took the issue to court (Pontius, 2000). Although the US EPA acknowledged that the best available science called for raising the MCLG above zero, it argued that retaining the zero MCLG was appropriate as an interim measure until it could obtain a review by its Science Advisory Board, which ultimately advised that use of a threshold for chloroform was justified by the scientific evidence.

On March 31 2000, the US District Court ruled that the US EPA had violated the SDWA by failing to use the best available science. The court found that the US EPA action of setting the MCLG of chloroform at zero to be "*arbitrary and capricious*" and in excess of statutory authority. The US EPA withdrew the zero MCLG in May 2000, ultimately replacing it with a MCLG of 70 μg/L for chloroform (US EPA,

2006). The four-fold lower (more cautious) MCLG was based, in part, upon assigning a lower proportion of total human chloroform exposure to ingestion of drinking water, recognizing the major exposures to chloroform from inhalation and dermal absorption associated with waterborne chloroform. This decision formalized the earlier withdrawal of a CSF for chloroform from the US EPA toxicology data base (Integrated Risk Information System, IRIS) by 2001, meaning that, based on considerable toxicological evidence and careful review, the US EPA no longer sanctioned the estimation of cancer risk for chloroform using a CSF. The World Health Organization had previously reached that conclusion in 1993, the Australian National Health and Medical Research Council in 1996 and Health Canada followed in 2006.

1.2 DAWN OF A NEW MILLENNIUM

Later in 2000, the other one of us (JWAC) began his PhD program in public health, investigating DBPs in drinking water, following what might be considered the most influential event for a generation of water quality researchers – Canada's Walkerton disaster.

1.2.1 Consequences of a disinfection failure

In May 2000 drinking water in Walkerton, Ontario became contaminated with *Escherichia coli* O157:H7 as well as *Campylobacter jejuni* (O'Connor, 2002). Without a properly maintained and operated water treatment plant, there was neither adequate disinfection nor chlorine residual to protect against consuming pathogenic microorganisms in the drinking water. The consequences of failed disinfection officially accounted for seven deaths and 2300 related illnesses (O'Connor, 2002), with children and the elderly taking the brunt of the effects.

The name "Walkerton" became synonymous with attributes of incompetence, failure, deceit, and mistrust. Unfortunately, this redefinition of "Walkerton" takes the focus away from what many people have too quickly forgotten or have failed to recognize altogether; that Walkerton is a real place where real people had their lives changed forever.

Justice Dennis O'Connor documented a few select personal stories of Walkerton town residents, providing a record of experiences and highlighting different aspects of the community's suffering. A short passage is included below (O'Connor, 2002) as a testament to the terrible personal suffering and loss that those in Walkerton experienced, not only in May of 2000 but for the lifelong pain many continue to endure. The personal account of suffering is a reminder that providing safe drinking water is a critical activity that requires not only proper technical training but also an understanding of the responsibility this activity carries with it, when not performed correctly, the consequences can be catastrophic (Hrudey and Walker, 2005). Walkerton serves as a constant reminder of what happens when disinfection goes wrong and why disinfection should never be secondary to DBPs.

The following is a description from a mother, Tracey Hammell, about her two-year-old son Kody who became infected with *E. coli* O157:H7 and developed hemolytic uremic syndrome (HUS) a serious kidney ailment:

> On Friday, Kody suddenly began to vomit. Mrs. Hammell took him inside to change his diaper and noticed that it was bloody. "I couldn't even tell what it was," she said. I had never seen that before."
>
> On Saturday morning Mrs. Hammell woke up to find Kody violently ill. At 10:00 a.m., she phoned the hospital but was told that the hospital was "backed up" and that she should not come in just yet. "Well, when can I come?" she asked. "He's really sick." The hospital staff told her to wait until 4:00 p.m. and to get fluids into her son in order to prevent dehydration. "Do whatever you have to do to get it in him," they said: "get a syringe."
>
> Mrs. Hammell followed their advice. She got a syringe and "shoved water down his throat." The water may still have been contaminated, but she did not know that. Finally, she phoned the hospital again around 12:30 p.m. and said, "You've got to see him. He's lifeless. His eyes are rolling in the back of his head … [He has] diarrhea every two minutes. He can't take it anymore." (O'Connor, 2002, pp. 43–44).

1.2.2 New horizons: Emerging DBPs and alternative disinfectants

The beginning of the 21st century also marked some major transitions on a number of fronts related to DBP research. Firstly after 25 years (since THMs were discovered) the water industry's preoccupation with halogenated DBPs was as strong as ever. Most drinking water research was still focused on the two most abundant classes of DBPs, THMs and haloacetic acids (HAAs). However, at the suggestion of Dr. Richard Bull (to JWAC), investigation of a novel class of emerging DBPs was pursued. The non-halogenated *N*-nitrosamines were a particularly interesting class of DBPs because as a chemical class, *N*-nitroso compounds had produced malignant tumours in every vital tissue tested (Shank and Magee, 1981) as well as the fact *N*-nitrosodimethylamine (NDMA) was beginning to be detected in North American drinking- and waste-waters. *N*-nitrosamine research was only being conducted by a few research groups, and thus a major opportunity presented itself for a PhD thesis (Charrois, 2006). Indeed, many parallels can be seen with *N*-nitrosamines and the early days of THM discovery.

Alternative disinfectants, such as chloramines, ozone, or chlorine dioxide, were (and continue to be) used in place of, or in combination with chlorination because of the tendency to produce less of the regulated chlorinated DBPs (Diehl *et al.* 2000). Compliance with drinking water limits (e.g. Stage 2 Disinfection By-products Rule) was a major regulatory driver causing water utilities to consider incorporating alternative disinfectants in hopes of achieving lower regulated DBP levels. Ironically, history has shown that attempts to reduce regulated DBP concentrations using alternative disinfection processes have, in some cases, been found to generate higher concentrations of unregulated and more toxic DBPs compared to chlorination alone (Charrois, 2011, Chapters 4, 5, 6, 7 & 10), posing uncertain levels of risk to human health.

Also around 2000 a US EPA led study – the Nationwide DBP Occurrence Study, which also included researchers from the Metropolitan Water District of Southern California and the University of North Carolina was underway. The main objective of this study was to reduce uncertainty over the identity and levels of DBPs that people were exposed to in their drinking water, with a focus on DBPs that had undergone a prior toxicological prioritization for the potential to cause adverse health effects (Weinberg *et al.* 2002). This pioneering study surveyed utilities using four types of disinfection methods including alternative disinfectants (chlorine, chloramines, ozone, and chlorine dioxide) and reported many emerging DBP species.

1.3 BALANCING MICROBIAL AND DBP HEALTH RISK

Because it is clear that a failure to disinfect drinking water adequately will guarantee a substantial risk of waterborne disease, whatever the human health risk posed by disinfection, DBP risk must be viewed as a trade-off. In principle, this reality suggests there should be merit in performing comparative risk assessment to judge what an appropriate human health risk trade-off between the risks of microbial pathogens and DBPs should be. Not surprisingly, there have been attempts to do this (Ashbolt, 2004; Havelaar *et al.* 2000).

Havelaar *et al.* (2000) applied the disability adjusted life-year (DALY) approach to demonstrate a basis for comparing microbial pathogen and DBP risks. While the DALY approach has attracted attention for comparing acute and chronic risks, the Havelaar *et al.* (2000) analysis does not address the considerable uncertainty associated with whether DBPs cause cancer from drinking water exposure (or if they do, which DBPs are responsible for cancer) compared with the absolute certainty that pathogen exposure via drinking water can cause human disease. Ashbolt (2004) notes the reliance of cancer risk assessment for DBPs on conceptual models rather than verifiable evidence and concludes there is clear justification for the often repeated requirement to avoid compromising disinfection in any attempt to address DBP risk. This caution is necessary because the regulatory agenda, which has been dominated by THMs and HAAs has driven water utilities to take action to achieve compliance (Chapter 3). As should become clear from the coverage in this book, measures to achieve simple compliance with THM or HAA requirements may generate unintended consequences because of the complexity of DBP formation and the often higher toxicity of DBPs formed to a greater degree by some measures aimed at reducing THMs (e.g. chloramination, Chapters 4, 5, 6, 7 & 10).

DBPs are a complex interdisciplinary subject, which makes communications about them a challenge. Comprehension about whether DBPs cause human health effects such as cancer via drinking water exposure is particularly challenging. We have to consider what message we convey by simply telling our utility operators that they must not compromise disinfection when seeking to achieve DBP reductions. If we also lead operators to believe that DBPs inevitably cause cancer, a thinking operator may do her or his own risk analysis trade-off between whether she or he should be more concerned about causing cancer versus gastrointestinal illness, unless we can provide better context about our comparative state of knowledge on these competing risks. There is clearly a role for better communications from our regulators on this subject, something we will revisit in our concluding chapter (12).

1.4 PURPOSE OF THIS BOOK

Our objective with this book is to provide front-line drinking water personnel and management, whether from water utilities or public health agencies with a timely and relevant perspective on our current state of knowledge about DBPs and human health, including an opportunity to appreciate the nature of uncertainty which remains. The chapters in this book are organized to provide a number of relevant perspectives bearing on the question of whether DBPs cause human health effects via drinking water exposure.

Chapter 2 provides an overview of DBP research, their regulatory history and the current worldwide status of DBP regulations and guidelines. Chapter 3 provides a utility water quality manager's perspective about managing DBPs under challenging source water conditions. Chapter 4 provides an overview of recent findings on halogenated DBPs and associated emerging issues. Chapter 5 provides an overview of some major themes regarding nitrogenous DBPs, their formation and control. Chapter 6 provides a comprehensive review of nitrogenous DBPs in drinking water including their toxicity, occurrence and control. Chapter 7 provides an overview of health risk assessment and management of NDMA and other *N*-nitrosamines. Chapter 8 introduces a recently discovered class of DBPs with potential toxicological concern, the halogenated benzoquinones. Chapter 9 provides health risk assessment of MX (a halogenated furanone), which was long regarded as the most potent mutagen found in drinking water. Chapter 10 provides a toxicological evaluation of experimental data regarding the magnitude of cancer risk from DBPs. Chapter 11 provides a review of epidemiological inference and evidence concerning DBPs and human health. Chapter 12 provides some concluding thoughts on DBPs, drinking water quality and public health risks.

We introduced this book with the saga of chloroform as a DBP. Knowing and understanding this story is absolutely critical if we ever want to move beyond the DBP studies of the past. The USEPA's Weight-of-Evidence Characterization for Human Carcinogenicity and Chloroform (2004) document states that, "*chloroform is likely to be carcinogenic to humans by all routes of exposure under high-exposure conditions that lead to cytotoxicity and regenerative hyperplasia in susceptible tissues (U.S. EPA, 1998a,b). Chloroform is not likely to be carcinogenic to humans by any route of exposure under exposure conditions that do not cause cytotoxicity and cell regeneration.* In plain English, this means that at low exposure concentrations, such as those typical for THMs in drinking water, the evidence shows that chloroform is not carcinogenic.

The fact that chloroform is not considered a carcinogen at drinking water concentrations has been known for over a quarter of a century now, yet the current literature (Chowdhury and Hall, 2010; Chowdhury *et al.* 2011) and conference presentations are still all too often showing examples of water professionals not knowing or understanding the evidence about chloroform. This book seeks to synthesize relevant information on our current understanding of DBPs in drinking water, including those areas of on going debate. Perhaps with an understanding that is based on current research evidence, water and public health professionals will be empowered to make sensible cautionary decisions to assure safe drinking water.

In closing, we offer some sage advice from one of the most respected and insightful science writers of our age, Carl Sagan (1995, p. 304), published shortly before his untimely death:

> "… at the heart of science is an essential balance between two seemingly contradictory attitudes
> – an openness to new ideas, no matter how bizarre or counterintuitive,
> – and the most ruthlessly sceptical scrutiny of all ideas, old and new.
> This is how deep truths are winnowed from deep nonsense. The collective enterprise of creative thinking and sceptical thinking, working together, keeps the field on track."
> [emphasis added]

REFERENCES

Ashbolt N. J. (2004). Risk analysis of drinking water microbial contamination versus disinfection by-products (DBPs). *Toxicology*, **198**, 255–262.

Bellar T. A., Lichtenberg J. J. and Kroner R. C. (1974). The occurrence of organohalides in chlorinated drinking waters. *Journal of the American Water Works Association*, **66**(12), 703–706.

Bull R. J., Brown J. M., Meierhenry E. A., Jorgenson T. A., Robinson M. and Stober J. A. (1986). Enhancement of the hepatoxicity of chloroform in B6C3F1 mice by corn oil: implications for chloroform carcinogenesis. *Environmental Health Perspectives*, **69**, 49–58.

Charrois J. W. A. (2006). Detection, Occurrence and Risk Assessment of N-Nitrosamines in Drinking Water. PhD thesis, University of Alberta, Edmonton, Alberta, Canada.

Charrois J. W. A. (2011). Analysis of emerging disinfection by-products in drinking water. In: Encyclopedia of Analytical Chemistry, Supplementary Volumes S1-S3, R. A. Meyers (ed.), John Wiley & Sons, Ltd., Chichester, UK, pp. 205–220.

Chowdhury S. et al. (2011). Disinfection byproducts in Canadian provinces: associated cancer risks and associated medical expenses. *Journal of Hazardous Materials*, **187**, 574–584.

Chowdhury S. and Hall K. (2010). Human health risk assessment from exposure to trihalomethanes in Canadian cities. *Environment International*, **36**, 453–460.

Diehl A. C., Speitel G. E., Symons J. M., Krasner S. W., Hwang S. J. and Barrett S. E. (2000). DBP formation during chloramination. *Journal of the American Water Works Association*, **92**, 76–90.

Golden R. J., Holm S. E., Robinson D., Julkunen P. H. and Reese E. A. (1997). Chloroform mode of action: implications for cancer risk assessment. *Regulatory Toxicology & Pharmacology*, **26**, 142–155.

Havelaar A. H., De Hollander A. E. M., Teunis P. F. M., Evers E. G., Van Kranen H. J., Versteegh F. M., Van Koten J. E. M. and Sob W. (2000). Balancing the risks and benefits of drinking water disinfection: disability adjusted life-years on the scale. *Environmetnal Health Perspectives*, **108**, 315–321.

Hrudey S. E. (2009). Chlorination disinfection by-products, public health risk tradeoffs and me. *Water Research*, **43**, 2057–2092.

Hrudey S. E. and Hrudey E. J. (2004). Safe Drinking Water – Lessons from Recent Outbreaks in Affluent Nations. IWA Publishing, London, 514.

ILSI (1997). An evaluation of EPA's Proposed Guideline for Carcinogenic Risk Assessment Using Chloroform and Dichloroacetic Acid as Case Studies, Report of an Expert Panel, International Life Sciences Institute, ILSI Press. Washington, DC.

Jorgenson T. A., Meierhenry J. R., Rushbook C. J., Bull R. J. and Robinson M. (1985). Carcinogenicity of chloroform in drinking water to male Osborne-Mendel rats and female B6C3F1 mice. *Fundamental & Applied Toxicology*, **5**, 760–769.

Larson J. L., Wolf D. C. and Butterworth B. E. (1994). Induced cytotoxicity and cell proliferation in the hepatotoxicity of chloroform in female B6C3F1 mice: comparison of administration by gavage in corn oil vs. ad libitum in drinking water. *Fundamental & Applied Toxicology*, **22**, 90–102.

Larson J. L., Wolf D. C., Morgan M. K. and Butterworth B. E. (1995). Toxicity and cell proliferation in the liver, kidneys and nasal passages of female F-344 rats, induced by chloroform administered by gavage. *Food & Chemical Toxicology*, **33**(6), 443–456.

NCI (1976). Carcinogenesis Bioassay of Chloroform. National Cancer Institute, Bethesda, MD.

Marx J. (1974). Drinking water: another source of carcinogens? *Science*, **186**, 809–811.

O'Connor D. R. (2002). Chapter 2: The Impact on Walkerton. In: Report of the Walkerton Inquiry. Part 1. The Events of May 2000 and Related Issues. Ontario Ministry of the Attorney General, Queens Printer for Ontario, Toronto, ON, pp. 42–54.

Pontius F. W. (2000). Chloroform: Science, Policy and Politics. *Journal American Water Works Association*, **92**(5), 12, 14, 16, 18.

Rook J. J. (1974). Formation of haloforms during chlorination of natural waters. *Water Treatment Examination*, **23**, 234–243.

Sagan C. (1995). Demon-Haunted World – Science as a Candle in the Dark. Random House, New York.

Shank R. C. (1981). Occurrence of N-nitroso compounds in the environment. In: Mycotoxins and N-Nitroso Compounds: Environmental Risks, R. C. Shank (ed.), Vol. 1, CRC Press, Boca Raton, FL, USA, pp. 155–183.

Symons J. M. (1975). National organics reconnaissance survey. *Preliminary Assessment of Suspected Carcinogens in Drinking Water (Appendices).* U.S. Environmental Protection Agency, Washington, DC, pp. 12–100.

Symons J. M. (2001a). The early history of disinfection by-products: a personal chronicle (Part I). *Environmental Engineer*, **37**(1), 20–26.

Symons J. M. (2001b). The early history of disinfection by-products: a personal chronicle (Part II). *Environmental Engineer*, **37**(2), 7–15.

US EPA (1998a). National Primary Drinking Water Regulations: Disinfectants and disinfection byproducts notice of data availability: Proposed rule. *Federal Register*, **63**(61), 15,673–15,692.

US EPA. (1998b). Health risk assessment/characterization of the drinking water disinfection byproduct chloroform. Prepared for Health and Ecological Criteria Division, Office of Science and Technology, Washington, DC, by Toxicology Excellence for Risk Assessment, Cincinnati, OH.

US EPA (2004). Integrated Risk Information System (Chloroform); Office of Research and Development (ORD), National Center for Environmental Assessment. http://www.epa.gov/iris/subst/0025.htm (accessed 15 February 2012).

US EPA (2006). 40 CFR Parts 9, 141, and 142: national primary drinking water regulations: stage 2 disinfectants and disinfection byproducts rule. Environmental protection agency. January 4, 2006, *Federal Register*, **71**(2), 388–494.

Vinten-Johansen P., Brody H., Paneth N., Rachman S. and Rip M. (2003). Cholera, Chloroform and the Science of Medicine – A Life of John Snow. Oxford University Press, Oxford.

Weinberg H. S., Krasner S. W., Richardson S. D. and Thruston A. D. Jr. (2002). The Occurrence of Disinfection By- Products (DBPs) of health concern in drinking water: Results of a nationwide DBP occurrence study, Report EPA/600/R-02/068, National Exposure Research Laboratory, Office of Research and Development, U.S. Environmental Protection Agency, Athens, GA.

Chapter 2

Research overview, regulatory history and current worldwide status of DBP regulations and guidelines

I. Kristiana, J. W. A. Charrois and S. E. Hrudey

2.1 DISINFECTION BY-PRODUCT RESEARCH

2.1.1 Discovery of DBPs

Numerous substances are present in water, even after it has been treated to very high standards. Many of these substances are known to us and analytical methods exist to identify and measure their presence in waters. However, despite our extensive array of analytical instrumentation and methods as well as our knowledge of these compounds, many are still unknown and unidentifiable. There is no universal analytical method that is able to identify and measure all substances present in water. What we can analyse in water is constrained by the characteristics of the analytical method that we choose. Every analytical method is selective and sensitive for a specific group of substances, also known as 'target compounds', which is defined by the sample preparation approach and the instrumentation used. Thus, the discovery of trace organic compounds in water has paralleled advances in analytical methods (Charrois, 2011).

Drinking water disinfection by-products (DBPs) are a group of compounds present in water resulting from disinfection practices in water treatment. The beginnings of trace organic analysis, made possible by the invention of gas chromatography (GC), laid the foundation for the discovery of DBPs in water. In the 1950s and 1960s, gas chromatographs were initially equipped with a universal, but relatively insensitive, detector, the thermal conductivity detector (TCD). Further developments in detector technology saw the conception of more selective detectors such as the flame ionisation detector (FID), which is selective for organic compounds that can be ionised by air/hydrogen flame; and the electron capture detector (ECD) which is selective for organic compounds that are capable of capturing low energy electrons to form negatively charged ions. In the late

1960s and early 1970s, ECD became popular as it provides far more sensitive and selective detection of compounds during GC analysis. The ECD is especially sensitive for halogenated materials and at the time, this advance led to a focus on halogenated contaminants (e.g. chlorinated pesticides such as DDT).

At the same time, there was a shift in sample extraction methods, from the use of solvents, such as dichloromethane, ether, and chloroform, to alternative means of extraction such as purge-and-trap and headspace extraction. The combination of the sensitivity of the ECD and the use of alternative sample preparation methods led to the discovery of chloroform and other trihalomethanes (THMs) in drinking water in the early 1970s. Using headspace sampling and GC-ECD, Rook (1974) found that THMs were present at higher concentrations in chlorinated drinking water compared to raw surface water supplies. Around the same time, Bellar *et al.* (1974) also found higher levels of THMs with increasing chlorine contact time during disinfection. They used a combination of purge and trap and GC-ECD to analyse their samples.

Further improvements in detector technology saw the development of the mass spectrometer (MS) as a GC detector. This allowed for the identification of "unknown" compounds present in water by interpretation of their mass spectra. The use of ECD and MS as GC detectors began to improve dramatically analytical sensitivity and selectivity, resulting in the detection of numerous trace organic compounds in treated drinking water supplies (Marx, 1974). However, due to the limitations of GC, these organic compounds are mostly volatile, relatively non-polar, and low molecular weight.

With advances in analytical instrument, there were also breakthroughs in sample preparation methods. Derivatisation methods were used to change the chemical properties of some target compounds so they could be analysed by GC. This advance allowed for the analysis and discovery of more polar compounds in water. Using derivatisation as a part of the sample preparation procedure, other DBPs were discovered and identified. Christman *et al.* (1983) found haloacetic acids (HAAs) in chlorinated solutions of fulvic acid isolated from a natural surface water by derivatising the ether-extracted samples using diazomethane. Hemming *et al.* (1986) found a chlorinated furanone, 3-chloro-4-(dichloromethyl)-5-hydroxy-2(5H)-furanone, also known as MX, in chlorinated drinking waters. They used acidic methanol to derivatise their samples. Further information on the formation and occurrence of MX is given in Chapter 9.

The establishment of selective and sensitive analytical methods for the measurement of DBPs in water led to several surveys of the occurrence of DBPs in treated drinking waters. For example, the US Environmental Protection Agency (US EPA) conducted the National Organics Reconnaissance Survey (NORS) in 1974. These early surveys revealed that THMs and HAAs are the most abundant DBPs in chlorinated waters. As such, the focus of DBP research in the 1970s and 1980s was on the formation occurrence and removal of THMs and HAAs in drinking water.

Further improvements in the sensitivity and selectivity of GC-MS instrumentation led to the discovery and identification of many other organic DBPs. Likewise, other developments in analytical instrumentation led to the discovery and identification of inorganic DBPs by ion chromatography. Many chemical classes of DBPs have been identified and more recent surveys have included some of these DBPs groups as well as the "traditional" THMs and HAAs. For example, the Urban Water Research Association of Australia conducted a survey of 16 drinking water distribution systems around Australia in 1994-1995 to determine the occurrence of 23 species of DBPs. In the US, the Information Collection Rule (ICR) resulted in the collection of data on the occurrence of DBPs across the US over an 18-month period from 1997 to 1998. Moreover, the use of alternative disinfectants to chlorine, as well as combination of disinfectants, have contributed to the large variety of DBPs. Alternative disinfectants have been used by water utilities because of the increasingly stringent regulations with respect to the most abundant chlorination DBPs, THMs and HAAs. Currently, there are over 600 DBPs that have been identified (Richardson, 2011). Some of the more commonly found DBPs in treated drinking waters are shown in Table 2.1. Initially, there was an emphasis on studies and identification of halogenated DBPs (i.e. chlorinated and brominated DBPs), since they are closely associated with chlorine-based disinfectants that are commonly used for drinking water treatment (i.e. chlorine and chloramine). However, there are numerous non-halogenated DBPs that have been identified (e.g. aldehydes, carboxylic acids, ketones), which can be produced by chlorination, chloramination, and other disinfection processes. The disinfectants that are associated with a particular class of DBPs are also listed in Table 2.1.

Table 2.1 Classes of DBPs identified in drinking waters with their corresponding species and the disinfectants that are associated with their formation in drinking waters (adapted from Hrudey, 2009; Richardson, 2011).

DBPs	Generic structure	Disinfectant
Trihalomethanes (THMs) Trichloromethane (Chloroform) Bromodichloromethane Chlorodibromomethane Tribromomethane (Bromoform)	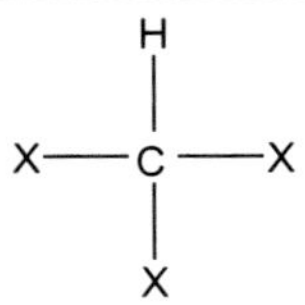	Chlorine Chloramine
Iodinated trihalomethanes (I-THMs) Dichloroiodomethane Bromochloroiodomethane Dibromoiodomethane		Chloramine

(*Continued*)

Table 2.1 Classes of DBPs identified in drinking waters with their corresponding species and the disinfectants that are associated with their formation in drinking waters (adapted from Hrudey, 2009; Richardson, 2011) (*Continued*).

DBPs	Generic structure	Disinfectant
Chlorodiiodomethane	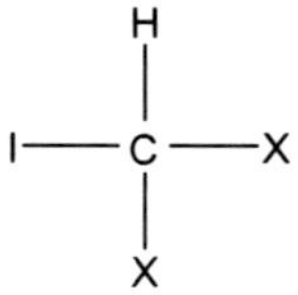	
Bromodiiodomethane		
Triiodomethane (Iodoform)		
Haloacetic acids (HAAs)	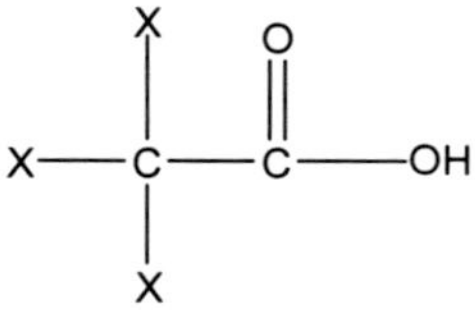	Chlorine Chloramine
Chloroacetic acid (MCAA)		
Dichloroacetic acid (DCAA)		
Trichloroacetic acid (TCAA)		
Bromoacetic acid (MBAA)		
Dibromoacetic acid (DBAA)		
Tribromoacetic acid (TBAA)		
Bromochloroacetic acid (BCAA)		
Bromodichloroacetic acid (BDCAA)		
Chlorodibromoacetic acid (CDBAA)		
Iodoacids		Chloramine
Iodoacetic acid		
Bromoiodoacetic acid		
(*Z*)-3-Bromo-3-iodopropenoic acid		
(*E*)-3-Bromo-3-iodopropenoic acid		
(*E*)-2-Iodo-3-methylbutenedioic acid		
Haloacetonitriles (HANs)	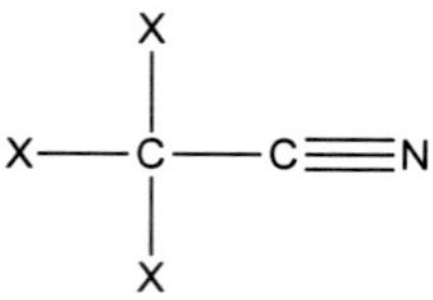	Chlorine Chloramine
Chloroacetonitrile		
Bromoacetonitrile		
Dichloroacetonitrile		
Dibromoacetonitrile		
Bromochloroacetonitrile		
Bromodichloroacetonitrile		
Dibromochloroacetonitrile		
Tribromoacetonitrile		
Trichloroacetonitrile		
Haloketones (HKs)	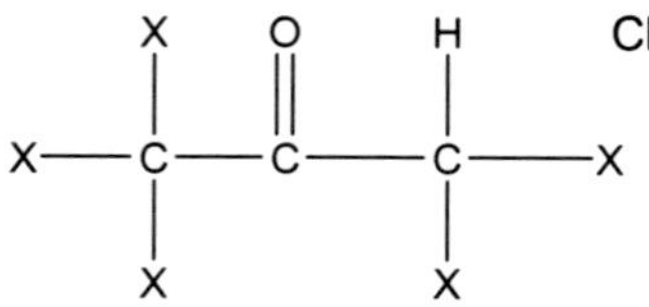	Chlorine
Chloropropanone		
1,1-Dichloropropanone		
1,3-Dichloropropanone		
1,1-Dibromopropanone		
1,1,1-Trichloropropanone		
1,1,3-Trichloropropanone		
1-Bromo-1,1-dichloropropanone		

(*Continued*)

Table 2.1 Classes of DBPs identified in drinking waters with their corresponding species and the disinfectants that are associated with their formation in drinking waters (adapted from Hrudey, 2009; Richardson, 2011) (*Continued*).

DBPs	Generic structure	Disinfectant
1,1,1-Tribromopropanone		
1,1,3- Tribromopropanone		
1,1,3,3-Tetrachloropropanone		
1,1,1,3-Tetrachloropropanone		
1,1,3,3-Tetrabromopropanone		
Halonitromethanes (HNMs)	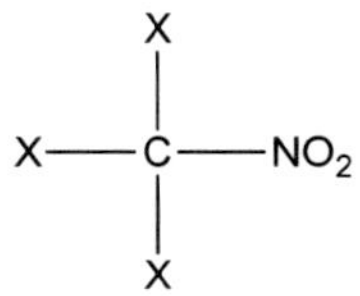	Chlorine Chloramine
Chloronitromethane		
Bromonitromethane		
Dichloronitromethane		
Dibromonitromethane		
Bromochloronitromethane		
Bromodichloronitromethane		
Dibromochloronitromethane		
Tribromonitromethane (bromopicrin)		
Trichloronitromethane (chloropicrin)		
Haloacetamides (HAMs)	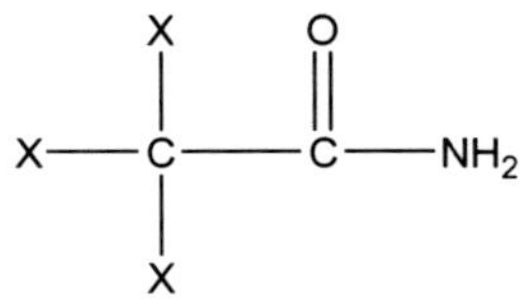	Chlorine Chloramine
Chloroacetamide		
Bromoacetamide		
Dichloroacetamide		
Dibromoacetamide		
Bromoiodoacetamide		
Trichloroacetamide		
Haloacetaldehydes (HALs)	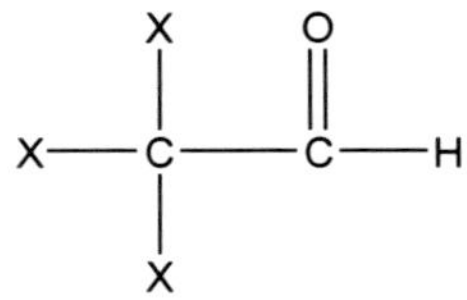	Chlorine
Chloroacetaldehyde		
Dichloroacetaldehyde		
Dibromoacetaldehyde		
Bromochloroacetaldehyde		
Trichloroacetaldehyde (chloral hydrate)		
Bromodichloroacetaldehyde		
Tribromoacetaldehyde		
Cyanogen halides (CNX)	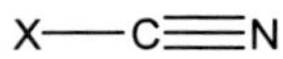	Chlorine Chloramine
Cyanogen chloride		
Cyanogen bromide		
***N*-Nitrosamines**	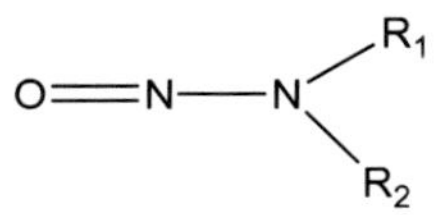	Chloramine Chlorine
N-Nitrosodimethylamine (NDMA)		
N-Nitrosodiethylamine (NDEA)		
N-Nitroso-*n*-propylamine (NDPA)		
N-Nitrosoethylmethylamine (NEMA)		

(*Continued*)

Table 2.1 Classes of DBPs identified in drinking waters with their corresponding species and the disinfectants that are associated with their formation in drinking waters (adapted from Hrudey, 2009; Richardson, 2011) (*Continued*).

DBPs	Generic structure	Disinfectant
N-Nitrosodi-*n*-butylamine (NDBA)		
N-Nitrosopiperidine (NPIP)		
N-Nitrosopyrrolidine (NPYR)		
N-Nitrosomorpholine (NMOR)		
N-Nitrosodiphenylamine (NDPhA)		
Oxyhalides		Ozone Chlorine dioxide
Bromate		
Chlorate		
Chlorite		
Aldehydes	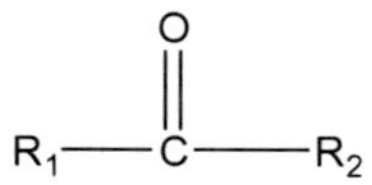	Ozone
Formaldehyde		
Acetaldehyde		
Cyanoformaldehyde		
Glyoxal		
Methylglyoxal		
Propanal		
Butanal		
Pentanal		
5-Keto-1-hexanal		
trans-2-Hexenal		
Carboxylic acids	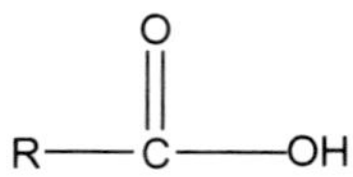	Ozone
2-Methlypropanoic acid		
Butanoic acid		
3-Methylbutanoic acid		
Pentanoic acid		
Hexanoic acid		
Heptanoic acid		
Octanoic acid		
Oxalic acid		
Malonic acid		
Succinic acid		
Halogenated furanones (MX compounds) (see Chapter 9)		Chlorine
3-Chloro-4-(dichloromethyl)-5-hydroxy-2(5H)-furanone (MX)		
3-Chloro-4-(dichloromethyl)-2(5H)-furanone (Red-MX)		
(E)-2-Chloro-3-(dichloromethyl)-butenedioic acid (Ox-MX)		

(*Continued*)

Table 2.1 Classes of DBPs identified in drinking waters with their corresponding species and the disinfectants that are associated with their formation in drinking waters (adapted from Hrudey, 2009; Richardson, 2011) (*Continued*).

DBPs	Generic structure	Disinfectant
(E)-2-Chloro-3-(dichloromethyl)-4-oxobutenoic acid (EMX)		
2,3-Dichloro-4-oxobutenoic acid (mucochloric acid)		
3-Chloro-4-(bromochloromethyl)-5-hydroxy-2(5H)-furanone (BMX-1)		
3-Chloro-4-(dibromomethyl)-5-hydoxy-2(5H)-furanone (BMX-2)		
3-Bromo-4-(dibromomethyl)-5-hydroxy-2(5H)-furanone (BMX-3)		
(E)-2-Chloro-3-(bromochloromethyl)-4-oxobutenoic acid (BEMX-1)		
Chloro-3-(dibromomethyl)-4-oxobutenoic acid (BEMX-2)		
(E)-2-Bromo-3-(dibromomethyl)-4-oxobutenoic acid (BEMX-3)		
Halobenzoquinones (see Chapter 8)		Chlorine
2,6-Dichloro-1,4-benzoquinone		Chloramine
2,6-Dichloro-3-methyl-1,4-benzoquinone		
2,3,6-Trichloro-1,4-benzoquinone		
2,6-Dibromo-1,4-benzoquinone		

As MS technology continues to advance, significant improvements in the sensitivity of GC-MS analysis have been evident in the extremely low detection limits, down to sub ng L^{-1}, that are commonly achievable. The discovery of *N*-nitrosodimethylamine (NDMA) and other nitrosamines as drinking water DBPs is largely due to the improved sensitivity in the analytical instrumentation that has allowed its determination at low ng L^{-1} levels (Chapters 5, 6 and 7).

Complementary to GC-based methods, liquid-chromatography (LC) based methods provide a means to analyse for compounds that are not amenable for GC separation. These include polar, high molecular weight, and thermally labile compounds. Prior to the development of LC-based methods, these compounds were undetected and thus unknown. The powerful combination of LC and MS has led the to the discovery of new DBPs, such as highly polar carbonyl DBPs: phenazine, *N*-chlorophenazine, and halobenzoquinones (Qin *et al.* 2010; Zhao *et al.* 2010; Zhou *et al.* 2009; Chapter 8) as well as the thermally labile DBP: *N*-nitrosodiphenylamine (NDPhA).

The majority of known, identified DBPs in drinking waters are halogenated compounds, which is logical given that chlorine and chloramine are the most commonly used disinfectants in drinking water treatment. Halogenated DBPs can be measured as individual species using GC- and LC-based methods, or as a bulk parameter such as total organic halogen (TOX). TOX provides a measure of all the halogenated species present in a sample. Technically speaking, TOX should be referred to as "adsorbable organic halogen" (AOX), since the analytical method used to measure TOX can only measure halogenated organic carbon that are adsorbed on to the activated carbon during the sample preparation procedure. For the purposes of this Chapter, the term AOX will be used instead of TOX, as it is the correct operationally defined term. The measurement of AOX is an alternative to the measurement of individual DBPs, which cannot be done routinely because of the diversity and the large range of molecular weights of DBPs. AOX is especially useful to evaluate the overall formation of halogenated DBPs in waters disinfected using halogen-based disinfectants such as chlorine and chloramines, or in waters containing high concentrations of halide ions. Conventionally, AOX has been measured as chloride by microcoulometric titration. However, current analytical instrument development has allowed for the measurement of halogen-specific AOX, which gives a separate measure of the formation of chlorinated (AOCl), brominated (AOBr), and iodinated (AOI) by-products (Kristiana *et al.* 2009). In chlorinated waters, the measured individual DBPs account for only 40–70% of AOX (Pressman *et al.* 2010; Richardson 2011), while in chloraminated waters, less than 20% of TOX can be assigned to individual species of DBPs (Hua and Reckhow, 2007a). Further information and discussions on the formation of TOCl, TOBr, and TOI are provided in Chapter 4.

2.1.2 DBP formation mechanisms

Investigation of DBP formation mechanisms follows the discovery of DBPs in drinking water, and is supported by a variety of analytical methods. These studies are critical in understanding the factors affecting the formation of DBPs as well as their reaction pathways, and ultimately in developing better strategies to control the formation of DBPs in drinking water. Organic and inorganic precursor materials are involved in the formation of DBPs.

2.1.2.1 Organic precursors of DBPs

Natural organic matter (NOM), algal organic matter (AOM), anthropogenic organic materials, and in some cases effluent organic matter (EfOM) all contribute to the formation of DBPs. As the names imply, NOM is of natural origin while AOM is specifically derived from algal blooms. Anthropogenic organic materials originate from human activities, and EfOM originates from wastewater treatment plant discharges.

NOM is an extremely complex mixture of organic compounds, which have been reported to exist mainly as macromolecules, in the form of large, polymeric-type structures (Croué *et al.* 1999). In aquatic environments, typically 90% of NOM occurs as dissolved organic matter (Amy, 1993). Knowledge of the characteristics of NOM is important, since they determine the extent of its removal during water treatment and its propensity to form DBPs. During coagulation and flocculation, the efficiency of the process depends on the nature of the NOM in the water (Boyer and Singer, 2005). The proteinaceous material within NOM has been associated with membrane fouling (Makdisy *et al.* 2004), while the biodegradable fraction of NOM is responsible for bacterial regrowth and biofilm formation in distribution systems (Croué *et al.* 2000). During disinfection, NOM reacts with disinfectant, resulting in an increase in disinfectant demand and the formation of DBPs. The more hydrophobic and more acidic fractions of NOM have been reported to be the most reactive DBP precursors (Croué *et al.* 2000). DBP formation has been found to be proportional to the aromatic carbon content of NOM (Reckhow and Singer, 1990). Westerhoff *et al.* (2004) further specified that hydroxy and amino functional groups on aromatic rings (activated aromatic compounds) increase the reactivity of NOM towards disinfectants.

AOM and EfOM have significant organic nitrogen content (e.g. amines and amino acids) that have been known as precursors for nitrogenous DBPs (N-DBPs) (Mitch *et al.* 2009). A recent study showed that chlorination of AOM produced more N-DBPs and haloaldehydes and less carbonaceous DBPs than did chlorination of NOM (Fang *et al.* 2010). The presence of NDMA precursors has been reported to be significantly higher in waters impacted by EfOM, resulting in higher concentrations of NDMA (Mitch *et al.* 2009, Chapters 5 and 6).

Anthropogenic organic materials include pharmaceuticals, personal care products, herbicides, pesticides, and surfactants that have entered source waters from human activities. Some of these materials have been implicated in the formation of DBPs, since many of them have activated aromatic rings that can readily react with oxidants. For example, the antibacterial agent triclosan, which is used in many hand soaps, can form chloroform and other chlorinated DBPs when reacted with chlorine under drinking water treatment conditions (Rule *et al.* 2005). The pharmaceutical acetaminophen can produce chlorinated and non-chlorinated products when treated with chlorine (Bedner and MacCrehan, 2006). Pharmaceuticals containing dimethylamine (DMA) functional groups can produce NDMA (Richardson, 2010). The fungicide tolylfluanide has also been identified, where present, as a major NDMA precursor (Schmidt and Brauch, 2008). Hair protein and urea from urine can provide precursor materials for the formation of HANs, especially in swimming pools (Kim *et al.* 2002). Further discussion on anthropogenic materials that serve as precursors to DBPs, especially N-DBPs, are presented in Chapter 6.

The formation of different groups of DBPs has been associated with specific functional groups within the organic matter. The formation mechanisms of THMs and HAAs have been reported to be similar, involving *m*-dihydroxybenzene (resorcinol) structures as major precursors (Gallard and von Gunten, 2002; Reckhow and Singer, 1985). Secondary aliphatic amines have been identified as important precursors of *N*-nitrosamines (Mitch *et al.* 2003). Amino acids have been associated with the formation of HANs, CNX, and HNMs (Lee *et al.* 2007; Yang *et al.* 2010).

2.1.2.2 Inorganic precursors of DBPs

Bromide and iodide ions have been identified as inorganic precursors of DBPs since they provide the source of halogens in the formation of brominated and iodinated DBPs. Bromide and iodide ions are naturally present in many source waters. Locations that can have high bromide and iodide levels include coastal cities where the groundwaters and surface waters are impacted by saltwater intrusion, and some inland locations where the surface waters are impacted by natural salt deposits from ancient seas or oil-field brines (Richardson, 2003). Naturally occurring bromide and iodide can be rapidly oxidised to hypobromous acid (HOBr) and hypoiodous acid (HOI) during disinfection, as shown by:

$$Br^- + \text{oxidant (e.g. HOCl)} \longrightarrow Cl^- + HOBr \quad (2.1)$$

$$I^- + \text{oxidant (e.g. HOCl)} \longrightarrow Cl^- + HOI \quad (2.2)$$

HOBr and HOI can then react with NOM to form brominated and iodinated DBPs in a way that is analogous to reactions with hypochlorous acid (HOCl). The concentration and the species distribution of DBPs are influenced by the ratio of bromide/iodide ion concentration to the concentration of reactive NOM, as well as the ratio of bromide/iodide ion concentration to the concentration of disinfectant/oxidant.

In the formation of THMs during chlorination, bromine incorporation into NOM, and thus the speciation of THMs, have been consistently shown to increase with increasing Br^-/DOC ratio (Liang and Singer, 2003; Westerhoff *et al.* 2004).

Iodinated DBPs have primarily been found in drinking waters disinfected with chloramines (Bichsel and von Gunten, 2000; Hua and Reckhow, 2007b; Krasner *et al.* 2006; Weinberg *et al.* 2002). Iodinated DBPs have been reported to be more toxic than brominated and chlorinated DBPs (Richardson *et al.* 2007). Iodinated THMs (I-THMs) have been identified as compounds that generate taste and odour issues in drinking waters. Bichsel and von Gunten (2000) reported that the concentrations of I-THMs were higher in chloraminated waters than in chlorinated waters, and that I-THMs were not formed in ozonation. Further information and discussion about I-DBPs are presented in Chapter 4.

2.1.2.3 Disinfectants

The type of disinfectant affects the type and the concentration of DBPs formed in drinking waters (Table 2.1). Chlorine produces the highest concentration of halogenated DBPs, with THMs and HAAs being the most abundant species of the identified chlorinated DBPs.

In general, chloramine produces significantly lower concentrations of halogenated DBPs than chlorine (3–20% of the concentrations produced in chlorination), especially the regulated DBPs THMs and HAAs (Hua and Reckhow, 2007a; Richardson, 2003). In chloramination, some nitrogen-containing DBPs were reported to be at higher concentrations than in chlorination systems (Richardson, 2011). Moreover, the nitrogen in NH_2Cl can be incorporated into some N-DBPs for example, NDMA, HANs, and CNX, thus contributing to the nitrogen source in the formation of N-DBPs (Yang *et al.* 2010). Chloropicrin is usually found at approximately the same concentrations as in chlorination, while cyanogen chloride (CNCl), NDMA, and haloacetamides are commonly present at higher concentrations in chloraminated waters compared to chlorinate sources (Choi and Valentine, 2002; Dotson *et al.* 2009; Richardson, 2011).

Chlorine dioxide does not produce organic halogenated DBPs to any significant degree, but it reacts with NOM to produce oxidation by-products that are most likely similar to those produced by ozone (Richardson, 1998). The major concern in the use of chlorine dioxide as a drinking water disinfectant is the formation of chlorite and chlorate as inorganic DBPs (Krasner, 1999). At low concentrations, chlorite and chlorate have been reported to cause haemolytic anaemia through oxidative damage of red blood cell membranes, while higher concentrations may result in an increase in methemoglobin and induce adverse effects on nervous systems in infants and young children (Veschetti *et al.* 2005). The production of chlorite and chlorate can be minimised by using high purity chlorine dioxide for drinking water treatment (White, 1999).

Ozone reacts with NOM to produce a variety of oxidation by-products, including aldehydes (e.g. formaldehyde, acetaldehyde, glyoxal, methylglyoxal), acids (e.g. oxalic acid, succinic acid, formic acid, acetic acid), ketones (e.g. acetone, butanone, 3-methylcyclopentanone), and hydrogen peroxide (Richardson, 2011). In bromide-containing waters, ozone reacts with NOM and bromide to produce brominated DBPs including bromate, bromoform, and bromopicrin (Koechling *et al.* 1996). The formation of bromate in ozonation is a potential public health concern because bromate has been shown to be a carcinogen in laboratory animals (Koechling *et al.* 1996), and causes chromosome damage *in vitro* and *in vivo* (Fawell, 1999).

UV irradiation promotes the formation of $^{\bullet}OH$ radicals, which are known to react with NOM, splitting large NOM molecules and generating lower molecular weight organic compounds including low molecular weight acids (Frimmel, 1998). These low molecular weight compounds can then react with the secondary disinfectant used in conjunction with UV irradiation (usually chlorine or chloramine), thus producing DBPs commonly associated with the disinfectant, for example, THMs and HAAs when chlorine or chloramine is used as the secondary disinfectant (Liu *et al.* 2006). Gallard and von Gunten (2002) found that pre-treatment using UV irradiation followed by chlorination does not affect THM formation (i.e. similar concentrations of THMs were formed in water treated with and without UV irradiation), but causes an increase in chlorine demand.

With increasingly stringent regulations and guidelines on the concentrations of THMs and HAAs in drinking waters, there has been an increasing use of disinfectants other than chlorine ('alternative disinfectants') for example, chloramines, ozone, chlorine dioxide, and combinations of these alternative disinfectants. However, the identities of DBPs associated with these alternative disinfectants, as well as their formation mechanisms, must also be considered for their potential public health risk.

2.1.2.4 Kinetics of DBP formation

The rate at which DBPs are formed is also an important factor that needs to be considered in developing better strategies to control the formation of DBPs in drinking water. Kinetic studies on the formation of DBPs provide this critical information. Fast and slow phases of DBP formation have been reported in various kinetic studies and these have been attributed to the presence of different types of DBP precursors and the different mechanisms involved in the formation of DBPs. For example, the kinetics of THM formation consists of an initial phase of rapid formation followed by a subsequent phase of slower formation that eventually levels off as any reactant becomes depleted (e.g. Gallard and von Gunten, 2002; Westerhoff *et al.* 2004). As a result of the initial, fast formation of THMs, a large proportion of THMs is formed in the early stages of chlorination. Some N-DBPs including HANs, CNX and NDMA have been reported to degrade in aqueous systems, especially in the presence of free chlorine residual (Charrois and Hrudey, 2007; Mitch and Schreiber, 2008; Reckhow and Singer, 1990). In this case, the concentrations of these DBPs in drinking water are a balance between their formation and degradation rates. In general, DBPs continue to be produced in the treatment plant and throughout the drinking water distribution systems as long as disinfectant residual and precursor materials are present.

2.1.3 Fate of DBPs in drinking water distribution system

Since the discovery of DBPs, numerous surveys of DBPs in drinking water have been carried out at various jurisdiction levels. The monitoring of the concentrations of specific DBPs in drinking waters has become an important part of water quality control and management.

Monitoring of DBPs in drinking water distribution systems has revealed temporal and spatial variations of DBP concentrations. These variations have been associated with changes in raw and treated water quality, variations in operational parameters related to disinfection (e.g. temperature, disinfectant dose and pH), and variations in distribution system structure and hydraulics (Rodriguez and Sérodes, 2001; Symanski *et al.* 2004). For example, the concentrations of THMs have been found to be higher in summer than in winter (Krasner *et al.* 1989; Sérodes *et al.* 2003; Symanski *et al.* 2004), and highest at the extremity of the distribution system (Guay *et al.* 2005; Sérodes *et al.* 2003). The concentrations of HAAs were found to be highest in spring and tend to be lower at the extremity of the distribution system possibly due to microbiological degradation (Rodriguez *et al.* 2004). Hydrodynamic factors and temperature have also been reported to play significant roles in the degradation of HAAs in the distribution system, by causing conditions that are favourable or unfavourable to biological activity (Speight and Singer, 2005). NDMA has been reported to continue to form in the distribution system (Charrois *et al.* 2004).

These findings have prompted further studies on temporal and spatial variations of DBPs in drinking water distribution systems to better understand the system and the behaviour of DBPs in the system. These studies have also been supported by laboratory-scale studies investigating the chemical and physical characteristics of DBPs, providing information on the stability and persistence of DBPs under various conditions. For example, HAAs have been reported to be susceptible to biodegradation (Baribeau *et al.* 2005) and chemical decomposition (Zhang and Minear, 2002). HANs are also susceptible to chemical degradation, especially at pH greater than 5 and in the presence of chlorine residual. Depending on the pH conditions, HANs may degrade to their corresponding HAAs or haloacetamides (Glezer *et al.* 1999).

The US EPA has acknowledged the importance of temporal and spatial variations in the concentrations of DBPs and their implications for compliance to the Disinfectants and DBP Rule. In Stage 1 of the Disinfectants and DBP Rule, the maximum concentrations of THMs and HAAs were regulated based on running annual averages (RAA), where a system-wide RAA at a minimum of four sampling locations in the distribution system must meet these regulations. In Stage 2 of the Disinfectants and DBP Rule, the requirement of running annual average (RAA) for the concentrations of DBPs was modified to locational RAA (LRAA), where utilities are required to select sampling locations that capture the peak concentrations of THMs and HAAs.

2.1.4 Evaluation of DBP exposure

The evaluation of human exposure to DBPs needs to consider the formation and occurrence of DBPs as well as all the possible routes of exposure to DBPs. The fate of DBPs in drinking water distribution systems influences the amount of DBPs available to enter a human body; while considerations of the routes of DBPs exposure determines the possible total amount of DBPs that may enter the human body. Considering the temporal and spatial variability in the concentrations of DBPs in drinking water distribution systems, measurement of DBPs at the outlet of a drinking water treatment plant can underestimate the concentrations of DBPs and thus underestimate exposure to DBPs. Therefore, understanding of the fate of DBPs in drinking water distribution systems and at the consumer tap is critical to provide accurate DBP exposure data. DBP exposure assessment underpins epidemiological studies on human health effects of DBPs. There are multiple routes of exposure to DBPs including ingestion, inhalation, and dermal absorption. The complexity of the multiple routes of human exposure to DBPs is discussed in Chapter 11.

Cantor *et al.* (2010) reported that dermal and inhalation exposures from showering, bathing, and swimming are significant factors in DBP exposure. Dermal absorption has been reported to be a significant route of exposure of THMs, where a 10-minute shower resulted in a higher exposure to THMs than ingestion of 1 L of water in 10 minutes, based on the concentrations of THMs in blood samples (Backer *et al.* 2000). Evaluation of DBP exposure assessment is discussed in more detail in Chapter 11.

2.1.5 Toxicological assessments of DBPs

The primary focus of toxicological assessment of DBPs was originally directed towards THMs and HAAs because they were, and remain, the dominant DBPs identified in drinking water on a concentration basis. One by one, the cases for individual DBPs among the THMs and HAAs to be plausible (genotoxic, no-threshold) carcinogens from drinking water exposure has been undermined over the years. The most complete case against the likelihood of a DBP contributing to human cancer via drinking water exposure having been made for chloroform (Bull *et al.* 2011; Pontius, 2000; SAB, 2000). More recently, the Health Canada (2009) decision to withdraw its 2006 guideline for BDCM based on an expert panel assessment (Health Canada, 2008a) of the 2006 cancer bioassay for BDCM (NTP, 2006) is also consistent with BDCM having a threshold, or at least not posing a cancer risk at drinking water exposure levels.

The default regulatory position in the US for possible carcinogens is to treat them as no-threshold (genotoxic) carcinogens (i.e. use of a cancer slope factor, CSF) until there is clear consensus on a mode of action to explain the mechanism for tumour formation observed in experimental animals. Consequently, it is unlikely that databases like the US EPA IRIS database will have their default no-threshold carcinogenic status (i.e. listing of CSF) of DBPs like BDCM, bromoform, dichoroacetic acid or trichloroacetic acid clarified in the foreseeable future. The inability of the dominant DBPs (THMs and HAAs) to explain cancer risk via drinking water exposure is a major issue addressed in Chapter 10.

As noted, a large number of DBPs beyond THMs and HAAs have been identified in drinking waters, thus it is not feasible to conduct toxicological, epidemiological, and occurrence studies on all DBPs reported in the literature. In order to address this issue, a multidisciplinary group of experts, including toxicologists, structure-activity specialists, and chemists met to carry out a toxicology review of the DBPs reported in the literature, to prioritise DBPs according to their predicted adverse health effects (Richardson, 2003). An in-depth mechanism-based structural activity relationship analysis resulted in 50 high-priority DBPs, which were then selected for a US nationwide DBP occurrence study. Yet, despite the large number of DBPs identified and measured in drinking waters, there are still many more unidentified DBPs whose potential toxicity and health effects on humans are not known (Richardson, 2003).

Moreover, the human body is exposed to a mixture of DBPs rather than single species or specific groups of DBPs. This also needs to be considered in DBP exposure assessment since the interactions between DBPs species may result in additive, synergistic, or antagonistic effects. Unfortunately, there are only a few studies investigating the toxicological effects of a mixture of DBPs, largely because it is so resource intensive to conduct such studies.

Recently, scientists from the US EPA Office of Research and Development collaborated on a multidisciplinary research project, "Integrated Disinfection By-Products Mixtures Research: Toxicological and Chemical Evaluation of Alternative Disinfection Treatment Scenarios". The project focused on the integration of toxicological and chemical evaluation of complex mixtures of DBPs (Simmons *et al.* 2011), and included: comprehensive chemical characterisation of water concentrates prepared from chlorinated and ozonated/post-chlorinated drinking water (Richardson *et al.* 2011), assessment of the developmental toxicity in Sprague-Dawley rats exposed to the concentrates (Narotsky *et al.* 2011), assessment of the *Salmonella* mutagenicity of the concentrates (Claxton *et al.* 2011), and large-scale differential gene expression analysis (Crosby *et al.* 2011). The study results to date found no adverse developmental effects on the Sprague-Dawley rats (Narotsky *et al.* 2011), demonstrated that the volatile and non-volatile organics of the chlorinated waters were mutagenic, while the ozonated/post-chlorinated waters were less mutagenic (Claxton *et al.* 2011). Crosby *et al.* (2011) showed that the water concentrates were cytotoxic and provided evidence for the causes of the observed cytotoxicity (e.g. cell cycle arrest, metabolic stasis, and oxidative stress).

As noted in Chapter 10, more research into DBPs that may be responsible for adverse human health effects is needed because those with plausible cause-effect relationships for adverse human health impacts at the exposure levels plausibly found in drinking water remain unknown.

2.1.6 Epidemiological studies of DBPs

The findings of epidemiological studies on DBPs are discussed in Chapter 11. Most epidemiological studies have utilized one drinking water source (surface vs. groundwater), disinfection method (chlorination vs. alternative) or THMs data as the exposure metric. The underlying logic for any epidemiology study involves seeking a statistical relationship between a specified adverse outcome (e.g. cancer or adverse reproductive effect) and a hypothesized exposure. As will be explained in Chapter 11, the findings of epidemiologic studies must be interpreted with considerable caution because, on their own, they cannot prove causation, only correlation. Evidence of a causal relationship, such as exposure to drinking water DBPs and a disease like cancer, requires a weight of evidence interpretation and judgement that relies on a wide range of contributing factors.

2.2 REGULATORY HISTORY

2.2.1 Regulations and guidelines for THMs

As explained in Chapter 1, the discovery of THMs by Rook (1974) and Bellar *et al.* (1974) was followed by a rapid expansion in the regulatory and research attention directed at DBPs in drinking water. THMs were the first group of DBPs regulated and included: chloroform, bromodichloromethane, dibromochloromethane, and bromoform. This combination was typically identified as "total THMs" or more accurately (given the later discovery of iodinated THMs) as "THM4". There have been some variations in the way the regulations were established and in the evolution of these regulations around the world.

The early health concerns associated with chloroform and THMs led to the establishment of guidelines and standards for these compounds. Canada was the first to set a guideline value for a "not to exceed" maximum total THMs at 350 $\mu g\ L^{-1}$ in 1978. In 1979, under the Safe Drinking Water Act, the US set a regulatory standard for maximum total THMs at 100 $\mu g\ L^{-1}$, as a RAA (running annual average) of four quarterly samples. The concept of using RAA was based on the knowledge that THMs varied substantially from summer (higher) to winter (lower) in many water systems and the health concern was for lifetime cancer risk, making short-term elevated exposure a lesser concern. In 1984, the WHO proposed a guideline for chloroform, as a maximum, of 30 $\mu g\ L^{-1}$ based on estimation that this concentration would assure less than a 1 in 100,000 lifetime cancer risk, based on interpreting chloroform as a no-threshold (genotoxic) carcinogen and a cancer slope factor (CSF) derived from the original cancer bioassay (NCI, 1976).

Further research on the formation, occurrence, and human health effects of THMs as well as the data collected from monitoring of THMs in drinking water distribution systems have prompted revisions and amendments of the previously established guidelines and standards. In 1993, the WHO published a revised version of the Guidelines for Drinking Water Quality (2nd edition) in which the guideline value for chloroform was revised to 200 $\mu g\ L^{-1}$ based on recognizing a threshold for chloroform causing cancer so that a tolerable daily intake was used rather than a CSF (WHO, 1993). Guideline values for the other species of THMs were established individually (bromodichloromethane 60 $\mu g\ L^{-1}$, chlorodibromomethane 100 $\mu g\ L^{-1}$, and bromoform 100 $\mu g\ L^{-1}$). Unlike the US and Canada, the WHO did not set a guideline value for total THMs.

In 1996, amendments to the US Safe Drinking Water Act were made and Stage 1 of the Disinfectant and Disinfection By-Product Rule was promulgated. Under this rule, the maximum total THM concentration was reduced, as a result of multi-stakeholder negotiations, informed by the US EPA risk assessments, to 80 $\mu g\,L^{-1}$, as a RAA. In 2005, Stage 2 of the Disinfectant and Disinfection By-Product Rule was signed by the US EPA administrator. Under this new rule, the maximum contaminant level (MCL) for total THMs remained at 80 $\mu g\,L^{-1}$, however, this value is now based on a locational running annual average, LRAA (USEPA, 2006).

In Canada, the guidelines for THMs were revised in 1996. The maximum concentration of total THMs was reduced to 100 $\mu g\,L^{-1}$, based on a RAA but retaining the use of CSF for chloroform on the assumption of it being a no-threshold carcinogen. In 2006, the guideline value for total THMs was re-affirmed at 100 $\mu g\,L^{-1}$ since there was essentially negligible difference in public health risk between maximum concentration of 80 $\mu g\,L^{-1}$ vs. 100 $\mu g\,L^{-1}$. But the 2006 guideline was based on treating chloroform as a threshold carcinogen, so it was based on a TDI and uncertainty factors, similar to the rationale previously adopted by WHO in 1993 for chloroform. As in the US, the Canadian guideline value for total THMs is based on LRAA of quarterly samples. However, unlike the situation in the US, these guideline values are not nationally enforced nor legally binding. Adoption of the guideline as an enforceable standard with legal consequences is at the discretion of the provincial government. Interestingly, the revision of the Guidelines for Canadian Drinking Water Quality in 2006 included an individual guideline value of 16 $\mu g\,L^{-1}$ for bromodichloromethane (BDCM). This guideline value was based on a cancer endpoint, using a CSF, but the value was designated as a "not-to-exceed" value as a precaution for potential adverse reproductive effects (Health Canada, 2008a). Ironically, the same year as an individual guideline for BDCM was adopted in 2006, a new National Toxicology Program cancer bioassay (NTP, 2006) found that BDCM in drinking water may not cause cancer in laboratory animals. New epidemiological studies on the reproductive effects of BDCM also found no association between exposure to BDCM and adverse reproductive outcomes (Health Canada, 2008a). In 2008, Health Canada convened a panel of experts to provide advice and recommendations on the BDCM guideline. Based on the assessment of the overall weight of scientific evidence and the recommendations of the expert panel, Health Canada rescinded the individual guideline for BDCM (Health Canada, 2008a). In April 2009, the guideline statement for total THMs in drinking water was modified to remove the separate guideline for BDCM (Health Canada, 2009).

In Australia, the approach to DBPs has been quite different to those in the US and Canada. Where the concentrations of THMs in Canada and the US have been regulated since 1978 and 1979, respectively, national attention to DBPs in Australia was first drawn only in 1987, with a one-paragraph section on 'total

THMs' in the 1987 Guidelines for Drinking Water Quality in Australia. This 1987 guideline was the result of the revision of the 1980 guidelines, taking into consideration the 1984 WHO Guidelines for Drinking Water Quality. There was no guideline value for any DBPs until 1996, when the guideline was revised and became one of the Australian Drinking Water Guidelines (ADWG). The 1996 ADWG included information (fact sheets) on 26 organic and inorganic species of DBPs and guideline values for some of these DBPs (11 species plus total THMs), reflecting an increasing concern on DBPs in Australia (NHMRC/NRMMC, 1996). The guideline value for total THMs was set at a maximum of 250 μg L^{-1} (not RAA) which was more comparable with the 1993 WHO guideline of 200 μg L^{-1} for chloroform alone (increased from a 1984 WHO guideline of 30 μg L^{-1} for chloroform) or the 1978 maximum (not RAA) of 350 μg L^{-1} for TTHM in Canada. The guideline values in the 1996 ADWG were based primarily on the latest WHO recommendations, which were based on working papers and assessments prepared by WHO expert panels, and reflected improvements in understanding of water quality problems and THM toxicology. Referenced materials were included in this publication, and included scientific papers, Guidelines published by overseas agencies, issues papers prepared by Australian water authorities, and assessments made by the NHMRC (National Health and Medical Research Council). The 1996 ADWG was revised in 2004, but there was no change with respect to guidelines for DBPs (NHMRC/NRMMC, 2004). In 2011, further revisions of the ADWG were published as the 2011 ADWG (NHMRC/NRMMC, 2011). With respect to DBPs, only two changes took place: the guideline value for chlorite was raised to 0.8 mg L^{-1} from 0.3 mg L^{-1} and NDMA was added to the list of DBPs, with a guideline value of 100 ng L^{-1}. Similar to the situation in Canada, the ADWG is not nationally enforced in Australia. Implementation of the Guidelines is subject to the policy established by each state and territory in Australia, which is determined by the Health Department in consultation with water authorities and suppliers in each state and territory.

Other countries have also established standards and guideline values for total THMs. In Europe, the European Council Directive set a standard of 100 μg L^{-1} for total THMs as an absolute maximum. Japan and China have also adopted the guideline value of 100 μg L^{-1} for total THMs as a maximum concentration, but they also have individual guideline values for each species of THMs (Karanfil *et al.* 2008).

2.2.2 Regulatory guidelines for HAAs

HAAs are often found as the second most abundant DBP class in chlorinated drinking waters. There are nine species of chlorinated and brominated HAAs that can be formed under conventional disinfection conditions (see Table 2.1). However, only five species of HAAs are regulated by the US EPA (HAA5: MCAA, MBAA, DCAA, DBAA, TCAA), and the WHO only provides guideline values for three species of HAAs (i.e. MCAA, DCAA, and TCAA).

Following the amendments to the US Safe Drinking Water Act and the establishment of the Stage 1 of the Disinfectant and Disinfection By-Product Rule, HAAs were added to the list of DBPs regulated in the US. Five species of HAAs (HAA5) were regulated at a total concentration of 60 μg L^{-1}, as a RAA. Stage 2 of the Disinfectant and Disinfection By-Product Rule did not bring any change to the MCL for HAA5, however the requirement for RAA was changed to LRAA, as in the case of THMs.

A guideline value for HAA5 also exists in Canada, though at a higher concentration of 80 μg L^{-1}. This guideline value was affirmed in 2008, based on RAA. This guideline was derived based on best available technology, rather than health-based targets, considering the technological limitations associated with reducing the concentrations of individual HAAs in drinking waters while maintaining effective disinfection (Health Canada, 2008b). DCAA and TCAA are the dominant (by concentration) HAAs found in Canadian drinking water supplies. TCAA was not judged to be a no-threshold carcinogen, so it was evaluated on non-cancer effects, resulting in a health-based limit for TCAA estimated at 300 μg L^{-1} (Health Canada, 2008b). The evidence for DCAA being genotoxic (i.e. no threshold) is not compelling, but the lack of a consensus, non-genotoxic mode of action for DCAA causing tumours led to a cautionary approach of estimating DCAA limits based on a no-threshold model.

The WHO provided guideline values for only three species of HAAs: MCAA, DCAA, and TCAA. The guideline values for DCAA (50 μg L^{-1}) and TCAA (200 μg L^{-1}) were published in 1993 while the guideline for MCAA (20 μg L^{-1}) was established in 2004. In the latest edition of the WHO Guidelines for Drinking Water Quality (4th edition), the guideline values for the three species of HAAs have remained the same (WHO, 2011). Australia has followed the WHO's example by setting guideline values for only three species of HAAs since the 1996 ADWG. However, the guideline values for MCAA (150 μg L^{-1}) and DCAA (100 μg L^{-1}) were higher than those established by the WHO, while the guideline value for TCAA (100 μg L^{-1}) was lower. Currently there is no guideline value for HAAs in the UK and Europe.

2.2.3 Regulations and guidelines considerations for emerging DBPs

Since the discovery of DBPs in the early 1970s, drinking water guidelines and regulations have focused on a proportion of DBPs that represent the most abundant and readily analysed DBPs, namely THMs and HAAs. Recent research on DBPs has shown that the regulated DBPs are not responsible for the observed adverse human health effects associated with DBPs (Chapters 10 and 11). For example, chloroform is not expected to cause human cancer at concentrations observed in drinking waters (Bull *et al.* 2001; Bull *et al.* 2011; Hrudey, 2009). The recent collaborative European study on human health impacts from DBPs showed that there was little evidence of any effects of THMs on reproductive outcomes including semen quality, congenital anomalies, stillbirth, pre-term delivery, and birth weight, and there is also little evidence for an epidemiologic relationship between THMs and colon cancer (Niewuwenhuijsen, 2010). A pooled analysis of bladder cancer (Chapter 11) found an exposure-response relationship that was marginally statistically significant, but with a shallower slope than found in previous North American studies. Because health effects that may be associated with DBPs in drinking waters cannot be explained on the basis of currently identified DBPs, further health-related research on DBPs other than THMs and HAAs, for example, emerging DBPs, is needed.

In 1998, the US EPA initiated a study that aimed to prioritise DBPs based on known/predicted health effects, to focus a new strategic initiative on the occurrence of emerging DBPs of health concern in drinking water. In the study, the DBPs reported in the literature were prioritized according to predicted cancer effects using a mechanism-based, structure-activity relationship analysis (Woo *et al.* 2002). This resulted in the identification of 50 "high priority" unregulated DBPs. Following this study, the US EPA conducted a "Nationwide Disinfection By-Product Occurrence Study" in 2002 which involved monitoring of these "high-priority" DBPs of toxicological significance in drinking water distribution systems managed by 12 water utilities across 6 EPA regions. This was the first systematic screening effort of DBPs to be based on a prior toxicological prioritization of DBPs. From this study, N-DBPs and I-DBPs were identified as having greater public health significance than the regulated DBPs, as they are more toxic and carcinogenic. Other emerging DBPs classes of interest were also identified: halonitromethanes, haloamides, halogenated furanones, and haloaldehydes (Weinberg *et al.* 2002). The strategic development of this study has helped refocus today's DBP research agenda toward a new generation of emerging DBPs of health significance. Among the 50 priority, new generation DBPs, varying intensity in research activities has been observed, with the rapid growth of research and regulatory activities on nitrosamines (Chapter 7) and I-DBPs (Chapter 4).

2.3 CURRENT STATUS OF DBP REGULATIONS AND GUIDELINES

Over 600 species of DBPs have been identified but only few DBPs are currently regulated or have associated guideline values. Where regulations require water utilities to comply with the regulated MCLs to avoid legal consequences, guidelines are enforced at the discretion of the local authorities, usually at state or provincal levels. The current regulations and guideline values for DBPs in drinking waters are presented in Table 2.2.

The current state of knowledge on DBPs is not accurately reflected in existing guidelines and regulations. Therefore, current guidelines and regulations need to be evaluated based on sound research findings that demonstrate a good understanding of the relative concentrations and toxicities of DBPs found in drinking waters. Specifically, there is a need to develop new guidelines for emerging DBPs since the current guidelines have driven the shift to alternative disinfectants, some of which have been shown to produce more toxic, emerging, unregulated DBPs (e.g. N-DBPs and I-DBPs from chloramination; halonitromethanes from pre-ozonation; dihaloaldehydes from combination of chloramination and ozonation). The evaluation of current regulations and guidelines and the development of new ones requires the involvement of interdisciplinary teams that include chemists, toxicologists, epidemiologists, statisticians, engineers and regulators. The utility of regulations or guidelines must be judged in terms of the relative occurrence of identified DBPs and the toxic potency for relevant human health effects at concentrations (drinking water exposure levels) that are relevant.

Table 2.2 Current regulations and guidelines for DBPs in drinking waters.

Country/Agency	DBPs	Guideline value ($\mu g\ L^{-1}$)
World Health Organization[a]	Chloroform	300
	Bromodichloromethane	60
	Chlorodibromomethane	100
	Bromoform	100
	Monochloroacetic acid	20[b]
	Dichloroacetic acid	50[b]
	Trichloroacetic acid	200[b]
	Bromate	10
	Chlorate	700
	Chlorite	700
	Dichloroacetonitrile	20
	Dibromoacetonitrile	70
	N-Nitrosodimethylamine	0.1
United States of America (US Environmental Protection Agency)[c]	Total THMs	80 (LRAA[d])
	HAA5	60 (LRAA[d])
	Bromate	10
	Chlorite	1000
Canada (Health Canada)[e]	Total THMs	100
	HAA5	80
	Bromate	10
	Chlorate	1000
	Chlorite	1000
	N-Nitrosodimethylamine[f]	0.04

(*Continued*)

Table 2.2 Current regulations and guidelines for DBPs in drinking waters (*Continued*).

Country/Agency	DBPs	Guideline value (μg L^{-1})
Australia (National Health and Medical Research Council and Natural Resource Management Ministerial Council)[g]	Total THMs	250
	Monochloroacetic acid	150
	Dichloroacetic acid	100
	Trichloroacetic acid	100
	Trichloroacetaldehyde	20
	Cyanogen chloride	80
	Bromate	20
	Chlorite	800
	Cyanogen chloride	80
	N-Nitrosodimethylamine	0.1
European Union[h]	Total THMs	100
	Bromate	10
Japan (Ministry of Health, Labour, and Welfare)[i]	Total THMs	100
	Chloroform	60
	Bromodichloromethane	30
	Chlorodibromomethane	100
	Bromoform	90
	Bromate	10
	Monochloroacetic acid	20
	Dichloroacetic acid	40
	Trichloroacetic acid	200
	Formaldehyde	80

[a]WHO 2011.
[b]as acetate – monochloroacetate, dichloroacetate, trichloroacetate.
[c]USEPA 2006.
[d]Locational Running Annual Average.
[e]Health Canada 2010.
[f]Health Canada 2011.
[g]NHMRC/NRMMC 2011.
[h]European Council 1998.
[i]Japanese Ministry of Health, Labour, and Welfare 2010.

2.4 CONCLUSIONS

Despite over 35 years of research on DBPs, there are still many issues that need to be further investigated. The formation and occurrence of DBPs, their toxicological significance and potential human health effects, and the need to regulate their concentrations in drinking waters are complex, inter-related issues that require an interdisciplinary understanding. Over the 35 years of research on DBPs, the following trends have been observed:

- *Chemistry of DBP formation*: there has been a shift from studies on regulated DBPs (THMs and HAAs) associated with chlorination to studies on emerging

DBPs (N-DBPs, I-DBPs and other new toxicologically-relevant DBPs, such as haloquinones, as well as transformation products of ozone, UV, and AOPs (i.e. alternative disinfectants).

- *Analytical chemistry for the identification of DBPs*: an expansion from the almost exclusive use of GC-MS based methods to include LC-MS based methods to determine and identify unknown DBPs, especially polar DBPs and high molecular weight DBPs.
- *Water treatment technology*: increasing understanding of the potential importance of emerging DBPs has been leading to technological research aimed to reducing these DBPs in finished water.
- *Toxicological studies on DBPs*: an expansion from studies on single compounds to mixtures; from animal studies to the use of mammalian cells to screen for toxic effects; and increasing use of modeling software to predict toxicity of DBPs.
- *Epidemiological studies on DBPs*: a better understanding of routes of exposure and temporal and spatial variations of DBPs informing DBP exposure assessment has raised concerns about the ability of epidemiology to assess accurately associations between DBPs and specific adverse health effects.

Key overview findings:

- Analytical chemistry realities dictate that you find what you look for, not necessarily what is important. In the context of all DBPs, there is no single analytical method that can be used to analyse for all DBPs in drinking waters, since DBPs vary in molecular weight, volatility, and polarity.
- Despite the relatively low human health risks associated with THMs and HAAs, they still dominate the published water quality literature and many DBP research agendas, likely because they are regulated and have associated guideline values worldwide.
- Alternative disinfectants also produce DBPs and these DBPs can be more individually toxic than the currently regulated DBPs, although their concentrations are generally, correspondingly lower. Therefore, DBP formation needs to be evaluated and risk trade-offs considered.
- Understanding the chemistry of DBP formation is crucial for developing strategies to minimize DBP formation in drinking water treatment plants.
- Numerous individual DBPs demonstrate serious toxicological effects in experimental animals at doses much higher than they are found in drinking water. Regulatory risk assessment based on these findings has been and continues to be precautionary because of the inherent uncertainties and the pervasive exposure of the public to DBPs.
- Epidemiological findings are often difficult to interpret because of the numerous limitations to such studies and their inherent inability to demonstrate causation without considerable additional supporting evidence from exposure assessment and toxicological evaluation.

REFERENCES

Amy G. L. (1993). Using NOM characterization for the evaluation of treatment. In: Proc. of Natural Organic Matter in Drinking Water: Origin, Characterisation, and Removal Conf., Chamonix, France, 19–22 September.

Backer L. C., Ashley D. L., Bonin M. A., Cardinali F. L., Kieszak S. M. and Wooten J. V. (2000). Household exposures to drinking water disinfection by-products: whole blood trihalomethane levels. *J. Expo. Anal. Environ. Epidemiol.*, **10**, 3321–3326.

Baribeau H., Krasner S. W., Chinn R. and Singer P. C. (2005). Impact of biomass on the stability of HAAs and THMs in a simulated distribution system. *J. Am. Water Works Assoc.*, **97**, 69–81.

Bedner M. and MacCrehan W. A. (2006). Transformation of acetaminophen by chlorination produces the toxicants 1,4-benzoquinone and *N*-acetyl-*p*-benzoquinone imine. *Environ. Sci. Technol.*, **40**, 516–522.

Bellar T. A., Lichtenberg J. J. and Kroner R. C. (1974). The occurrence of organohalides in chlorinated drinking water. *J. Am. Water Works Assoc.*, **66**, 703–706.

Bichsel Y. and von Gunten U. (2000). Formation of iodo-trihalomethanes during disinfection and oxidation of iodide-containing waters. *Environ. Sci. Technol.*, **34**, 2784–2791.

Boyer T. H. and Singer P. C. (2005). Bench-scale testing of a magnetic ion exchange resin for removal of disinfection by-product precursors. *Water Res.*, **39**, 1265–1276.

Bull R. J., Fawell J., Cotruvo J. and Hrudey S. E. (2011). Critique of Chowdhury *et al.* – Disinfection byproducts in Canadian provinces: associated cancer risks and medical expenses. *J. Hazard Mat.*, **187**, 574–584.

Bull R. J., Krasner S. W., Daniel P. A. and Bull R. D. (2001). Health Effects and Occurrence of Disinfection By-Products. AWWA Research Foundation, Denver, Colorado, USA.

Cantor K. P., Villanueva C. M., Silverman D. T., Figueroa J. D., Real F. S., Garcia-Closas M., Malats N., Chanock S., Yeager M., Tardon A., Garcia-Closas R., Serra C., Carrato A., Castano-Vinyals G., Samanic C., Rothman N. and Kogevinas M. (2010). Polymorphisms in GSTT1, GSTZ1, and CYP2E1, disinfection by-products, and risk of bladder cancer in Spain. *Environ. Health Perspect.*, **118**, 1545–1550.

Charrois J. W. A. (2011). Analysis of emerging disinfection by-products in drinking water. In: Encyclopedia of Analytical Chemistry, Supplementary Volumes S1-S3, R. A. Meyers (ed.), John Wiley & Sons, Ltd., Chichester, UK, pp. 205–220.

Charrois J. W. A., Arend M. W., Froese K. L. and Hrudey S. E. (2004). Detecting *N*-nitrosamines in drinking water at nanogram per liter levels using ammonia positive chemical ionization. *Environ. Sci. Technol.*, **38**, 4835–4841.

Charrois J. W. A. and Hrudey S. E. (2007). Breakpoint chlorination and free-chlorine contact time: implications for drinking water *N*-nitrosodimethylamine concentrations. *Water Res.*, **41**, 674–682.

Choi J. and Valentine R. L. (2002). Formation of *N*-nitrosodimethylamine (NDMA) from reaction of monochloramine: a new disinfection by-product. *Water Res.*, **36**, 817–824.

Christman R. F., Norwood D. L., Millington D. S., Johnson J. D. and Stevens A. A. (1983). Identity and yields of major halogenated products of aquatic fulvic acid chlorination. *Environ. Sci. Technol.*, **17**, 625–628.

Claxton L. D., Pegram R., Schneck K. M., Simmons J. E. and Warren S. H. (2011). Integrated disinfection by-products research: salmonella mutagenicity of water concentrated disinfected by chlorination and ozonation/postchlorination. *J. Toxicol. Environ. Health*, Part A **71**(17), 1187–1194.

Crosby L. M., Simmons J. E., Ward W. O., Moore T. M., Morgant K. T. and DeAngelo A. B. (2011). Integrated disinfection by-products (DBP) mixtures research: gene expression alteration in primary rat hepatocyte cultures exposed to DBP mixtures formed by chlorination and oxonation/postchlorination. *J. Toxicol. Environ. Health*, Part A **71**(17), 1195–1215.

Croué J.-P., Debroux J.-F., Amy G. L., Aiken G. R. and Leenheer J. A. (1999). Natural organic matter: structural characteristics and reactive properties. In: Formation and Control of Disinfection By-Products in Drinking Water, P. C. Singer (ed.), American Water Works Association, Denver, Colorado, USA, pp. 65–93.

Croué J.-P., Korshin G. V. and Benjamin M. (2000). Characterisation of Natural Organic Matter in Drinking Water. American Water Works Association Research Foundation, Denver, Colorado, USA.

Dotson A., Westerhoff P. and Krasner S. W. (2009). Nitrogen enriched dissolved organic matter (DOM) isolates and their affinity to form emerging disinfection by-products. *Water Sci. Technol.*, **60**, 135–143.

European Council (1998). Drinking Water Directive, Council Directive 98/83/EC on the quality of water intended for human consumption, European Commission. http://ec.europa.eu/environment/water/water-drink/index_en.html (accessed 07 February 2012).

Fang J., Yang X., Ma J., Shang C. and Zhao Q. (2010). Characterization of algal organic matter and formation of DBPs from chlor(am)ination. *Water Res.*, **44**, 5897–5906.

Fawell J. (1999). The risks of DBPs in perspective. In: Disinfection By-Products in Drinking Water: Current Issues, M. Fielding and M. Farrimond (eds), The Royal Society of Chemistry, Cambridge, UK, pp. 157–164.

Frimmel F. H. (1998). Characterisation of natural organic matter as major constituents in aquatic systems. *J. Contam. Hydrol.*, **35**, 201–216.

Gallard H. and von Gunten U. (2002). Chlorination of natural organic matter: kinetics of chlorination and of THM formation. *Water Res.*, **36**, 65–74.

Glezer V., Harris B., Tal N., Iosefzon B. and Lev O. (1999). Hydrolysis of haloacetonitriles: linear free energy relationship, kinetics, and products. *Water Res.*, **33**, 1938–1948.

Guay C., Rodriguez M. and Sérodes J. (2005). Using ozonation and chloramination to reduce the formation of trihalomethanes and haloacetic acids in drinking water. *Desalination*, **176**, 229–240.

Health Canada (2008a). Findings and recommendations of the BDCM expert panel meeting. Available by request to: water_eau@hc-sc.gc.ca

Health Canada (2008b). Guidelines for Canadian Drinking Water Quality: Guideline Technical Document – Haloacetic Acids. Water, Air and Climate Change Bureau, Healthy Environments and Consumer Safety Branch, Health Canada, Ottawa, Ontario.

Health Canada (2009, April addendum). Guidelines for Canadian Drinking Water Quality: Guideline Technical Document – Trihalomethanes. Water Quality and Health Bureau, Healthy Environments and Consumer Safety Branch, Health Canada, Ottawa, Ontario.

Health Canada (2010). Guidelines for Canadian Drinking Water Quality, Ontario, Canada. http://www.hc-sc.gc.ca/ewh-semt/pubs/water-eau/2010-sum_guide-res_recom/index-eng.php (accessed 07 February 2012).

Health Canada (2011). Guideline for Canadian Drinking Water Quality: Guideline Technical Document – *N*-Nitrosodimethylamine. Water, Air and Climate Change Bureau, Healthy Environments and Consumer Safety Branch, Health Canada, Ottawa, Ontario.

Hemming J., Holmbom B., Reunanen M. and Kroberg L. (1986). Determination of the strong mutagen 3-chloro-4-(dichloromethyl)-5-hydroxy-2(5H)-furanone in chlorinated drinking and humic waters. *Chemosphere*, **15**(5), 549–556.

Hrudey S. E. (2009). Chlorination disinfection by-products, public health risk tradeoffs and me. *Water Res.*, **43**, 2057–2092.

Hua G. and Reckhow D. A. (2007a). Comparison of disinfection byproduct formation from chlorine and alternative disinfectants. *Water Res.*, **41**, 1667–1678.

Hua G. and Reckhow D. A. (2007b). DBP formation during chlorination and chloramination: effect of reaction time, pH, dosage, and temperature. *J. American Water Works Assoc.*, **100**, 82–95.

Japanese Ministry of Health, Labour, and Welfare (2010). Drinking Water Quality Standards. http://www.mhlw.go.jp/english/policy/health/water_supply/4.html (accessed 07 February 2012).

Karanfil T., Krasner S. W., Westerhoff P. and Xie Y. (2008). Recent advances in disinfection by-produc formation, occurrence, control, health effects, and regulations. In: Disinfection By-Products in Drinking Water Occurrence, Formation Health Effects, Control, T. Karanfil, S. W. Krasner, P. Westerhoff and Y. Xie (eds), American Chemical Society, Washington, DC, USA, pp. 2–19.

Kim H., Shim J. and Lee S. (2002). Formation of disinfection by-products in chlorinated swimming pool water. *Chemosphere*, **46**, 123–130.

Koechling M. T., Shukairy H. M. and Summers R. S. (1996). Effect of ozonation and biotreatment on molecular size and hydrophilic fractions of natural organic matter. In: Water Disinfection and Natural Organic Matter: Characterisation and Control, R. A. Minear and G. L. Amy (eds), American Chemical Society, Washington, DC, pp. 196–210.

Krasner S. W. (1999). Chemistry of disinfection by-product formation. In: Formation and Control of Disinfection By-Products in Drinking Water, P. C. Singer (ed.), American Water Works Association, Denver, Colorado, USA, pp. 27–51.

Krasner S. W., McGuire M. J., Jacangelo J. G., Patania N. L., Reagan K. M. and Aieta E. M. (1989). The occurrence of disinfection by-products in US drinking water. *J. Am. Water Works Assoc.*, **81**, 41–53.

Krasner S. W., Weinberg H. S., Richardson S. D., Pastor S. J., Chinn R., Sclimenti M. J., Onstad G. D. and Thruston A. D. Jr. (2006). Occurrence of a new generation disinfection byproducts. *Environ. Sci. Technol.*, **40**, 7175–7185.

Kristiana I., Croué J.-P., Gallard H. and Joll C. A. (2009). The formation of halogen-specific TOX from chlorination and chloramination of natural organic matter isolates. *Water Res.*, **43**, 4177–4186.

Lee W., Westerhoff P. and Croue J.-P. (2007). Dissolved organic nitrogen as a precursor for chloroform, dichloroacetonitrile, *N*-nitrosodimethylamine, and trichloronitromethane. *Environ. Sci. Technol.*, **41**, 5485–5490.

Liang L. and Singer P. C. (2003). Factors influencing the formation and relative distribution of haloacetic acids and trihalomethanes in drinking water. *Environ. Sci. Technol.*, **37**, 2920–2928.

Liu W., Cheung L.-M., Yang X. and Shang C. (2006). THM, HAA, and CNCl formation from UV irradiation and chlor(am)ination of selected organic waters. *Water Res.*, **40**, 2033–2043.

Makdisy G., Croué J. P. and Buisson A. H. (2004). Fouling of a polyethersulfone ultrafiltration membrane by natural organic matter. *Water Sci. Technol.*, **4**, 205–212.

Marx J. (1974). Drinking water: another source of carcinogens? *Science*, **186**, 809–811.

Mitch W. A., Krasner S. W., Westerhoff P. and Dotson A. (2009). Occurrence and Formation of Nitrogenous Disinfection By-Products. Water Research Foundation, Denver, Colorado, USA.

Mitch W. A. and Schreiber I. M. (2008). Degradation of tertiary alkylamines during chlorination/chloramination: implications for formation of aldehydes, nitriles, halonitroalkanes, and nitrosamines. *Environ. Sci. Technol.*, **42**, 4811–4817.

Mitch W. A., Sharp J. O., Trussell R. R., Valentine R. L., Alvarez-Cohen L. and Sedlak D. L. (2003). *N*-Nitrosodimethylamine (NDMA) as a drinking water contaminant: a review. *Environ. Eng. Sci.*, **20**, 389–404.

Narotsky M. G., Best D. S., Rogers E. H., McDonald A., Sey Y. M. and Simmons J. E. (2011). Integrated disinfection by-products mixtures research: assessment of developmental toxicity in Sprague-Dawley rats exposed to concentrates of water disinfected by chlorination and ozonation/postchlorination. *J. Toxicol. Environ. Health,* Part A **71**, 1216–1221.

NCI (1976). Report on Carcinogenesis Bioassay of Chloroform. NTIS PB-264018. National Cancer Institute, US Government Printing Office, Washington, DC, USA.

NHMRC/NRMMC (1996). Australian Drinking Water Guidelines. http://www.nhmrc.gov.au/publications/synopses/eh19syn.htm (accessed 07 February 2012).

NHMRC/NRMMC (2004). Australian Drinking Water Guidelines. http://www.nhmrc.gov.au/guidelines/publications/eh34 (accessed 07 February 2012).

NHMRC/NRMMC (2011). Australian Drinking Water Guidelines. http://www.nhmrc.gov.au/guidelines/publications/eh52 (accessed 07 February 2012).

Nieuwenhuijsen M. J. (2010). Health Impacts of Long-Term Exposure to Disinfection By-Products in Drinking Water. Project No. 036224. HiWATE Publishable Final Report. http://www.hiwate.eu/files/Publishable%20Final%20Report%20V2.pdf (accessed 07 February 2012).

NTP (2006). Toxicology and Carcinogenesis Studies of Bromodichloromethane (CAS No. 75-27-4) in Male F344/N Rats and Female B6C3F1 Mice (Drinking Water Studies). Research Triangle Park, NC, National Toxicology Program.

Pontius F. W. (2000). Chloroform: science, policy and politics. *J. Am. Water Works Assoc.*, **92**, 12–18.

Pressman J. G., Richardson S. D., Speth T. F., Miltner R. J., Narotsky M. G., Hunter E. S. III, Rice G. E., Teuschler L. K., McDonald A., Parvez S., Krasner S. W., Weinberg H. S., McKague A. B., Parrett C. J., Bodin N., Chinn R., Lee C.-F. T. and Simmons J. E. (2010). Concentration, chlorination, and chemical analysis of drinking water for disinfection byproduct mixtures health effects research: U.S. EPAs four lab study. *Environ. Sci. Technol.*, **44**, 7184–7192.

Qin F., Zhao Y. Y., Zhao Y. L., Boyd J. M., Zhou W. J. and Li X.-F. (2010). A toxic disinfection by-product, 2,6-dichloro-1,4-benzoquinone, identified in drinking water. *Angew. Chem. Int. Ed.*, **49**, 790–792.

Reckhow D. A. and Singer P. C. (1985). Mechanisms of organic halide formation during fulvic acid chlorination and implications with respect to preozonation. In: Water Chlorination: Chemistry, Environmental Impact and Health Effects, Volume 5, R. L. Jolley, R. J. Bull, W. P. Davis, S. Katz, M. H. Roberts and V. A. Jacobs (eds), Lewis Publishers Inc., Chelsea, Michigan, USA, 1229–1257.

Reckhow D. A. and Singer P. C. (1990). Chlorination by-products in drinking waters: from formation potentials to finished water concentrations. *J. Am. Water Works Assoc.*, **82**, 173–180.

Richardson S. D. (1998). Drinking water disinfection by-products. In: The Encyclopedia of Environmental Analysis and Remediation, F. A. Meyers (ed.), John Wiley and Sons Inc., New York, USA, 1398–1421.

Richardson S. D. (2003). Disinfection by-products and other emerging contaminants in drinking water. *Trends Anal. Chem.*, **22**, 666–684.

Richardson S. D. (2010). Environmental mass spectrometry: emerging contaminants and current issues. *Anal. Chem.*, **82**, 4742–4774.

Richardson S. D. (2011). Disinfection by-products: formation and occurrence in drinking water. In: The Encyclopedia of Environmental Health, Vol. 2, J. O. Nriagu (ed.), Elsevier, Burlington, pp. 110–136.

Richardson S. D., Plewa M. J., Wagner E. D., Schoeny R. and DeMarini D. M. (2007). Occurrence, genotoxicity, and carcinogenicity of regulated and emerging disinfection by-products in drinking water: a review and roadmap for research. *Mutat. Res.*, **636**, 178–242.

Richardson S. D., Thruston A. D. Jr, Krasner S. W., Weinberg H. S., Miltner R. J., Schenck K. M., Narotsky M. G., McKague A. B. and Simmons J. E. (2011). Integrated disinfection by-products mixtures research: comprehensive characterisation of water concentrates prepared from chlorinated and ozonated/postchlorinated drinking water. *J. Toxicol. Environ. Health,* Part A **71**, 1165–1186.

Rodriguez M. J. and Sérodes J.-B. (2001). Spatial and temporal evolution of trihalomethanes in three water distribution systems. *Water Res.*, **35**, 1572–1586.

Rodriguez M. J., Sérodes J.-B. and Levallois P. (2004). Behaviour of trihalomethanes and haloacetic acids in a drinking water distribution system. *Water Res.*, **38**, 4367–4382.

Rook J. J. (1974). Formation of haloform during chlorination of natural waters. *Water Treat. Exam.*, **23**, 234–243.

Rule K. L., Ebbett V. R. and Vikesland P. J. (2005). Formation of chloroform and chlorinated organics by free-chlorine-mediated oxidation of triclosan. *Environ. Sci. Technol.*, **39**, 3176–3185.

SAB (2000). Review of the Darft Chloroform Risk Assessment. Letter to Carol Browner, Administrator of the US Environmental Protection Agency, EPA-SAB-EC-00-009, April 28, 2000, Choroform Risk Assessment Review Subcommittee of the US EPA Science Advisory Board (SAB).

Schmidt C. K. and Brauch H.-J. (2008). *N*, *N*-dimethylsulfamide as precursor for *N*-nitrosodimethylamine (NDMA) formation upon ozonation and its fate during drinking water treatment. *Environ. Sci. Technol.*, **42**, 6340–6346.

Sérodes J.-B., Rodriguez M. J., Li H. and Bouchard C. (2003). Occurrence of THMs and HAAs in experimental chlorinated waters of the Quebec City area (Canada). *Chemosphere*, **51**, 253–263.

Simmons J. E., Richardson S. D., Teuschler L. K., Miltner R. J., Speth T. F., Schenck K. M., Hunter S. E. III and Rice G. (2011). Research issues underlying the four-lab study: integrated disinfection by-products mixtures research. *J. Toxicol. Environ. Health,* Part A **71**, 1125–1132.

Speight V. L. and Singer P. C. (2005). Association between residual chlorine loss and HAA reduction in distribution systems. *J. Am. Water Works Assoc.*, **97**, 82–91.

Symanski E., Savitz D. A. and Singer P. C. (2004). Assessing spatial fluctuations, temporal variability, and measurement error in estimated levels of disinfection by-products in tap water: implications for exposure assessment. *Occup. Environ. Med.*, **61**, 65–72.

US EPA (2006). 40 CFR Parts 9, 141, and 142, National primary drinking water regulations: stage 2 disinfectants and disinfection byproducts rule; final rule. *Fed. Regist.*, **71**(2), 388–493.

Veschetti E., Cittadini B., Maresca D., Citti G. and Ottaviani M. (2005). Inorganic by-products in waters disinfected with chlorine dioxide. *Microchem. J.*, **79**, 165–170.

Weinberg H. S., Krasner S. W., Richardson S. D. and Thruston A. D. Jr (2002). The Ooccurrence of Disinfection By-Products (DBPs) of Health Concern in Drinking Water: Results of a Nationwide DBP Occurrence Study EPA/600/R02/068. US Environmental Protection Agency, National Exposure Research Laboratory, Athens, Georgia, USA.

Westerhoff P., Chao P. and Mash H. (2004). Reactivity of natural organic matter with aqueous chlorine and bromine. *Water Res.*, **38**, 1502–1513.

White C. (1999). Handbook of Chlorination and Alternative Disinfectants. John Wiley and Sons Inc., New York, USA.

WHO (1993). Guidelines for Drinking Water Quality. 2nd edn, World Health Organization, Geneva. http://www.who.int/water_sanitation_health/dwq/gdwq2v1/en/index.html (accessed 07 February 2012).

WHO (2011). Guidelines for Drinking Water Quality. 4th edn, World Health Organization, Geneva. http://www.who.int/water_sanitation_health/publications/2011/dwq_guidelines/en/index.html (accessed 07 February 2012).

Woo Y. T., Lai D., McLain J. L., Manibusan M. K. and Dellarco V. (2002). Use of mechanism-based structure-activity relationships analysis in carcinogenic potential ranking for drinking water disinfection by-products. *Environ. Health Perspect.*, **110**, 75–87.

Yang X., Fan C., Shang C. and Zhao Q. (2010). Nitrogenous disinfection byproducts formation and nitrogen origin exploration during chloramination of nitrogenous organic compounds. *Water Res.*, **44**, 2691–2702.

Zhang X. and Minear R. A. (2002). Decomposition of trihaloacetic acids and formation of the corresponding trihalomethanes in drinking water. *Water Res.*, **36**, 3665–3673.

Zhao Y., Qin F., Boyd J. M., Anichina J. and Li X.-F. (2010). Characterization and determination of chloro- and bromo-benzoquinones as new chlorination disinfection byproducts in drinking water. *Anal. Chem.*, **82**, 4599–4605.

Zhou W. J., Boyd J. M., Qin F., Hrudey S. E. and Li X.-F. (2009). Formation of *N*-nitrosdiphenylamine and two new *N*-containing disinfection byproducts from chloramination of water containing diphenylamine. *Environ. Sci. Technol.*, **43**, 8443–8448.

Chapter 3

Managing DBPs under challenging conditions – a utility water quality manager's perspective

R. Walker

3.1 INTRODUCTION

As a water quality manager in a utility, I have come to understand my job is all about risk management – spot the hazard, manage the risk. Sounds easy, but the dilemma is there are so many hazards and so many potential risks. I am responsible for the water quality delivered to customers in Perth (with a population of 1.8 M) as well as 220 small and midsized towns spread across one third of continental Australia. One hundred and sixty towns have a population of less than 1000 people and are up to 3000 km from Perth. Sources of water include surface-, ground- and seawater. With over 200 sources, the quality of source water is indeed diverse. Australia is a dry country and providing adequate domestic water quantity has been a concern throughout Australia's recorded history, while water quality has been a more recent focus. I only have limited resources, so where do I rank disinfection by-products (DBPs)? My approach is to accept DBPs (essentially THMs for the purposes of this Chapter) as a hazard like any other, which requires preventive risk management based on knowing my own systems.

In addition, DBPs are easy to measure in the distribution system but I am trying to manage water quality as a proactive, risk-based operation. I do not want my first indication of a problem to be water supplied to customers being in excess of the maximum concentration recommended in the Australian Drinking Water Guidelines (ADWG). That said, allowing inadequately disinfected water to cause illness in any of our customers is totally unacceptable, making the management of DBPs produced by disinfection, of challenging source waters, even more difficult.

This chapter provides a practical approach to addressing the issues associated with utility management of DBPs with a particular focus on THMs because these are the DBPs that have had ADWG limits since 1987 (Chapter 2).

3.2 PATHOGENS, CHLORINE, AND DBPS

The ADWG make it clear that pathogens represent the greatest risk to human health and disinfection should never be compromised to meet DBP targets (NHMRC/NRMMC, 2004; 2011). Nevertheless, attention is often concentrated on those sections of the guidelines devoted to chemical parameters and maximum recommended concentrations. Chemicals assume an unwarranted priority for most – especially operators and (dare I say it) some regulators. Maybe it is because they are easy to measure and manage – you either pass or fail!

Many stakeholders are unaware that there are very few chemicals (including DBPs), which are known with confidence to cause ill health in humans as a result of drinking water consumption (Hrudey *et al.* 2011; WHO, 2007). The guideline values for DBPs are all inherently precautionary (Chapter 12). On the other hand, there is no dispute that pathogens cause illness and death. Likewise, it is certain that chlorine inactivates most of the waterborne pathogens (essentially all of the bacterial and viral pathogens) known to infect humans. It is important to understand that for all the inadequacies of the Walkerton water supply system, if the operators had just maintained or at least monitored for a chlorine residual at the head-works, that tragedy, which killed 7 and made more than 2000 ill, could have been avoided (Hrudey and Walker, 2005).

Thus, water quality managers must never ever allow disinfection to be compromised to meet a DBP target. In Western Australia we must also not allow chlorine residuals in the reticulation system to be compromised as the residual is an essential barrier to the free-living, lethal amoebic pathogen *Naegleria fowleri*, which can colonise pipe networks. I know DBP exceptions always trigger a "please explain" request from the utility Executive, the Health Department and often the media can run a story that will create public outcry. But the reality is that there will be no effective increase in public health risk due to short-term DBP guideline exceedences. Supplying inadequately disinfected water, on the other hand, potentially exposes a large number of people to pathogens and almost immediate illness. This must never be allowed to happen!

The truth is that in the 15 years I have been managing water quality, chlorine has become my best friend. Whenever there is a hint of trouble such as an *Escherichia coli (E. coli)* or *Naegleria* detection in the reticulation – what do I do? I dose the tank with liquid chlorine and flush the reticulation system until I have a measureable chlorine residual. If it rains and my source has a high *E. coli* count or rising turbidity – what do I do? I increase the dose rate on my chlorinator of course. When operators of rural and indigenous water supplies ask for my advice, it is always the same. Ensure you implement reliable chlorination and you will be 99% of the way to supplying safe drinking water.

Somehow it seems ironic that so many have spent so long trying to prove chlorinated drinking water produces ill effects but I recognise the nature of precaution when it comes to drinking water. Accordingly, I understand that my obligation is to meet all the requirements of the ADWG, which includes DBPs. This means I should always keep DBPs below the maximum specified in the ADWG and minimise DBP concentrations where practical.

3.3 UTILITY-WIDE RISK ASSESSMENT

Most utilities are managing a multitude of risks. We have developed a "whole-of-utility" risk assessment model in order to prioritise the allocation of scarce resources. Typically this model plots the risk on a consequence vs. likelihood matrix (Figure 3.1). At the utility level, consequence is not just the public health outcomes, but also the financial, reputational, political and social consequences of an adverse event.

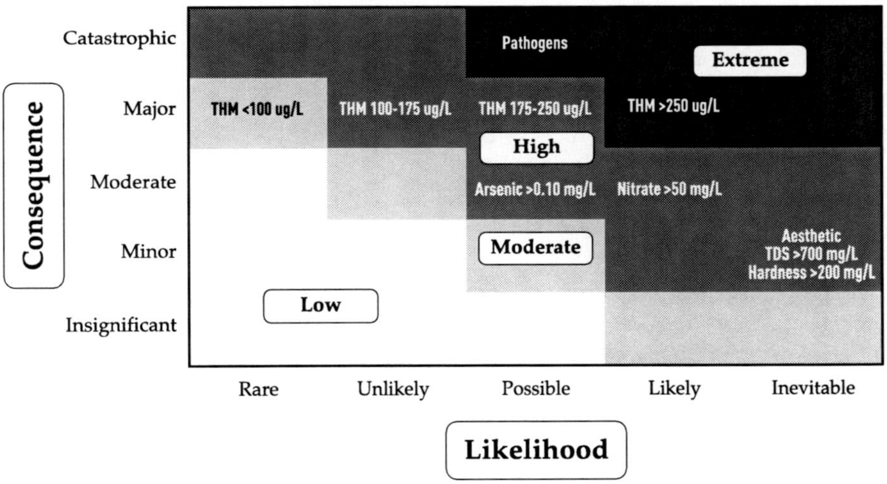

Figure 3.1 Typical water quality risk ratings

Our utility has had several occasions where DBP issues have been covered in the media (Guest, 2008). Our experience is that when cancer and birth defects are raised as concerns we can understandably expect a strong reaction. We see the biggest risk from DBPs is that they become politicised and the utility is directed to install treatment to reduce DBPs whether or not there is a rational case for such investments. The consequence of such an event would be "major" because treatment investments can be extremely expensive. We have a responsibility to all of our customers to insure that they are not asked to pay additional treatment costs that may offer them no public health benefit.

The likelihood of DBPs becoming politicised is dependent on the actual DBP concentrations in our systems. If the guideline for trihalomethanes (THMs) is exceeded (0.25 mg/L) it is "inevitable" DBPs will become politicised. If THM concentrations are maintained at less than 0.100 mg/L we think it would be "rare" for the issue to attract media and/or political attention. Between these "bookends" the likelihood is assessed as "unlikely" if THMs are between 0.1–0.175 mg/L and "possible" if between 0.175–0.250 mg/L. Each utility needs to undertake its own risk assessment based on the local conditions and environment. This will influence the priority given to DBP operational management and investment in long-term interventions.

3.4 THE CHALLENGE OF DBP MANAGEMENT

This section will mainly deal with THMs as this is the most common indicator of DBPs in our systems. The THMs are difficult to manage because they are one of the few chemicals where the concentration changes within the distribution system. Utilities know that the concentration of THMs is influenced by (Singer, 1999):

- Total organics in the source water
- Character of organics (humic, wastewater, algal) in the source water
- Treatment plant performance (where treatment exists)
- Chlorine dose
- Time since chlorination (water age)
- Water temperature
- pH of water

The problem is that operators are often only able to exercise limited control over many of these factors. For example, unlike pathogen management, source protection generally cannot change the organic loading in the source water. Often water temperature and water age are functions of climate, asset configuration and water demand characteristics – matters that an operator usually inherits and cannot readily change.

3.4.1 Know your system

The fundamental rule of water quality management is "know your system" and this also applies to DBP management because there are so many variables. Unless you understand the significance of each factor and can monitor how they are changing, you will always be playing "catch up" after you experience high THMs in the distribution system. Consider, how do you "get to know your system".

3.4.2 Source water

The concentration of organics and their character in source water correlates to the THM formation potential of the system (i.e. if you continue to apply chlorine then THMs will continue to form until all the reactive organics are consumed). We have found the best way to characterise, initially, the source water organics, is to sample the raw water and undertake laboratory determinations of dissolved organic carbon (DOC) and UV absorbance (at 254 nm). Regular monitoring of UV absorbance (either *in situ* or by laboratory) then informs us on changes in source water quality. Some sources can be highly variable. Figure 3.2 illustrates the seasonal variability of UV absorbance for a large surface water source in the south west of Western Australia. The peaks in UV absorbance correspond to peaks in DOC and therefore THM formation potential. We believe the DOC variability is influenced by the quantum and pattern of rainfall together with the volume of water in the dam at the start of winter.

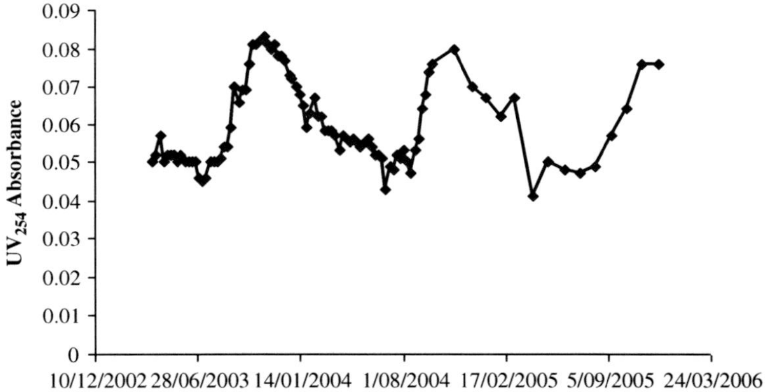

Figure 3.2 Temporal variability of UV_{254} absorbance of a source water in the south west of Western Australia

3.4.3 Chlorination

Source water is dosed with chlorine to provide a disinfection barrier. Typically, it is necessary to dose only modestly to achieve effective disinfection (e.g. Ct 15–30 mg/L-min) (NHMRC/NRMMC, 2011; Appendix A1.8). Usually utilities are dosing chlorine at the source to achieve a chlorine residual throughout the reticulation system. This is where we run into trouble with THMs. As mentioned before, THMs will be at their maximum at the end of the reticulation system where the chlorine has had longer to react with the organics in the source water. To understand how these formation processes play out, we have undertaken 7-day chlorine decay and THM formation potential (THMFP) tests on all of our source waters. We chose seven days because this is a typical maximum water-age in a distribution system. However, for some regional schemes we have extended the period to 30 days because that reflects their reality. Figure 3.3 is an example illustrating THMFP test results, which can be generated in laboratory experiments. These tests provide some useful information on scheme performance.

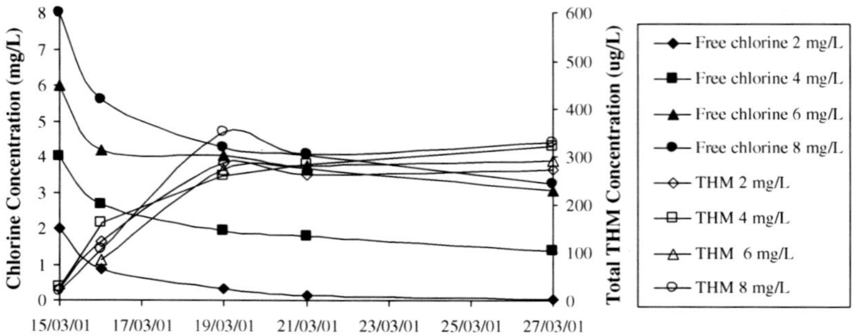

Figure 3.3 THM formation potential concentrations in relation to chlorine dose and contact time

Note the chlorine decay profile is similar regardless of the chlorine dose rate. It is possible to calculate the dose rate at the chlorinator to provide the required chlorine residual at the end of the system, if we can estimate water age. Alternatively, by measuring the chlorine residual at the start and end of the system it is possible to estimate water age using the chlorine decay curves. We have found a good correlation between laboratory and field results and that it is unnecessary to introduce a "pipe factor" to allow for additional chlorine decay as a result of biofilm in the pipe network. The THMs are formed over four days following chlorination and then "flat line" after all the reactive organics are consumed. Within experimental error, the maximum THM concentrations formed are independent of the chlorine dose rate.

Figure 3.3 is worth closer examination, particularly the part where THMs are being formed.

We have found there is a straight-line relationship between THMs formed and chlorine consumed (i.e. chlorine dosed minus chlorine residual). This is true for all sources, with the slope of the line changing for different sources. Combining the information in Figures 3.3 and 3.4 allows us to conclude:

(1) Maintaining high chlorine residuals does not produce higher THMs compared to maintaining lower residuals (i.e. THMs in the reticulation will be the same whether the chlorine residual is 0.5 mg/L or 1.5 mg/L because it is chlorine used (not chlorine dosed), which determines THM concentration).

(2) The only way to reduce THM concentration (by adjusting chlorination) is to allow chlorine residuals to drop to zero (i.e. stop the THM formation reaction). However, this choice is generally not acceptable because it allows other contamination risks such as backflow, ingress and in our case, *Naegleria*, to arise.

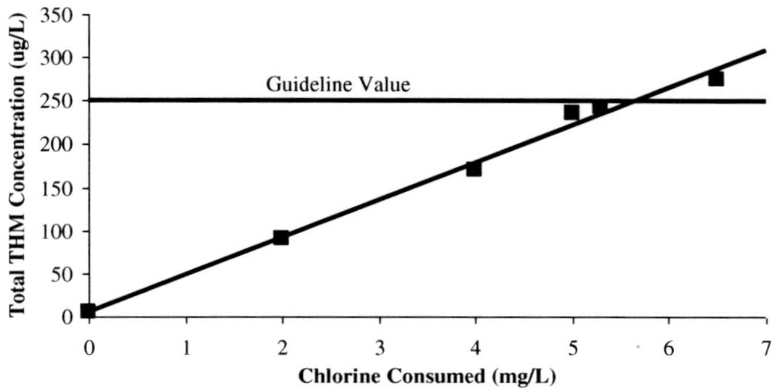

Figure 3.4 THM formation as a function of chlorine consumed

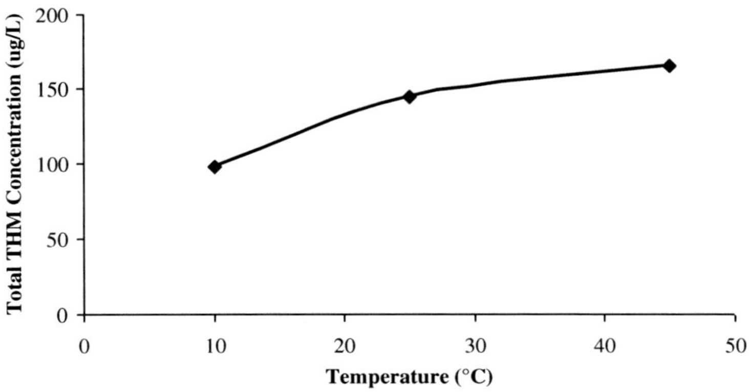

Figure 3.5 The influence of temperature on THM formation

3.4.4 Temperature and pH effects

Laboratory experiments can also determine the effect of pH and temperature on THM formation. In south western Australia surface source waters can be mildly aggressive and we have observed the pH can rise from about 7 to 8 as water reacts with the cement lining in the extensive pipe network. This increase in pH may result in the THM concentration increasing by about 10%.

The water from this source is typically about 15°C in winter. However, as it travels through sections of above ground main in summer considerable thermal heating occurs, and temperatures may rise to 40°C. The effect on THM formation is marked and may cause a 40% elevation in THM concentration (Figure 3.5).

3.4.5 Management opportunities

The knowledge, described above, about our systems can be used to better manage THMs.

Case 1. This is as easy as it gets! No matter how much chlorine I dose into this water I cannot exceed the target THM concentration. The only thing that can cause grief is if the organic load in the source water increases. So I just need to check UV absorbance data regularly and make sure organic loading does not start to increase. The THM formation characteristic of this scheme is shown in Figure 3.6.

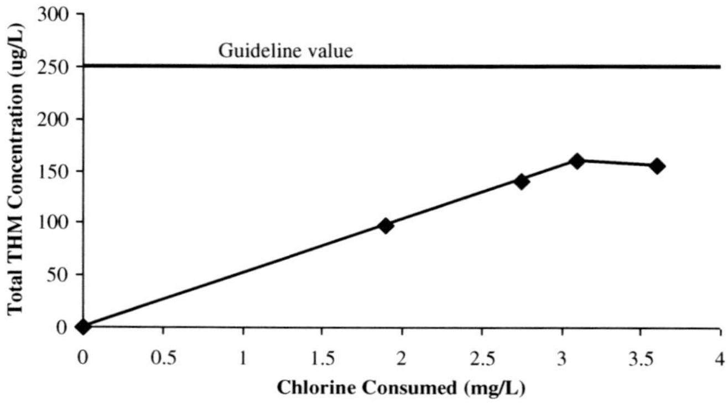

Figure 3.6 Peak THM concentration as a function of free chlorine consumed, guideline value not exceeded

Case 2. This is not so easy! If I dose too much chlorine I will exceed my target. The THM formation characteristic of this scheme is shown in Figure 3.7. I can, however, set a band in which I can dose enough chlorine to achieve effective disinfection and provide a chlorine residual in the reticulation and not exceed my THM target. Figure 3.8 illustrates how this can be achieved. This mode of operation will be upset if there is an increase in source organic load. Therefore, as in Case 1, I need to regularly monitor UV absorbance in the raw water and the chlorine dose rate.

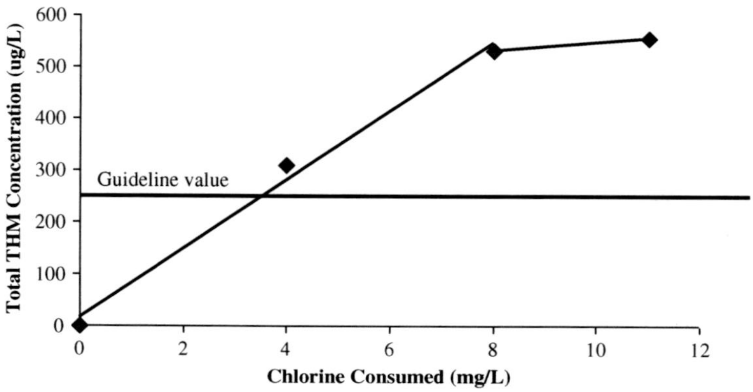

Figure 3.7 Peak THM concentration as a function of free chlorine consumed, guideline value exceeded

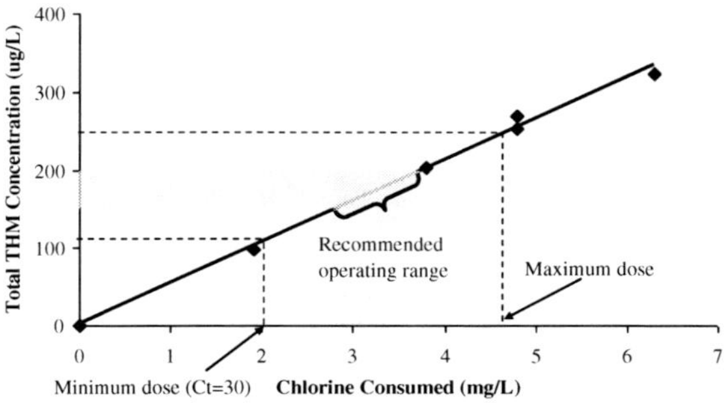

Figure 3.8 Setting an operation band to limit THM formation

3.4.6 Management interventions

The previous section explained how desired end points for chlorine residual and THMs could be achieved by improved management of chlorination based on knowledge of chlorine decay and THM formation. However, in some schemes this knowledge will indicate it is not possible to maintain disinfection and minimum chlorine residuals without exceeding the THM target. In this case, additional interventions are required. Some of the possible options are discussed below.

3.4.6.1 Water age

One way to reduce THMs is to reduce the time the water is held in the water supply system (i.e. the time for the THM formation to occur). The THM concentrations will continue to increase until either:

- The chlorine is entirely consumed
- The reactive organics are consumed
- The water is consumed, by customers

In Australia our systems are designed for very hot days and high demands. In winter the demand may be three or four times lower than summer. Excessive water age in winter can be a problem for THM management and a simple measure to alleviate the problem is to run tanks and reservoirs half full or lower.

Another problem occurs with poorly designed systems where tanks do not have a separate outlet. Short-circuiting can occur and by-passed water may stagnate in a tank. In such cases, THMs will continue to form. These tanks may be dosed regularly with liquid chlorine or tablets to maintain a chlorine residual, allowing THM formation to continue. Then one day, there will be a power failure or similar incident and the water from the tank will be discharged to customers and contain very high THM concentration. If a THM sample is collected at this time, the result will appear anomalous. This problem results from a deficiency in the system design. Each tank should have a separate inlet and outlet pipe designed to avoid short-circuiting to ensure control can be exercised over water age.

3.4.6.2 pH control

Stabilising pH is not a trivial exercise; usually involving dosage of chemicals such as lime and carbon dioxide. This can be operationally difficult and relatively expensive. Given the modest impact on THM formation pH control should be lower down the list in terms of value for money interventions unless corrosion problems dictate otherwise.

3.4.6.3 Water temperature

Temperature can be an important factor and may explain "unexpected" results in the reticulation. However, temperature is often difficult to address for existing assets. We have had some success with putting shades on above ground pipes but not on a large scale. The real lesson is to take temperature into consideration when planning new assets and replacement of old assets (i.e. avoid above ground pipes in hot climates and consider cooling deep ground water before introduction into the system).

3.4.6.4 Aeration

During our initial laboratory experiments into chlorine decay and THM formation we noticed that aeration was effective in reducing the THM concentration in treated water. For example, passing water from one beaker to another ten times reduced THM concentration by 40%. The challenge was to convert this potential into field reality. Immediate success was found by installing "splash plates" on the inlet to small tanks (say less than 250 kL). However, a different technique was required for larger tanks and reservoirs. A submersible pump was placed in the bottom of the reservoir to pump continuously water into a 100 mm section of PVC pipe. Slots were cut in the PVC pipe with an angle grinder. The system cost less than $5000 and produced outstanding results as shown in Figure 3.9.

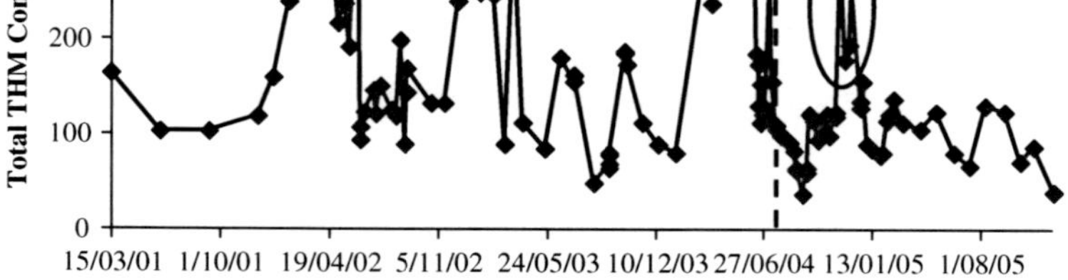

Figure 3.9 Successful THMs reductions achieved with recirculation aeration in a service reservoir

Since then, we have refined the system for tanks using overhead nozzles (Trolio *et al.* 2007). The results of this set up in a 4.5 ML tank are shown in Figure 3.10.

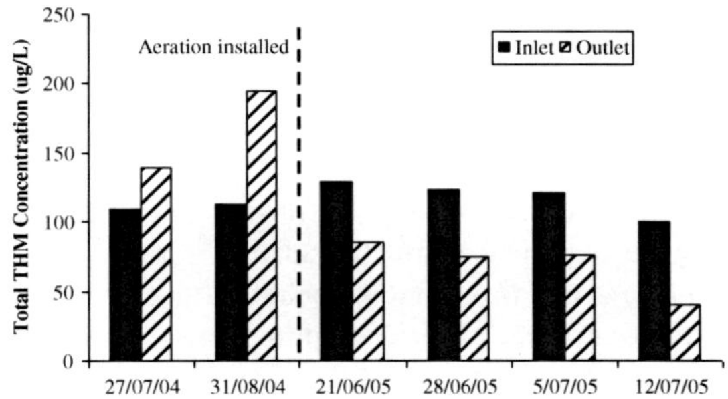

Figure 3.10 THM reduction with recirculating aeration in a service tank

3.4.6.5 Limitations of aeration

Aeration has been very effective at reducing THM concentrations at many locations. If aeration had not been installed it is likely THM concentrations would have regularly exceeded the maximum set in the ADWG or operators would have "tinkered" with chlorination to avoid exceedences. In this respect introduction of aeration has been highly beneficial. However, the limitations of aeration need to be recognised. Firstly, it is only effective for volatile DBPs. We have found the four species of THMs that we measure (chloroform, bromodichloromethane, dibromochloromethane, bromoform) were all reduced by a similar percentage by aeration. Aeration will not reduce haloacetic acids or chloral hydrate concentrations.

Aeration is regarded as an effective short-term measure to reduce the risk of exceeding volatile DBP guidelines prior to the longer-term solution, which is to choose another source (often not possible) or introduce treatment to remove DBP precursors.

3.5 TREATMENT FOR ORGANICS REMOVAL

This chapter will not attempt to address the complicated science of water treatment plant design and operation to remove DBP precursors. Suffice to say that our experience is that different source waters contain not only different total organic loadings but also different organic "fractions" (i.e. distribution of molecular weights of natural organic matter). This means a treatment process that is effective for one source may be ineffective for another. Water treatment is an expensive exercise and "pilot plants" are essential to prove treatment process effectiveness. The laboratory testing outlined earlier to determine chlorine decay and THM formation characteristics on raw waters needs to be repeated on the treated water from the pilot treatment plants. Testing needs to demonstrate that sufficient chlorine can be applied to the treated water to achieve effective disinfection, maintain sufficient chlorine residual throughout the reticulation system and maintain DBPs below targets for all conditions (i.e. high temperatures, low demand etc.). Some case studies for use of water treatment to remove organics follow.

3.5.1 Treatment case study 1

This study refers to a surface water source in the south of Western Australia; a highly coloured source. The treatment plant was struggling to remove enough colour as the water demand increased because of population growth. The existing treatment plant was replaced and during the design of the new treatment process THM exceptions were experienced in the scheme. Our understanding of THM precursors was limited, at that time, and we believed that treating the colour (using enhanced coagulation) to meet ADWG requirements would also deal with the THM issue. Unfortunately, there was still a THM problem after the treatment process was commissioned. A review of treatment effectiveness occurred with the results shown in Table 3.1.

Table 3.1 Treatment effectiveness.

Water quality parameter	Raw	Enhanced coagulation alum	Alum & PAC
Colour (TCU)	375	5	1
UV Abs_{254} (cm^{-1})	1.35	0.13	0.02
DOC (mg/L)	24.6	7.0	2.3
SUVA (L/mg m)	5.5	1.8	N/A
THMFP (μg/L)	810	382	173
Cl_2 demand (mg/L)	extreme	7.5	3.2

While the enhanced coagulation process effectively removed colour, sufficient organics remained after treatment to yield a high chlorine demand and THM formation. Aeration was used as a short-term THM mitigation strategy and powdered activated carbon (PAC) subsequently was added to the treatment train. PAC was highly effective but comes at great cost and is not an easy process to operate. We use UV absorbance on the outlet of the WTP to confirm organics removal is within specification to manage THM formation.

3.5.2 Treatment case study 2

Another surface water source in the south west of Western Australia presented another situation where enhanced coagulation was not achieving sufficient organics reduction to meet the THM target. In this case, granular activated carbon (GAC) was added to the treatment process. The history of THMs in the reticulation before and after GAC introduction is shown in Figure 3.11. There was a rapid reduction in THMs initially when the carbon was operating in a simple carbon adsorption mode. As the activated carbon transformed to biological mode performance deteriorated. Gradually, operators have been able to optimise the WTP so that its performance is more consistent and acceptable, but this has required intensive and ongoing input from expert water treatment engineers. We use UV absorbance on the outlet of the water treatment plant to confirm acceptable removal of organics.

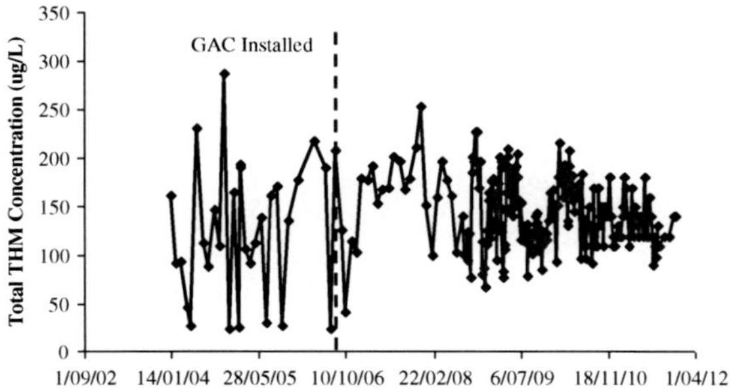

Figure 3.11 Treatment effectiveness and performance of granular activated carbon associated with THM removal

3.6 SPATIAL AND TEMPORAL VARIATIONS

THMs are sampled in the reticulation to meet the requirements of the ADWG. Sampling usually occurs well into the distribution system and at frequencies varying between 4 to 12 times per year. This sampling is part of Element 5 of the Framework for Management of Drinking Water Quality (i.e. verification) (NHMRC/NRMMC, 2004, 2011). We are verifying the quality of water supplied to our customers and therefore also verifying that the upstream processes and barriers are operating effectively.

We need to give this concept a bit of thought when applying it to THMs. As THMs will tend to increase with water age we need to remember that customers upstream of the sample point will probably experience lower THMs (i.e. most of the zone) but some downstream may see higher THMs. Also, because THM formation is affected by so many factors, variability is a natural consequence. In reality, the THM sampling acts as a coarse form of "process control". Generally it receives little attention unless guideline values are exceeded. To understand these exceedences requires an understanding of variability of THM formation in the distribution system. Another reason for acquiring this understanding is to allow the verification monitoring to be used in health studies (Chapter 11).

Epidemiology researchers may take the sampling results and ascribe them uniformly to all customers in the supply zone and, in so doing, fail to appreciate the error band associated with each sample result and the spatial, temporal and seasonal variability in any supply zone. We investigated these effects in several of our supply zones to address this concern (Hazell *et al.* 2009). The results for one investigation are presented in Figure 3.12 for interest.

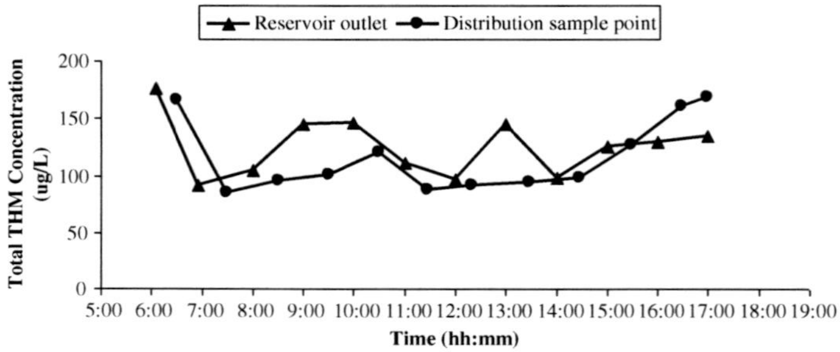

Figure 3.12 Spatial and temporal variations in THM concentration in a large reservoir zone

Because the water for this zone is held in a large service reservoir for several days we expected most of the THM formation to occur prior to entering the distribution and therefore that the THMs at the reservoir outlet would be similar to the THMs at the normal sample point near the end of the distribution system. This in fact was the case but interestingly, the THM values at the reservoir outlet were higher than those at the end of the distribution system for much of the time. Of course, it is "different" water being sampled – the water at the end of the distribution system probably left the reservoir outlet a week earlier. But what we did not expect were the significant variations over the 12-hour period samples were taken. The THM results at the reservoir outlet varied from 91 to 176 μg/L while in the reticulation the variation was 85 to 168 μg/L. We need to do more work to truly "know this system".

3.7 WATER SAFETY PLANS – BRINGING IT TOGETHER

Water safety plans are a tool we use to manage all water quality risks. In this regard, DBPs are just another "hazard". Using this approach we set a "specification" for operators from catchment to consumer. This specification will be a combination of the level of challenge on the scheme and the performance of the barriers that are in place to meet the challenge. In this case the challenge is the organic loading of the raw water. The barrier performance will include chlorine dose, chlorine residual at the chlorinator and in the reticulation. Where treatment is in place it will also cover water treatment plant performance. The targets for these set points are derived from the work outlined earlier to "know your system". Water temperature and pH may also be monitored in the reticulation if they are critical to keeping water supplied to customers "in spec".

A deviation from the specification for any parameter is a warning to the operator that something has changed and an investigation is required with possible intervention necessary. The elements of a water safety plan concept are illustrated below (Figure 3.13, Table 3.2). Only parameters relevant to DBPs have been included for clarity.

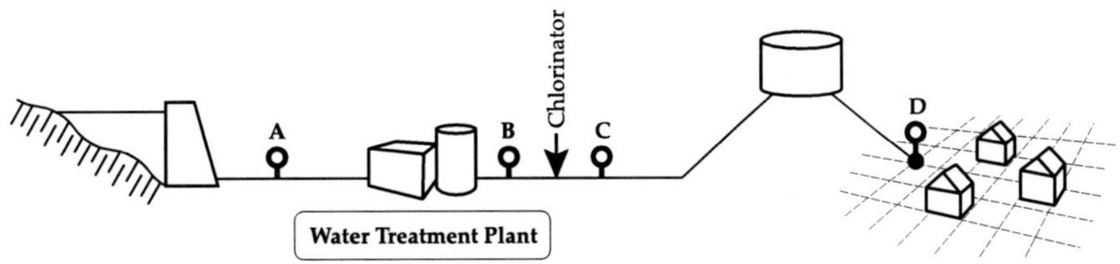

Figure 3.13 Drinking water system schematic illustrating various elements of a water safety plan

Table 3.2 Guidance for water safety plan elements in whole water system.

Sampling location	Measure	Frequency	Target	Alert	Corrective action
A Raw water	UV_{abs}	Monthly	<.07	.07	Ensure treatment process optimised. Consider short term options to increase organic removal (e.g. increase Alum dose)
B Treated water	UV_{abs}	Continuous/ weekly	<.02	>.02	Rework THM vs. chlorine used relationship to determine if exceptions likely
Chlorinator	Chlorine dose	Weekly	<2 mg/L	>2 mg/L	Check source and treated water UV_{abs}. Check chlorinator. Rework chlorine demand and THM formation potential in the lab
C Post chlorinator	Cl residual	Continuous	1.5 mg/L	1.2–2 mg/L	Check chlorinator for malfunction
D Reticulation	Cl residual	Daily	0.5 mg/L	<0.2 mg/L >1 mg/L	Check chlorinator Readjust dose rate if necessary
	pH	Weekly	7.5	>8	Assess impact on THM formation
	Temperature	Weekly	25°C	>30°C	Check chlorine decay across system

3.8 CONCLUSIONS

DBPs are not easy to manage because there are many factors that can influence their formation in a water supply system. Utilities can undertake laboratory studies to determine the relationship between chlorination and DBP formation as well as chlorine decay characteristics and the effect of other factors such as pH and temperature. This information can be used in a water safety plan to set a specification from catchment to consumer, which will provide confidence that disinfection, chlorine residual and DBP end points are satisfactory. Deviation away from specifications provide an early warning there is a change in the system status, which may result in unsatisfactory outcomes.

Interventions may be required to achieve satisfactory water quality. These could be short-term seasonal operational changes such as lowering operating levels in tanks to reduce water age. Alternatively it may be possible to reduce some DBPs after they have formed (but before delivery to customers) by introducing aeration into tanks and reservoirs. In some cases it may be necessary to provide treatment specifically to remove DBP precursors. In this case it is necessary to understand the nature of the organics in the source water and ensure the chosen treatment process will provide effective removal. The treatment process needs to be "piloted" and tests undertaken to provide confidence that chlorine decay and THM formation characteristics are satisfactory, i.e. for the system in question enough chlorine can be dosed under all conditions to provide adequate chlorine residual at the end of the reticulation while not breaching DBP targets.

We have been applying these principles for the last eight years. The Corporation supplies the city of Perth and 220 regional and remote supplies. Approximately 1000 THM samples are undertaken each year. Figure 3.14 plots the number of samples each year that have exceeded the recommended maximum concentration in the ADWG.

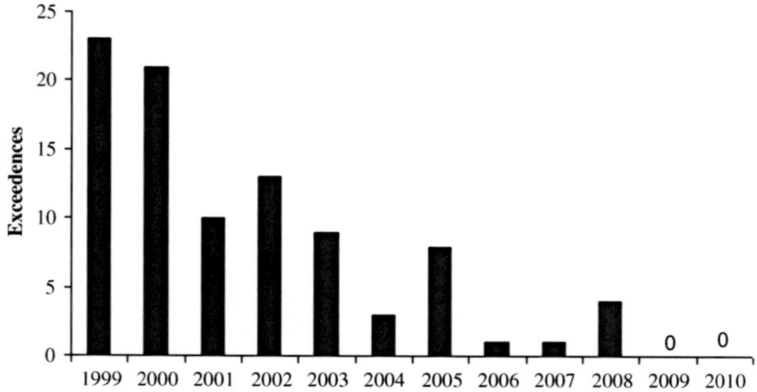

Figure 3.14 Decline in THMs exceedences based on the Australian Drinking Water Guidelines

Our objective is that operators should only have to worry about disinfection performance and not be "diverted" into THM management. We are not quite there but consider that THMs are now "under control".

In summary:

- Utilities should endeavour to minimise DBPs in water delivered to customers, but never at the expense of effective disinfection and system management to minimise microbial risk for customers.
- DBPs can be considered as another water quality "hazard" and are amenable to management by the water safety plan approach.
- Effective management requires good knowledge of chlorine decay and the chlorine used versus DBP formation relationship. The effect of temperature and pH on DBP formation should also be established.
- This knowledge can be deployed via the water safety planning process to establish an "operating window" for the water supply system to achieve acceptable disinfection and DBP end points. Appropriate operational monitoring of source water quality, treatment performance, chlorine dose, pH, temperature and chlorine residuals in the distribution system can give an early warning of changes in the system which signal possible trouble and allow timely interventions.
- Interventions can be operational, such as adjustment of tank levels to reduce water age or small capital works such as aeration in tanks or changes to inlet/outlet pipes. Sometimes the knowledge of system characteristics confirms longer-term works such as water treatment to reduce organics is justified. In this case it is necessary to confirm the proposed treatment process will remove sufficient of the appropriate organic fractions to achieve the required DBP outcomes under all operating conditions.

REFERENCES

Guest D. (2008). Tap water 'increases risk of birth defects'. *The West Australian*, 22 April, 1.

Hazell A., Trolio R., Walker R. and McNeil S. (2009). Temporal and spatial variations in THM and HAA concentrations in Perth metropolitan water supplies. In: Proc of OzWater 2009. Melbourne, Australia, 16–18 March.

Hrudey S. E., Conant B., Douglas I. P., Fawell J., Gillespie T., Hill D., Leiss W., Rose J. B. and Sinclair M. (2011). Managing uncertainty in the provision of safe drinking water. *Water Science & Technology: Water Supply*, **11**, 675–681.

Hrudey S. E. and Walker R. (2005). Walkerton – 5 years later. Tragedy could have been prevented. *Opflow*, **31**(1), 4–7.

NHMRC/NRMMC (2004). Australian Drinking Water Guidelines. http://www.nhmrc.gov.au/guidelines/publications/eh34 (accessed 07 February 2012).

NHMRC/NRMMC (2011). Australian Drinking Water Guidelines. http://www.nhmrc.gov.au/guidelines/publications/eh52 (accessed 07 February 2012).

P. C. Singer (ed.) (1999). Formation and Control of Disinfection By-Products in Drinking Water. American Water Works Association, Denver, CO.

Trolio R., Walker R., McNeil S., Huynh T. and Jismi N. (2007). Trihalomethane management in drinking water: a novel approach. *Water (Journal of the Australian Water Association)* **34**, 77–81.

WHO (2007). *Chemical Safety of Drinking-water: Assessing Priorities for Risk Management*. World Health Organization, Geneva. www.who.int/water_sanitation_health/dwq/dwchem_safety/en/index.html (accessed 07 February 2012).

Chapter 4

Halogenated DBPs and emerging issues

S. W. Krasner

4.1 INTRODUCTION

When drinking water is disinfected, a wide range of disinfection by-products (DBPs) of health and regulatory concern can form (Richardson *et al.* 2007). However, there are standards for only a limited number of DBPs (e.g. trihalomethanes [THMs], haloacetic acids [HAAs], bromate, and chlorite in the United States [U.S.]) (U.S. Environmental Protection Agency [USEPA] 1998). In the U.S. DBP Rule, the control of THMs, HAAs, and total organic carbon (TOC) was envisioned to also enable the control of other halogenated DBPs. In the U.S. and many developed countries, alternative disinfectants to the use of free chlorine (i.e. ozone, chlorine dioxide, chloramines, ultraviolet [UV] irradiation) have been used as the primary and/or secondary disinfectant to minimize the formation of THMs and HAAs. However, it has been shown that some emerging DBPs may increase in concentration when certain alternative disinfectants are used to control the formation of THMs and HAAs (Krasner *et al.* 2006). The latter research focused on the study of high-priority DBPs (e.g. iodine-containing THMs, haloacetonitriles [HANs], haloketones, halonitromethanes [HNMs], haloacetaldehydes, haloacetamines, halogenated furanones), as based on an expert determination of DBPs likely to exhibit greater toxicological concern (Woo *et al.* 2002). Thus, there has been a concern over possible risk/risk tradeoffs associated with the use of alternative disinfectants.

Recent research has shown that certain emerging DBPs may be of higher cytotoxicity and genotoxicity than that of the regulated THMs or HAAs (Plewa and Wagner, 2009). In the latter research, certain classes of nitrogenous (N) DBPs (i.e. HANs, HNMs, haloacetamides) and one class of carbonaceous (C) DBPs (i.e. haloacetaldehydes) were found to be comparatively high in toxicity.

Moreover, iodine-containing DBPs were found to be more toxic than bromine-containing species, with chlorine-containing DBPs least toxic. However, regulations for THMs and HAAs do not include iodine-containing species. In addition, many studies have examined the occurrence and formation of chloral hydrate (trichloroacetaldehyde) (McGuire *et al.* 2002), but have not examined dihalogenated and/or bromine-containing analogues that appear to be more toxic. Of course, in any consideration of comparative toxicity with other DBPs, comparative concentration must also be considered. This chapter will discuss the occurrence and formation of emerging C-DBPs, whereas Chapter 5 will focus on N-DBPs and Chapter 6 will present findings on novel new DBPs. Moreover, this chapter will present information on various ways to cost-effectively balance and minimize the formation of regulated and emerging DBPs.

4.1.1 Overview of formation of DBPs

The formation of DBPs involves the reaction of the disinfectant (i.e. chlorine, chloramines, chlorine dioxide, ozone) with the organic (i.e. natural organic matter [NOM]) and inorganic (i.e. bromide) precursors (Krasner, 1999). NOM includes decaying vegetation or humic substances (Reckhow *et al.* 1990), whereas bromide comes from sources such as: saltwater intrusion, connate water (geologically trapped ancient seawater), and oil-field brines (Amy *et al.* 1994). This reaction is impacted by pH, temperature, and time (Stevens *et al.* 1989). The formation of emerging DBPs follows a similar formula with some differences. In addition to NOM, sources of DBP precursors attracting more recent attention include algal organic matter (AOM) and effluent organic matter (EfOM) from wastewater treatment plant discharges (Dotson *et al.* 2009). In addition to bromide, other inorganic DBP precursors include iodide (Moran *et al.* 2002) and nitrite. An emerging disinfectant is UV irradiation, including medium-pressure systems, which can also impact DBP formation (Reckhow *et al.* 2010).

4.2 FORMATION OF IODINE-CONTAINING DBPS

4.2.1 Chemistry of formation

Ozone, chlorine or chloramines can oxidize iodide to hypoiodous acid (HOI). Ozone or chlorine can further oxidize HOI to iodate, which becomes a sink for the iodide, whereas chloramines do not (Bichsel and von Gunten, 2000). Chlorine dioxide in the dark can oxidize iodide to iodine, whereas it will not further oxidize it to iodate (Fabian and Gordon, 1997). HOI or iodine can react with NOM to form iodine-containing DBPs.

4.2.2 Formation of TOI

Total organic halogen (TOX) is a surrogate for the sum of the halogenated DBPs formed. It can be segmented into total organic chlorine (TOCl), bromine (TOBr), and iodine (TOI) (Hua and Reckhow, 2006a). Hua and Reckhow (2006b) studied the relative formation of TOCl and TOI for an iodide-spiked water using different disinfectants or combinations. As expected, chlorine produced a substantial amount of TOCl, whereas very little TOCl was produced with chloramines or chlorine dioxide. Alternatively, chlorine produced very little TOI, whereas TOI was produced by chloramines or chlorine dioxide. Moveover, pre-ozonation resulted in a substantial reduction in TOI formation during post-chloramination. These results are consistent with the chemistry.

4.2.3 Formation of iodine-containing THMs

Krasner and colleagues (2006) found the most iodine-containing THMs at a plant that treated a water source high in bromide (iodide was not measured) that used chloramines only or, on one occasion, used chlorine dioxide as a pre-oxidant. Plants that used chlorine only or ozone and chlorine typically produced some of the lowest levels of iodine-containing THMs. Plants with pre-chlorination and post-chloramination typically had intermediate levels of iodine-containing THMs. Table 4.1 summarizes the overall U.S. occurrence.

Table 4.1 Occurrence of iodine-containing THMs* in 12 U.S. plant effluents, where 10 plants used chloramines and most plants treated water with moderate-to-high levels of bromide (data from Weinberg *et al.* 2002).

Statistic	Concentration (μg/L)					
	DCIM	BCIM	DBIM	CDIM	BDIM	TIM
25th Percentile	ND†	ND	ND	ND	ND	ND
Median	0.3	ND	ND	ND	ND	ND
75th percentile	1.0	0.5	ND	ND	ND	ND
Maximum	11	3.0	4.0	2.0	0.7	2.0

*D = Di, T = tri, B = bromo, C = chloro, I = iodo, M = methane.
†ND = Not detected.

Karanfil and colleagues (2011) found that pre-chlorination before post-chloramination did not always result in lower formation of iodine-containing THMs and, in some cases (e.g. a high-humic water), resulted in more formation. These researchers speculated that the chlorine was resulting in the "oxidation and break down" of the NOM, forming more hydrophilic and/or lower molecular weight NOM components that reacted with chloramines to form iodine-containing THMs. Alternatively, the humic NOM may have been consuming chlorine, diverting chlorine from further oxidation of the HOI to iodate, where the HOI can form iodinated DBPs.

4.2.4 Formation of iodine-containing acids

Weinberg and colleagues (2011) studied the formation of iodinated acids at six U.S. plants that used chloramines and treated water with moderate-to-high levels of iodide (total iodine = 15–230 μg/L). (In a survey of major U.S., Canadian, and European rivers, iodine was 0.5–212 μg/L [median = 10 μg/L] [Moran *et al.* 2002].) Although only ng/L levels of iodoacids were formed (Table 4.2), Plewa and colleagues (2004) found that iodoacetic acid was 2 and 47 times more genotoxic than monobromo- or monochloroacetic acid (two HAAs currently regulated in the U.S. and Canada), respectively, as measured by the Chinese hamster ovary cell assay.

Table 4.2 Occurrence of iodoacids in six U.S. distribution systems that used chloramines and treated water with moderate-to-high levels of iodine (adapted from Weinberg *et al.* 2011).

Statistic	Concentration (ng/L)			
	Iodoacetic acid	Chloroiodo-acetic Acid	Bromoiodo-acetic Acid	Diiodoacetic acid
Minimum	40	8.9	32	10
Median	116	53	73	21
Maximum	376	275	168	48

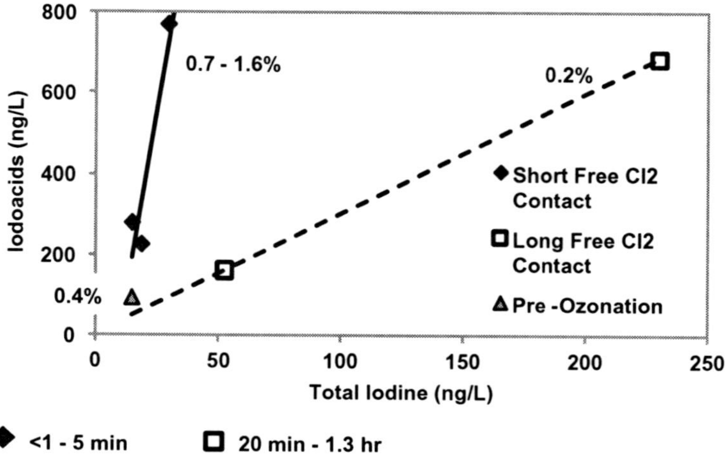

Figure 4.1 Impact of pre-oxidation on iodoacid formation in distribution systems of six U.S. plants that used chloramines (molar conversion of iodine to iodoacids, expressed as %) (data from Weinberg *et al.* 2011)

In the Weinberg *et al.* (2011) survey (Figure 4.1), plants that used relatively short free chlorine contact times (i.e. <1 to 5 min) had a higher molar conversion of iodine to iodoacids (0.7–1.6%) than plants that used longer free chlorine contact times (i.e. 20 min–1.3 hr; 0.2%). This is noteworthy for overall water treatment operations because a longer free chlorine contact time generally provides greater pathogen disinfection. The downside of longer free chlorine contact time is generally greater conventional THM formation, but a greater understanding of other DBPs formed helps to keep these issues in perspective. In addition, a plant that used pre-ozonation also had a relatively low molar conversion (0.4%).

4.3 FORMATION OF EMERGING CHLORINE- AND BROMINE-CONTAINING DBPS

4.3.1 Formation of TOCl and TOBr

In a survey of U.S. plants, Krasner and colleagues (2006) found that the percentage of TOCl in plant effluents accounted for by the individually measured halogenated DBPs ranged from 12 to 33 percent, with a median value of 24 percent. The percentage of TOBr accounted for by the halogenated DBPs ranged from 6 to 58 percent, with a median value of 39 percent. In bench-scale testing (Zhang *et al.* 2000), the percentage of TOX, TOCl or TOBr that was accounted for in chlorinated waters was higher than in samples treated with alternative disinfectants. The percentages accounted for in the full-scale survey (Krasner *et al.* 2006) compared favourably with the range of values observed in the bench-scale study (Zhang *et al.* 2000). Research continues to identify the unaccounted for TOX (e.g. Chapter 8).

4.3.2 Haloacetaldehydes

Trehy and colleagues (1986) found that chlorination of an amino acid could form an aldehyde and a nitrile, where additional chlorination could form a haloacetaldehyde and a HAN, respectively. Freuze *et al.* (2005) and Froese *et al.* (1999) also showed that nuisance, odorous aldehydes and chlorinated aldimines could be formed. Moreover, McKnight and Reckhow (1992) found that acetaldehyde, formed from the ozonation of NOM, could react with chlorine to form chloroacetaldehyde, which in the presence of excess chlorine could form the trihalogenated species (chloral hydrate). In other research, Krasner and colleagues (2006) found that dichloroacetaldehyde was formed in a water that used chloramines (no free chlorine contact time) and ozone.

Figure 4.2 shows the occurrence of di- and trihalogenated acetaldehydes in U.S. plant effluents. This survey focused on wastewater and algal-impacted waters, as such sources would tend to have more amino acids in their water supplies (Dotson and Westerhoff, 2009). In addition to chloral hydrate, di- and bromine-containing analogues of chloral hydrate were detected, where some of these other species are of higher toxicity than chloral hydrate (Plewa and Wagner, 2009).

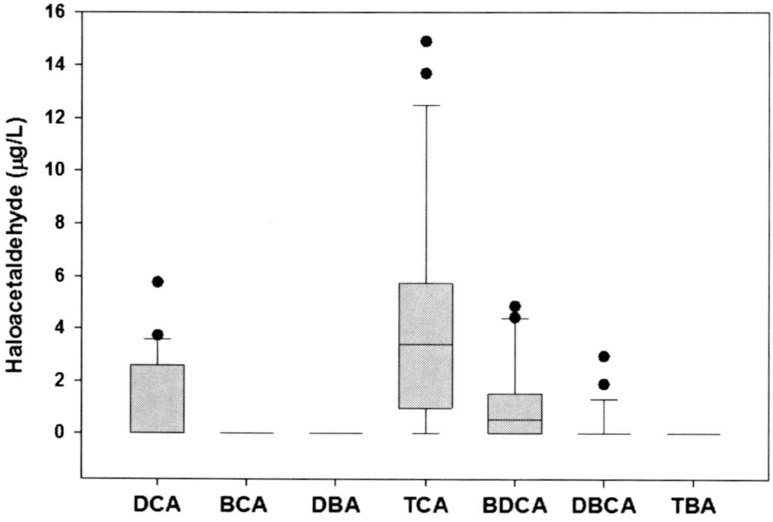

Figure 4.2 Occurrence of haloacetaldehydes in 11 U.S. plant effluents (wastewater and algal impacted waters; A = acetaldehyde; bottom and top of box = 25th and 75th percentiles, respectively, line inside box = median, bottom and top of whiskers = 10th and 90th percentiles, respectively, and points above whisker = outliers) (data from Mitch *et al.* 2009)

When there is a shift in speciation to more bromine incorporation in one class of DBPs, the same should be observed for other classes of DBPs (Krasner *et al.* 2008). The bromine incorporation factor (BIF) for THMs is defined as follows (Gould *et al.* 1983; Symons *et al.* 1993):

$$\begin{aligned} \text{BIF} &= \text{THM-Br/THMs} \\ &= \frac{0 \times CHCl_3 + 1 \times CHCl_2Br + 2 \times CHClBr_2 + 3 \times CHBr_3}{CHCl_3 + CHCl_2Br + CHClBr_2 + CHBr_3} \end{aligned} \quad (4.1)$$

where the THM concentrations are on a molar basis.

For the THMs, BIF values range from 0 (all chloroform [$CHCl_3$] to 3 (all bromoform [$CHBr_3$]). A BIF of 1.0 corresponds to a water in which the "average" species is bromodichloromethane ($CHCl_2Br$). A similar equation was used to determine the BIF for trihaloacetaldehydes (THAs). Figure 4.3 shows the relationship between the BIF for THMs and that of THAs in U.S. plant effluents impacted by wastewater or algae, where there was a strong correlation ($R^2 = 0.73$, $p < 0.0001$). The slope was less than one (0.61), which indicates somewhat more bromine incorporation into the THMs. A similar slope (0.69) was observed for samples from nitrified wastewater treatment plants (Krasner *et al.* 2008). There may have been less bromine incorporation into the THAs from steric interference from the carbonyl moiety. Alternatively, bromine-containing THAs may have formed but then degraded (Xie and Reckhow, 1996).

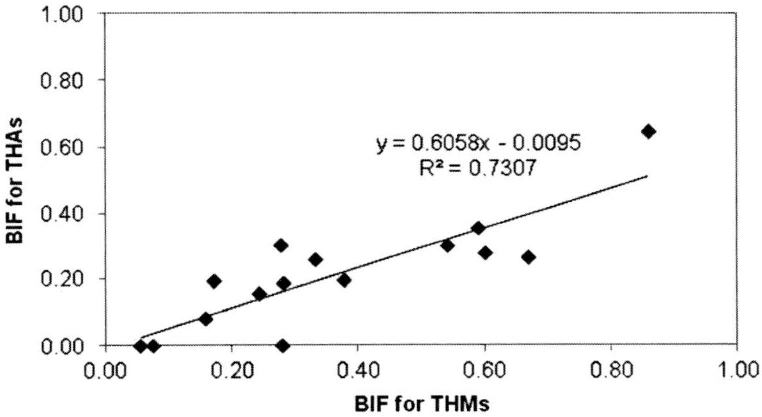

Figure 4.3 Comparison of bromine incorporation factors (BIFs) for trihalo-acetaldehydes (THAs) and trihalomethanes (THMs) in 11 U.S. plant effluents (wastewater and algal impacted waters) (data from Mitch *et al.* 2009)

Figure 4.4 shows the impact of the unit processes at an example treatment plant, which used softening, ozonation, and biologically active filtration (on granular activated carbon [GAC]) (Mitch *et al.* 2009). UV absorbing (UVA) NOM (e.g. humic substances) was better removed during the softening process than the bulk dissolved organic carbon (DOC). The removal of precursors (formation potential [FP]) measured in the presence of chlorine [Cl_2]) for THMs and THAs was in between that of the DOC and UVA. Ozone destroyed UVA, but had little impact on DOC or THM FP. However, ozone resulted in a substantial increase in THA FP. Biofiltration removed some DOC but little UVA or THM FP. However, biofiltration removed a significant amount of THA FP. This was due in part to the removal of acetaldehyde on the biofilter.

Medium-pressure UV has been shown to increase chloropicrin formation during post-chlorination (Reckhow *et al.* 2010). In recent research (Mitch *et al.* 2010), medium-pressure UV was also found to increase the formation of chloral hydrate during post-chlorination/chloramination (Figure 4.5). The latter research was part of a study on *N*-nitrosamines, in which waters impacted by wastewater or the coagulation polymer polyDADMAC were studied. Although polyDADMAC is a precursor for *N*-nitrosodimethylamine (NDMA) (Kohut and Andrews, 2003), it is not a precursor for chloral hydrate. The wastewater-impacted waters studied appeared to have more precursors for chloral hydrate than the polymer-impacted waters. In either case, an increase in UV fluence resulted in more chloral hydrate formation. However, such high fluences would not be used for germicidal applications of UV.

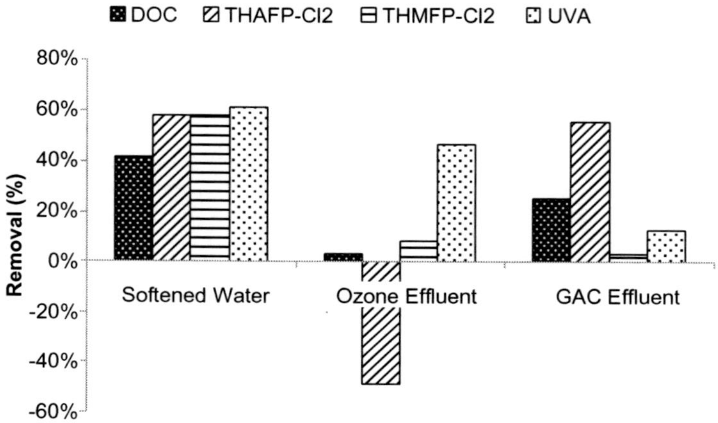

Figure 4.4 Impact of the unit processes at a U.S. ozone plant using biologically active filtration on DOC, UVA, and the precursors for THAs and THMs. Copyright Water Research Foundation 2009 (Mitch *et al.* 2009)

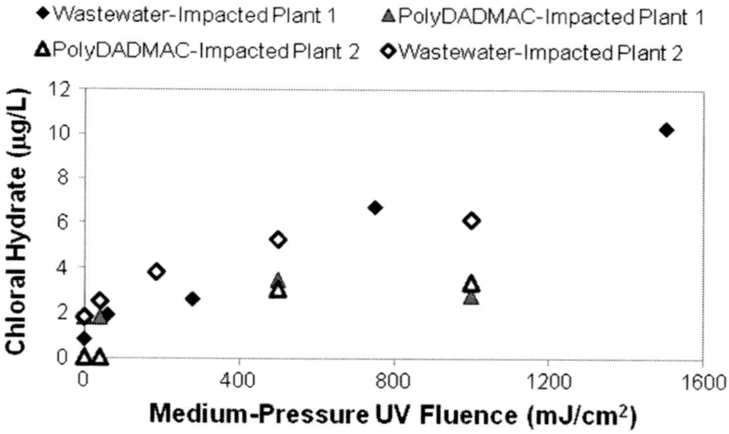

Figure 4.5 Impact of medium-pressure UV on chloral hydrate during post-chlorination/chloramination (data from Mitch *et al.* 2010)

4.3.3 Halogenated furanones

The health risks associated with halogenated furanones are discussed in Chapter 9. Although 3-chloro-4-(dichloromethyl)-5-hydroxy-2[5H]-furanone (MX) is the most commonly measured halogenated furanone, there are a number of important analogues of MX, which include an open ring form (EMX), bromine-containing analogues of MX (i.e. BMXs) and EMX (i.e. BEMXs), and reduced or oxidized MX (Onstad *et al.* 2008). Figure 4.6 shows the impact of the disinfection process on halogenated furanone formation at two plants that treated the same water (Weinberg *et al.* 2002). Because of the moderate amount of bromide in the water (0.12 mg/L), bromine-containing furanones were present (where 1, 2, and 3 signify 1, 2, and 3 bromine atoms in the MX or EMX analogue, respectively). Plant 1 used ozonation, biofiltration, and post-chloramination (sodium hypochlorite [NaOCl] was added before the ammonia [NH_3]), whereas Plant 2 added chlorine to the raw, settled, and filtered water, and ammonia was added at the end of the plant to form chloramines. Plant 1 formed very little halogenated furanones, as ozone may have destroyed some precursors and/or biofiltration may have removed some, and there was little free chlorine contact time before the ammonia was added. Plant 2 formed much more of the halogenated furanones because of the extensive free chlorine contact time before ammonia addition. Note, neither plant formed MX, so a survey based on only MX would have missed the formation of halogenated furanones at these plants.

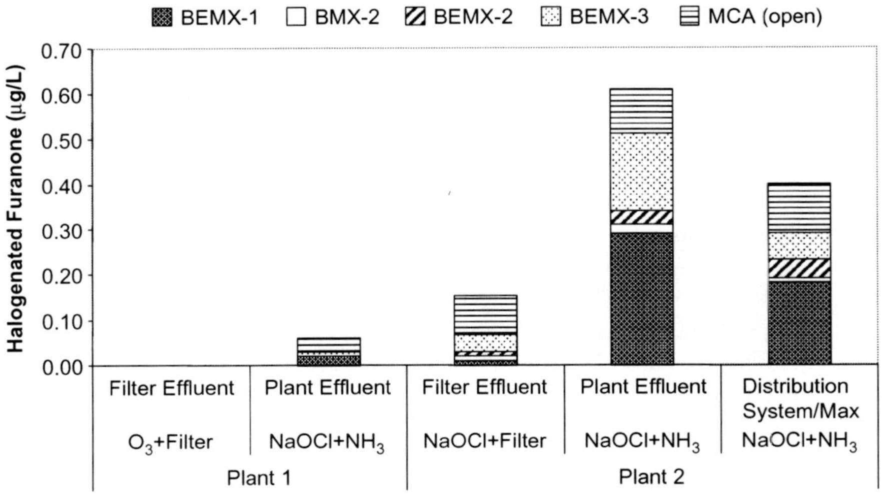

Figure 4.6 Impact of disinfection process on formation of halogenated furanones (MCA open = mucochloric acid, open form) at two plants that treated the same water (bromide = 0.12 mg/L) (Weinberg *et al.* 2002)

4.4 SUMMARY AND CONCLUSIONS

There are a wide range of DBPs of health concern in water, which include chlorine-, bromine-, and iodine-containing by-products. However, many studies have not examined all species within certain classes of DBPs and, thus, may underestimate the total formation of these by-product families.

The formation of DBPs is also impacted by the types of precursors present in the water supply (NOM, AOM, EfOM, bromide, iodide). For example, waters impacted by AOM or EfOM are expected to have more amino acids and may form more haloacetaldehydes. Waters containing bromide will result in a shift in formation to more bromine-incorporated species (e.g. THMs, THAs, halogenated furanones), some of which may be more toxic than their chlorine-containing species.

The formation of DBPs is impacted by the treatment process and disinfection scenario, where the impact on some emerging DBPs may be different than that of the regulated by-products. (For more information on emerging N-DBPs, see Chapters 5 and 6.)

- Although free chlorine results in the formation of bromine- and chlorine-containing THMs and HAAs, as well as haloacetaldehydes, it can oxidize iodide to iodate and may minimize the formation of iodine-containing DBPs (e.g. THMs, acids) during post-chloramination. Likewise, free chlorine can be used to destroy/transform NDMA precursors (Charrois and Hrudey, 2007), so that less of this *N*-nitrosamine is formed during post-chloramination. Of course, free chlorine contact time may also provide an important contribution to disinfection process credits.
- Although ozone can destroy to some extent the precursors of THMs and halogenated furanones, it can increase the formation of THAs, as well as that of chloropicrin (Hoigné and Bader, 1988). However, ozone can oxidize iodide to iodate and, thus, minimize the formation of iodine-containing DBPs during post-chloramination, and it can destroy/transform NDMA precursors (Lee *et al.* 2007).
- Chloramines can be used to minimize the formation of bromine- and chlorine-containing THMs and HAAs, as well as the formation of THAs and halogenated furanones. However, chloramines preferentially form iodine-containing DBPs and NDMA (Mitch *et al.* 2003), which can be mitigated (in part) with some degree of pre-chlorination or pre-ozonation.
- Medium-pressure UV, especially at high fluence, may increase the formation of certain DBPs (e.g. chloral hydrate, chloropicrin) during post-disinfection with chlorine and/or chloramines.
- Although biofiltration may have a small impact on THM precursors, it may decrease THA FP and that of halogenated furanones.

Not only is there a need to balance disinfection and DBP formation, there is a need to balance the control of the regulated with the emerging DBPs, where some of the latter by-products are believed to be of higher health concern. This may be accomplished by optimizing a series of various unit processes and disinfection scenarios to minimize the formation of a wide range of DBPs.

It should be clear that a water utility that seeks to manage the DBP issue needs to know and understand the dynamics of DBP formation with its own source water and specific treatment/disinfection practices. Although published evidence can provide a valuable guide, only studies of actual process conditions will provide meaningful site-specific answers.

REFERENCES

Amy G., Siddiqui M., Zhai W., Debroux J. and Odem W. (1994). Survey of Bromide in Drinking Water and Impacts on DBP Formation. American Water Works Association Research Foundation and AWWA, Denver, CO, USA.

Bichsel Y. and von Gunten U. (2000). Formation of iodo-trihalomethanes during disinfection and oxidation of iodide-containing waters. *Environ. Sci. Technol.*, **34**(13), 2784–2791.

Charrois J. W. A. and Hrudey S. E. (2007). Breakpoint chlorination and free-chlorine contact time: implications for drinking water *N*-nitrosodimethylamine concentrations. *Water Res.*, **41**(3), 672–682.

Dotson A. and Westerhoff P. (2009). Occurrence and removal of amino acids during water treatment. *J. AWWA*, **101**(9), 101–115.

Dotson A., Westerhoff P. and Krasner S. W. (2009). Nitrogen enriched dissolved organic matter (DOM) isolates and their affinity to form emerging disinfection by-products. *Water Sci. Technol.*, **60**(1), 135–143.

Fabian I. and Gordon G. (1997). The kinetics and mechanism of the chlorine dioxide-iodide ion reaction. *Inorg. Chem.*, **36**(12), 2494–2497.

Freuze I., Brosillon S., Laplanche A., Tozza D. and Cavard J. (2005). Effect of chlorination on the formation of odorous disinfection by-products. *Water Res.*, **39**, 2636–2642.

Froese K., Wolanski A. and Hrudey S. E. (1999). Factors governing odorous aldehyde formation as disinfection by-products in drinking water. *Water Res.*, **33**, 1355–1364.

Gould J. P., Fitchhorn L. E. and Urheim E. (1983). Formation of brominated trihalomethanes: extent and kinetics. In: Water Chlorination: Environmental Impact and Health Effects, R. L. Jolley, W. A. Brungs, J. A. Cotruvo, R. B. Cumming, J. S. Mattice and V. A. Jacobs (eds), Vol. 4, Ann Arbor Sci. Publ., Ann Arbor, MI, USA.

Hoigné J. and Bader H. (1988). The formation of trichloronitromethane (chloropicrin) and chloroform in a combined ozonation/chlorination treatment of drinking water. *Water Res.*, **22**(3), 313–319.

Hua G. and Reckhow D. A. (2006a). Determination of TOCl, TOBr, and TOI in drinking water by pyrolysis and off-line ion chromatography. *Anal. Bioanal. Chem.*, **384**(2), 495–504.

Hua G. and Reckhow D. A. (2006b). Comparison of total organic halogen formation from chlorine and alternative disinfectants. In: Proc. 2006 AWWA Ann. Conf. AWWA, Denver, CO, USA.

Karanfil T., Hu J., Jones D. B., Addison J. W. and Song H. (2011). Formation of Halonitromethanes and Iodo-Trihalomethanes in Drinking Water. Water Research Foundation, Denver, CO, USA.

Kohut K. D. and Andrews S. A. (2003). Polyelectrolyte age and *N*-nitrosodimethylamine formation in drinking water treatment. *Water Qual. J. Can.*, **38**, 719–735.

Krasner S. W. (1999). Chemistry of disinfection by-product formation. In: Formation and Control of Disinfection By-Products in Drinking Water, P. C. Singer (ed.), AWWA, Denver, CO, USA, pp. 27–52.

Krasner S. W., Lee C. F. T., Chinn R., Hartono S., Weinberg H., Richardson S. D., Pressman J. G., Speth T. F., Miltner R. J. and Simmons J. E. (2008). Bromine incorporation in regulated and emerging DBPs and the relative predominance of mono-, di-, and trihalogenated DBPs. In: Proc. 2008 AWWA Wat. Qual. Technol. Conf. AWWA, Denver, CO, USA.

Krasner S. W., Weinberg H. S., Richardson S. D., Pastor S. J., Chinn R., Sclimenti M. J., Onstad G. D. and Thruston A. D., Jr. (2006). Occurrence of a new generation of disinfection byproducts. *Environ. Sci. Technol.*, **40**(23), 7175–7185.

Lee C., Schmidt C., Yoon J. and von Gunten U. (2007). Oxidation of *N*-nitrosodimethylamine (NDMA) precursors with ozone and chlorine dioxide: kinetics and effect on NDMA formation potential. *Environ. Sci. Technol.*, **41**(6), 2056–2063.

McGuire M. J., McLain J. L. and Obolensky A. (2002). Information Collection Rule Data Analysis. American Water Works Association Research Foundation and AWWA, Denver, CO, USA.

McKnight A. and Reckhow D. A. (1992). Reactions of ozonation by-products with chlorine and chloramines. In: Proc. 1992 AWWA Ann. Conf. AWWA, Denver, CO, USA, pp. 399–409.

Mitch W., Krasner S., Lee C.-F.T., Wong T. M. and Bubnis B. (2010). Tradeoffs in DBP formation during chloramination following pre-oxidation for nitrosamine control. Presented at 2010 AWWA Wat. Qual. Technol. Conf., Savannah, GA, USA.

Mitch W. A., Krasner S. W., Westerhoff P. and Dotson A. (2009). Occurrence and Formation of Nitrogenous Disinfection By-Products. Water Research Foundation, Denver, CO, USA.

Mitch W. A., Sharp J. O., Trussell R. R., Valentine R. L., Alvarez-Cohen L. and Sedlak D. L. (2003). *N*-Nitrosodimethylamine (NDMA) as a drinking water contaminant: a review. *Environ. Eng. Sci.*, **20**(5), 389–404.

Moran J. E., Oktay S. D. and Santschi P. H. (2002). Sources of iodine and iodine 129 in rivers. *Water Resour. Res.*, **38**(8), 1149.

Onstad G. D., Weinberg H. S. and Krasner S. W. (2008). Occurrence of halogenated furanones in U.S. drinking waters. *Environ. Sci. Technol.*, **42**(9), 3341–3348.

Plewa M. J. and Wagner E. D. (2009). Quantitative Comparative Mammalian Cell Cytotoxicity and Genotoxicity of Selected Classes of Drinking Water Disinfection By-Products. Water Research Foundation, Denver, CO, USA.

Plewa M. J., Wagner E. D., Richardson S. D., Thruston A. D. Jr, Woo Y.-T. and McKague A. B. (2004). Chemical and biological characterization of newly discovered iodoacid drinking water disinfection byproducts. *Environ. Sci. Technol.*, **38**(18), 4713–4722.

Reckhow D. A., Linden K. G., Kim J., Shemer H. and Makdissy G. (2010). Effect of UV treatment on DBP formation. *J. AWWA*, **102**(6), 100–113.

Reckhow D. A., Singer P. C. and Malcolm R. L. (1990). Chlorination of humic materials: byproduct formation and chemical interpretations. *Environ. Sci. Technol.*, **24**(11), 1655–1664.

Richardson S. D., Plewa M. J., Wagner E. D., Schoeny R. S. and DeMarini D. M. (2007). Occurrence, genotoxicity, and carcinogenicity of emerging disinfection by-products in drinking water: a review and roadmap for research. *Mutat. Res.*, **636**(1–3), 178–242.

Stevens A. A., Moore L. A. and Miltner R. J. (1989). Formation and control of non-trihalomethane disinfection by-products. *J. AWWA*, **81**(8), 54–60.

Symons J. M., Krasner S. W., Simms L. A. and Sclimenti M. (1993). Measurement of THM and precursor concentrations revisited: the effect of bromide ion. *J. AWWA*, **85**(1), 51–62.

Trehy M. L., Yost R. A. and Miles C. J. (1986). Chlorination byproducts of amino acids in natural waters. *Environ. Sci. Technol.*, **20**(11), 1117–1122.

USEPA (1998). National primary drinking water regulations: disinfectants and disinfection byproducts; final rule. *Fed. Reg.*, **63**(241), 69,390–69,476.

Weinberg H. S., Krasner S. W., Richardson S. D. and Thruston A. D. Jr. (2002). The Occurrence of Disinfection By-Products (DBPs) of Health Concern in Drinking Water: Results of a Nationwide DBP Occurrence Study. EPA/600/R-02/068. USEPA, Athens, GA, USA. http://www.epa.gov/Athens/publications/reports/EPA_600_R02_068.pdf (accessed 13 Aug 2011).

Weinberg H. S., Kritsch K. and Krasner S. W. (2011). Iodoacids in Drinking Water Supplies: Methods and Occurrence. Water RF, Denver, CO, USA.

Woo Y.-T., Lai D., McLain J. L., Manibusan M. K. and Dellarco V. (2002). Use of mechanism-based structure–activity relationships analysis in carcinogenic potential ranking for drinking water disinfection by-products. *Environ. Health Perspect.*, **110** (suppl. 1), 75–87.

Xie Y. and Reckhow D. A. (1996). Hydrolysis and Dehalogenation of Trihaloacetaldehydes. In: Disinfection By-Products in Water Treatment: The Chemistry of Their Formation and Control, R. A. Minear and G. L. Amy (eds), CRC Lewis Publishers, Boca Raton, FL.

Zhang X., Echigo S., Minear R. A. and Plewa M. J. (2000). Characterization and comparison of disinfection by-products of four major disinfectants. In: Natural Organic Matter and Disinfection By-Products: Characterization and Control in Drinking Water, S. E. Barrett, S. W. Krasner and G. L. Amy (eds), American Chemical Society, Washington, DC, USA.

Chapter 5

Nitrogenous DBPs, formation, control and new frontiers

W. A. Mitch

5.1 INTRODUCTION

Disinfection by-product (DBP) research has traditionally focused on the reaction of free chlorine with natural organic matter (NOM), such as humic substances in pristine water supplies (i.e. those not significantly impacted by human activities). Humic substances are the long-term degradation products of carbon-rich plant polymers, such as cellulose and lignin. These reactions form the most commonly regulated DBPs, trihalomethanes (THMs) and haloacetic acids (HAAs), as well as a host of other emerging carbon-based by-products, as described in other chapters. However, because NOM is nitrogen-poor, and because analytical methods for many of these compounds were not prevalent, nitrogen-based by-products have not, until the past decade, been a focus of DBP research.

As a result of trends in population growth, climate change, and urbanization, pristine water supplies are increasingly rare, such that some utilities have increasingly needed to exploit source waters impacted by wastewater or algae. These types of waters feature relatively fresh biomolecules (e.g. human waste products or algal exudates) and a higher proportion of organic nitrogen than the NOM in pristine waters (Westerhoff and Mash, 2002). Accordingly, reactions of disinfectants with these materials may promote the formation of nitrogen-based DBPs.

Nitrogen-based by-products include several families of current concern (e.g. *N*-nitrosamines, halonitriles and halonitroalkanes) because of their higher cyto- and genotoxicity as shown by *in vitro* assays or estimated cancer potencies (Muellner *et al.* 2007; Plewa *et al.* 2004; US EPA, 2011). *N*-Nitrosamines include compounds with a nitroso (—N=O) functional group. *N*-nitrosodimethylamine (NDMA) is of particular importance (Figure 5.1a), because the US EPA's

Integrated Risk Information System (IRIS) database indicates that a drinking water concentration of 0.7 ng/L would be associated with a 10^{-6} upper bound lifetime cancer risk (US EPA, 2011). *N*-Nitrosamines are currently being considered for regulation in drinking water in the United States and NDMA has been assigned a guideline value in Australia and Canada.

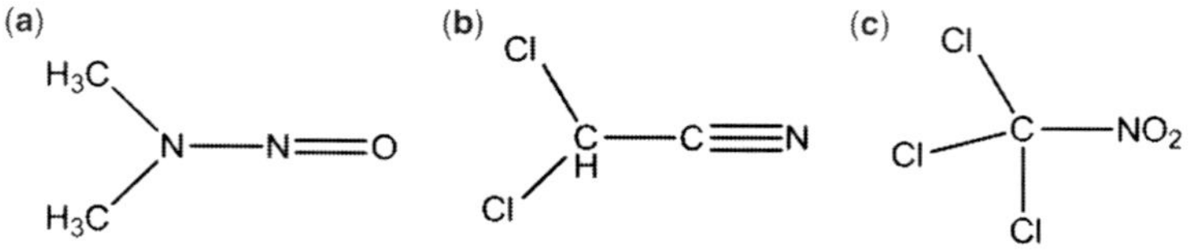

Figure 5.1 Structures of (a) *N*-nitrosodimethylamine (b) dichloroacetonitrile and (c) chloropicrin

Halonitriles constitute halogenated compounds with a nitrile (—C≡N) functional group, such as dichloroacetonitrile (Figure 5.1b). Halonitroalkanes consist of halogenated compounds with a nitro (-NO_2) functional group, including chloropicrin (Figure 5.1c). *In vitro* toxicological assays indicate that both families of by-products are potentially more cyto- and genotoxic than the regulated THMs (Muellner *et al.* 2007; Plewa *et al.* 2004).

The wide range of nitrogen-based DBPs precludes providing a comprehensive background on each grouping but more details are provided in Chapter 6. Instead, the purpose of this chapter is to provide some examples that serve to highlight four important major themes as we approach this emerging by-product field:

(1) All disinfection techniques may promote the formation of some unwanted by-products.
(2) Combining disinfectants may help, but tradeoffs in DBP formation must be balanced.
(3) Understanding how DBPs form may lead to inexpensive modifications to disinfection systems that minimize their formation.
(4) Current DBPs of interest may not be the most important for driving drinking water health risk, but future research may predict by-products of importance.

5.2 THEME 1: ALL DISINFECTION SCHEMES FORM DBPs

THMs and HAAs have long been associated with chlorine disinfection. As a result of regulations limiting their permissible concentrations in disinfected waters, utilities have been experimenting with alternative disinfectants. However, many of these disinfectants are associated with by-product formation of their own. For example, ozonation is associated with formation of bromate in bromide-containing waters, another DBP regulated in many countries. Additionally, ozonation followed by chlorination has been associated with promotion of the formation of chloropicrin (trichloronitromethane), a generally unregulated contaminant (Merlet *et al.* 1985). Chlorine dioxide is associated with chlorite, another by-product that is regulated in many countries. As discussed in Section 5.3, chloramination is associated with *N*-nitrosamine formation (Mitch *et al.* 2003).

The popularity of ultraviolet treatment has increased in recent years partly as a result of its effectiveness against protozoan pathogens, but also as a result of the assumption that, as a "chemical-free" disinfectant, it should reduce DBP formation. However, in many countries, ultraviolet treatment must be followed by the application of chlorine or chloramines to maintain a residual in the distribution system. Accordingly, it is important to understand whether ultraviolet treatment alters precursor materials, rendering them more susceptible to by-product formation upon post-chlorination or chloramination.

Recent research has indicated that treatment of authentic drinking waters by medium-pressure ultraviolet light at relevant germicidal fluence (i.e. up to ~200 mJ/cm^2) followed by application of chlorine or chloramines at concentrations relevant to distribution systems (i.e. ~2.5 mg/L as Cl_2), promoted the formation of chloropicrin at concentrations up to 2.5 μg/L (Reckhow *et al.* 2010; Shah *et al.* 2011). This concentration is nearly an order of magnitude higher than observed in the absence of the ultraviolet treatment. The increased formation was attributable to the photolysis of nitrate. However, this promotion was not observed with low-pressure mercury lamps, because of limited overlap between the lamp output spectrum (predominantly at 254 nm) and the nitrate absorption spectrum. These results indicate that it is likely that all disinfection schemes will promote the formation of some unwanted DBPs. Accordingly, rather than attempting to discover a "magic bullet" treatment scheme that will be by-product-free, it would likely be more fruitful to optimize individual disinfectants or combinations of disinfectants to minimize the overall toxicity of the by-products formed.

5.3 THEME 2: DISINFECTANT COMBINATIONS MUST BALANCE TRADEOFFS IN DBP FORMATION

Combining disinfectants may help achieve pathogen deactivation goals while avoiding elevated formation of the particular by-products associated with each individual disinfectant. Indeed, the popularity of the combination of pre-chlorination and post-chloramination is related to the effective deactivation of pathogens by chlorination, the stronger oxidant of the two chlorine-based disinfectants, with the reduced formation of regulated DBPs such as HAAs and THMs during chloramination. However, combining disinfectants will result in the formation of by-products associated with each disinfectant, and a balance must be struck to minimize overall by-product formation.

As a case study, utilities may wish to consider combinations of strong oxidants with post-chloramination. As noted previously, chloramination has been associated with *N*-nitrosamine formation (Schreiber and Mitch, 2006). Strong oxidants, including chlorine (Charrois and Hrudey, 2007), ozone and chlorine dioxide (Lee *et al.* 2007) have been noted to deactivate precursors for *N*-nitrosodimethylamine (NDMA). Combinations of these pre-oxidants with post-chloramination may reduce NDMA formation, but at the expense of the formation of the DBPs associated with these other disinfectants.

As an example, tradeoffs in by-product formation as a function of pre-oxidant contact time (CT) were evaluated after exposure of an authentic sample collected from the filter effluent of a drinking water utility to chloramines at 2.5 mg/L as Cl_2 and pH ~8 for 3 days; the utility used polyDADMAC, a cationic polymer that can serve as a NDMA precursor because it contains a dimethylamine functional group (Krasner *et al.* 2011). Increasing the chlorine CT from 0 to 178 mg · min/L decreased NDMA formation from 21 ng/L to 3 ng/L, but at the expense of an increase in the formation of trihalomethanes (THM4) from 17 to 39 μg/L, and haloacetic acids (HAA5) from 8.5 to 12 μg/L. Increasing the pre-ozone CT from 0 to 2.9 mg · min/L reduced NDMA formation from 11 ng/L to 5.5 ng/L, but at the expense of an increase in bromate formation from 1.5 to 10 μg/L. Increasing the pre-chlorine dioxide CT from 0 to 19 mg · min/L reduced NDMA formation from 11 to 5.6 ng/L, but at the expense of an increase in chlorite from non-detectable to 0.21 mg/L. Note that the zero CT values for NDMA differ for pre-ozone and pre-chlorine dioxide vs. pre-chlorination because the former were conducted on an aged sample, where some of the NDMA precursors may have decayed.

5.4 THEME 3: KNOWLEDGE OF DBP FORMATION PATHWAYS PERMITS DESIGN OF INEXPENSIVE MODIFICATIONS TO REDUCE DBP FORMATION

Although NDMA formation is associated with chloramination, it is important to note that monochloramine and dichloramine always coexist in chloraminated solutions according to the equilibrium:

$$2NH_2Cl + H^+ \longleftrightarrow NHCl_2 + NH_4^+ \qquad (5.1)$$

Although monochloramine is typically the dominant chloramine constituent during chloramination, dichloramine will always co-occur. Mechanistic research indicates that essentially all *N*-nitrosamine formation during chloramination could be attributed to the traces of dichloramine present in solution (Schreiber and Mitch, 2006).

Accordingly, one technique to reduce *N*-nitrosamine formation during chloramination would be to alter the disinfection scheme to minimize dichloramine formation. Chloramines are formed by mixing free chlorine with ammonia. These reagents are generally added with a slight molar excess of ammonia (i.e. ~1:1.1 chlorine to ammonia molar ratio or ~4.75 chlorine as Cl_2 to ammonia as N weight ratio). At circumneutral pH (5.5–7.4), these conditions favor the predominant formation of monochloramine according to the equilibrium described above, although ~5% of the total chloramines may be dichloramine. However, the ammonia and chlorine reagents can be added in either order. Inorganic chloramine formation occurs extremely rapidly. It was demonstrated that the order of reagent addition may affect the prevalence of dichloramine because inorganic chloramine formation may occur on the timescale of reagent mixing (Schreiber and Mitch, 2005). If ammonia is added upstream of chlorine, the ammonia will be well-mixed into solution at the point of chlorine addition. However, at this addition point, the relatively high chlorine: ammonia ratio associated with the injection of a concentrated chlorine stock may promote formation of dichloramine at levels higher than suggested by the equilibrium reaction described above. Accordingly, addition of chlorine prior to ammonia may minimize *N*-nitrosamine formation by minimizing dichloramine formation, in addition to any benefits associated with precursor deactivation prior to ammonia addition. There should be minimal additional THM/HAA formation with the brief free chlorine contact time required.

In some wastewater recycling operations, non-nitrified effluents are chlorinated upstream of membrane filtration to minimize biofouling. The inorganic chloramines formed by the reaction of the free chlorine with the ammonia are less harmful than free chlorine to the membranes, but may promote the formation of *N*-nitrosamines. Formation of *N*-nitrosamines, particularly NDMA, has been a concern for wastewater recycling facilities due to the prevalence of NDMA precursors in these waters. Because ammonia already occurs in non-nitrified wastewaters, it is not possible to add chlorine upstream of the ammonia. An alternative would be to pre-form chloramines under conditions that favour monochloramine over dichloramine (e.g. addition of free chlorine prior to ammonia to tap water at elevated pH), and apply the pre-formed monochloramine to the non-nitrified wastewater. This technique was successfully applied to a recycled wastewater to reduce NDMA formation (Mitch *et al.* 2005).

5.5 THEME 4: DBPs YET TO BE DISCOVERED SHOULD BE PURSUED IN A RATIONAL MANNER

Over 600 halogenated DBPs form during chlorine disinfection, yet these DBPs account for only ~30% of the Total Organic Halides (i.e. TOX) pool (Richardson, 2003). Although THMs and HAAs constitute the bulk of the fraction of TOX that has been characterized, the low fraction of the TOX pool constituted by these compounds raises the question whether the most important halogenated by-products are targeted by current regulations. In this context, it would be desirable to apply known reaction pathways of disinfectants to precursor structures to predict by-products likely to form in the highest concentrations. However, the humic substance precursors predominating in pristine water supplies represent highly degraded materials, such that it is likely that each molecule is different. The poor characterization of individual molecules has prevented the application of disinfectant reaction pathways to these structures to predict by-products likely to form in highest concentrations. As a result, most by-product research has focused on compounds identified by analytical chemists, whose overall importance to the toxicity of a disinfected water is unclear. In contrast, Woo *et al.* (2002) and Bull *et al.* (2011) have pursued a more focused approach of looking for plausible DBPs that are of toxicological interest. The results of the former are mentioned in Chapter 4 and the latter is elaborated in Chapters 8 and 10.

The US EPA developed Method 521 as an analytical method enabling the detection of seven *N*-nitrosamines in drinking waters (USEPA, 2004). Based upon the occurrence of these seven *N*-nitrosamines, the US EPA is considering whether or not to regulate *N*-nitrosamines in drinking waters. As regulatory agencies evaluate whether or not to regulate *N*-nitrosamines, it would be helpful to have a total *N*-nitrosamine (TONO) analytical method, akin to TOX, to understand to what extent the seven *N*-nitrosamines targeted by the US EPA capture the total *N*-nitrosamine pool. Following development of a TONO method, it was applied to disinfected recreational waters alongside Method 521 (Kulshrestha *et al.* 2010). The results indicated that NDMA was the most prevalent of the seven *N*-nitrosamines in recreational waters, but that NDMA generally constituted only ~10% of the TONO pool. Accordingly, Method 521 may not be targeting the most important *N*-nitrosamines.

One significant difference between humic substances in pristine water supplies and organic nitrogen precursors in algal- or wastewater-impaired water supplies is that the latter materials are relatively fresh biomolecules compared to more degraded humic substances. Although biopolymers, such as protein, represent complex assemblages in terms of the number and order of monomers, there is a limited array of these monomers and their structures are well-characterized. For example there are approximately 20 amino acids from which proteins are constructed. Known disinfectant reaction pathways could be applied to these monomer structures to predict by-products likely to form at high yield. Some research has been done on DBPs produced by disinfection of free amino acids with most notable findings suggesting that resulting DBPs may explain taste and odour problems that had previously gone unexplained (Freuze *et al.* 2005; Froese *et al.* 1999; Trehy *et al.* 1986); however, as free amino acids constitute ~5–10% of the total amino acids in source waters (Dotson and Westerhoff, 2009), DBPs produced from them are anticipated to be of minor importance to the toxicity of a disinfected water. Note that for proteins, exposure to the modified amino acid residues during ingestion is likely, as the amino acids are liberated from the protein within the digestive system.

As an example, mechanistic research indicated that chlorination of primary amines forms nitriles (Joo and Mitch, 2007). Because the side chain of the amino acid lysine contains a primary amine, it was hypothesized that a nitrile would form during chlorination of lysine-containing proteins (Figure 5.2). An analytical method was developed to achieve this detection, and lysine nitrile was detected in a chlorinated drinking water at 0.1 μg/L (Walse *et al.* 2009). Although the relative importance of this particular by-product to the toxicity of a disinfected water is unclear, the ability to predict precursors forming in high yield should enable researchers to target by-products likely to be more important, rather than to respond to the plethora of by-products remaining to be discovered by analytical chemists that may form at only trace yields.

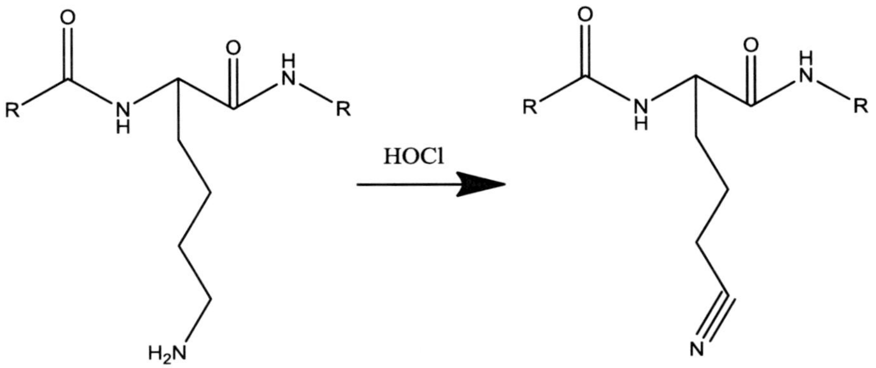

Figure 5.2 Formation of lysine nitrile from lysine. R refers to a continuation of the protein polymer.

5.6 SUMMARY AND CONCLUSIONS

While early research on DBPs focused on the discovery of THMs and HAAs produced by disinfectant reactions with NOM, in the past decade, an increasing number of nitrogenous DBPs have been identified as research has evaluated DBPs produced in source waters affected by domestic wastewaters and/or algal exudates, both of which are richer in organic nitrogen DBPs precursors than NOM. As a result, *N*-nitrosamines, halonitriles and halonitroalkanes have been identified as nitrogenous DBPs that occur widely, albeit at much lower concentrations than the most commonly regulated DBPs, the THMs and HAAs.

Considering the diversity and complexity that these additional DBP precursors introduce, some major themes are important considerations for water utilities including:

- It is highly unlikely that a disinfection scheme will be identified that is free of DBP formation.
- Disinfectant combinations should be optimized to minimize the overall exposure to DBPs. Such a balance often entails tradeoffs in the formation of the DBPs associated with each disinfectant.

- Knowledge of formation pathways permits the design of disinfection systems that minimize DBP formation by minimizing the most important precursors.
- DBP research should pursue a rational strategy involving prediction of DBPs likely to form at high yield and exhibit elevated toxicity to efficiently focus research efforts.

REFERENCES

Bull R. J., Reckhow D. A., Li X., Humpage A. R., Joll C. and Hrudey S. E. (2011). Potential carcinogenic hazards of non-regulated disinfection by-products: haloquinones, halo-cyclopentene and cyclohexene derivatives, *N*-halamines, halonitriles, and heterocyclic amines. *Toxicology*, **286**, 1–19.

Charrois J. W. A. and Hrudey S. E. (2007). Breakpoint chlorination and free-chlorine contact time: implications for drinking water *N*-nitrosodimethylamine concentrations. *Water Res.*, **41**, 674–682.

Dotson A. and Westerhoff P. (2009). Occurrence and removal of amino acids during drinking water treatment. *J. AWWA*, **101**, 101–115.

Freuze I., Brosillon S., Laplanche A., Tozza D. and Cavard J. (2005). Effect of chlorination on the formation of odorous disinfection by-products. *Water Res.*, **39**, 2636–2642.

Froese K., Wolanski A. and Hrudey S. E. (1999). Factors governing odorous aldehyde formation as disinfection by-products in drinking water. *Water Res.*, **33**, 1355–1364.

Joo S.-H. and Mitch W. A. (2007). Nitrile, aldehyde and halonitroalkane formation during chlorination/chloramination of primary amines. *Environ. Sci. Technol.*, **41**, 1288–1296.

Krasner S. W., Lee C. F. T., Mitch W. A. and von Gunten U. (2011). Development of a Bench-Scale Test to Predict the Formation of Nitrosamines. Water Research Foundation, Denver.

Kulshrestha P., McKinstry K. C., Fernandez B. O., Feelisch M. and Mitch W. A. (2010). Application of an optimized total *N*-nitrosamine (TONO) assay to pools: placing *N*-nitrosodimethylamine (NDMA) determinations into perspective. *Environ. Sci. Technol.*, **44**, 3369–3375.

Lee C., Schmidt C., Yoon J. and von Gunten U. (2007). Oxidation of *N*-nitroso-dimethylamine (NDMA) precursors with ozone and chlorine dioxide: kinetics and effect on NDMA formation potential. *Environ. Sci. Technol.*, **41**, 2056–2063.

Merlet N., Thibaud H. and Dore M. (1985). Chloropicrin formation during oxidative treatments in the preparation of drinking water. *Sci. Tot. Environ.*, **47**, 223–228.

Mitch W. A., Oelker G. L., Hawley E. L., Deeb R. A. and Sedlak D. L. (2005). Minimization of NDMA formation during chlorine disinfection of municipal wastewater by application of pre-formed chloramines. *Environ. Eng. Sci.*, **22**, 882–890.

Mitch W. A., Sharp J. O., Trussell R. R., Valentine R. L., Alvarez-Cohen L. and Sedlak D. L. (2003). *N*-nitrosodimethylamine (NDMA) as a drinking water contaminant: a review. *Environ. Eng. Sci.*, **20**(5), 389–404.

Muellner M. G., Wagner E. D., McCalla K., Richardson S. D., Woo Y. T. and Plewa M. J. (2007). Haloacetonitriles vs. regulated haloacetic acids: are nitrogen-containing DBPs more toxic? *Environ. Sci. Technol.*, **41**(2), 645–651.

Plewa M. J., Wagner E. D., Jazwierska P., Richardson S. D., Chen P. H. and McKague A. B. (2004). Halonitromethane drinking water disinfection byproducts: chemical characterization and mammalian cell cytotoxicity and genotoxicity. *Environ. Sci. Technol.*, **38**(1), 62–68.

Reckhow D. A., Linden K. G., Kim J., Shemer H. and Makdissy G. (2010). Effect of UV treatment on DBP formation. *J. Am. Water Works Assoc.*, **102**, 100–113.

Richardson S. D. (2003). Water analysis: emerging contaminants and current issues. *Anal. Chem.*, **75**, 2831–2857.

Schreiber I. M. and Mitch W. A. (2005). The influence of the order of reagent addition on *N*-nitrosodimethylamine formation. *Environ. Sci. Technol.*, **39**, 3811–3818.

Schreiber I. M. and Mitch W. A. (2006). Nitrosamine formation pathway revisited: the importance of dichloramine and dissolved oxygen. *Environ. Sci. Technol.*, **40**, 6007–6014.

Shah A. D., Dotson A. A., Linden K. G. and Mitch W. A. (2011). Impact of UV disinfection combined with chlorination/chloramination on the formation of halonitromethanes and haloacetonitriles in drinking water. *Environ. Sci. Technol.*, **45**, 3657–3664.

Trehy M. L., Yost R. A. and Miles C. J. (1986). Chlorination byproducts of amino acids in natural waters. *Environ. Sci. Technol.*, **20**(11), 1117–1122.

USEPA (2004). Method 521. Determination of Nitrosamines in Drinking Water by Solid Phase Extraction and Capillary Column Gas Chromatography with Large Volume Injection and Chemical Ionization Tandem Mass Spectrometry (MS/MS) EPA/600/R-05/054 National Exposure Research Laboratory, Office of Research and Development (ORD), United States Environmental Protection Agency, Cincinnati, OH. http://epa.gov/microbes/m_521.pdf (accessed 11 January 2011).

USEPA (2011). Office of Research and Development (ORD), National Center for Environmental Assessment. United States Environmental Protection Agency. Integrated Risk Information System. http://www.epa.gov/iris/ (accessed 11 January 2011).

Walse S. S., Plewa M. J. and Mitch W. A. (2009). Exploring amino acid side chain decomposition using enzymatic digestion and HPLC-MS: combined lysine transformations in chlorinated waters. *Anal. Chem.*, **81**, 7650–7659.

Westerhoff P. and Mash H. (2002). Dissolved organic nitrogen in drinking water supplies: a review. *J. Water Supply Technol. – Aqua*, **51**, 415–448.

Woo Y.-T., Lai D., McLain J. L., Manibusan M. K. and Dellarco V. (2002). Use of mechanism-based structure–activity relationships analysis in carcinogenic potential ranking for drinking water disinfection by-products. *Environ. Health Perspect.*, **110** (suppl. 1), 75–87.

Chapter 6

Nitrogenous DBPs in drinking water: toxicity, regulation, analysis, occurrence and control

D. Liew, K. L. Linge, A. Heitz and C. A. Joll

6.1 INTRODUCTION

Since the discovery of disinfection by-products (DBPs) in chlorinated drinking water over 35 years ago, there has been considerable research effort to understand how DBPs are formed and how they can be avoided (Richardson, 2003; Richardson, 2005). The formation of DBPs is influenced by the disinfectant used, source water quality and reaction environment. Most of the initial research was focused on investigating and minimising the trihalomethanes (THMs) and the haloacetic acids (HAAs), the two most abundant classes of DBPs found in chlorinated drinking water, and this has led to alternative disinfectants being used in place of chlorine (e.g. chloramine, ozone and chlorine dioxide). While these generally reduce THM and HAA concentrations, they can increase the concentrations of other potentially toxic DBPs. Over 600 DBPs have been identified from the treatment of water with the major disinfectants and their combinations (Krasner *et al.* 2006; Richardson, 2005). Little information is available on the toxicities of many of these DBPs and in fact, only a relatively small proportion have been quantified or surveyed for prevalence in drinking waters (Krasner *et al.* 2006; Richardson, 2005).

One particular group of DBPs that has not been studied extensively are nitrogen-containing, or nitrogenous, DBPs (N-DBPs). In water sources that are impaired by agricultural runoff or by wastewater effluents, the high content of dissolved organic nitrogen may serve as precursors for N-DBPs (Hwang *et al.* 2001). They include DBPs containing nitroso- (e.g. *N*-nitrosamines), nitro- (e.g. halonitromethanes), nitrile- (e.g. halonitriles, cyanogen halides) or amide- (e.g. haloacetamides) functional groups, and concerns have arisen due to the toxicities of these groups of compounds. The *N*-nitrosamines and halonitriles have been reported most frequently in the literature, but studies of other N-DBPs are rare.

While most of these N-DBPs are not yet regulated in drinking waters, the toxicities and carcinogenic potential of these compounds may pose greater health risks than the commonly regulated THMs and HAAs. At present, there is still limited information available on these emerging DBPs and there is a need to provide sufficient knowledge on the associated health risks, formation mechanisms and mitigation methods of these N-DBPs for appropriate regulation. Chapter 5 has introduced some major themes related to N-DBPs. This review summarizes currently available research findings for seven classes of N-DBPs and identifies important gaps in knowledge.

6.2 *N*-NITROSAMINES

6.2.1 Background, toxicity, and regulatory status

The *N*-nitrosamines are a group of emerging DBPs that have been detected in groundwater, recycled water used for groundwater recharge, and finished drinking water. The importance of *N*-nitrosamines as DBPs of health concern is illustrated by the publication of recent literature reviews describing their occurrence and control (Bond *et al.* 2011; Nawrocki and Andrzejewski, 2011). The occurrence, regulation, health risks and an overview of risk management options for *N*-nitrosamines are addressed in Chapter 7.

6.2.2 Analysis

Analysis of *N*-nitrosamines is challenged by the difficulties in selective separation and detection of these small, highly polar but uncharged molecules from water, combined with present guideline values in the range of ng/L. Several analytical methods have been developed, typically comprising preconcentration or extraction followed by chromatographic mass spectrometric analysis. Recent techniques achieve method detection limits (MDLs) down to sub- or low ng/L levels, as illustrated in Table 6.1. For preconcentration of samples, liquid-liquid extraction (LLE) (Gerecke and Sedlak, 2003; Raksit and Johri, 2001), solid-phase extraction (SPE) with carbonaceous adsorbents such as coconut charcoal (Munch and Bassett, 2004; Planas *et al.* 2008) or Ambersorb 572 (Charrois *et al.* 2004; Cheng *et al.* 2006; Ontario MoE, 2004; Zhao *et al.* 2006), and solid-phase microextraction (SPME) (Grebel *et al.* 2006) have been used, with variable recoveries. SPE is commonly utilized over conventional liquid-liquid extractions because of the shorter processing times, less sample handling, smaller volumes of environmentally toxic solvents, and ease of automation (Charrois *et al.* 2004; Jurado-Sanchez *et al.* 2007).

Gas chromatography (GC) provides good separation of *N*-nitrosamines, but challenges exist in attaining sensitive and specific detection for quantification.

Table 6.1 Methods for *N*-nitrosamine analysis and corresponding method detection limits (MDLs).

Extraction method	Instrumentation	NDMA[a] MDL (ng/L)	All other *N*-nitrosamines MDL (ng/L)	References
LLE	GC-EI-MS	3	N/A	(Raksit and Johri, 2001)
	Large volume injection and GC-EI-MS	0.52	N/A	(Yang *et al.* 2008)
	Methanol GC-CI-MS/MS	2.2	N/A	(Gerecke and Sedlak, 2003)
SPE	GC-EI-MS	10	7 *N*-nitrosamines 0.6–1.5 ng/L	(Jurado-Sanchez *et al.* 2007)
	Ammonia GC-CI-MS	0.7	8 *N*-nitrosamines 0.4–1.6 ng/L	(Charrois *et al.* 2004)
	Large volume injection and Methanol GC-CI-MS/MS	0.28	7 *N*-nitrosamines 0.17–0.66 ng/L	(Munch and Bassett, 2004)
	Methanol/ Acetonitrile GC-CI-MS/MS	0.7	8 *N*-nitrosamines 0.3–1.4 ng/L	(Cheng *et al.* 2006)
	GC-HRMS	0.8	9 *N*-nitrosamines 0.08–1.7 ng/L	(Planas *et al.* 2008)
	GC-HRMS	0.4		(Ontario MoE, 2004)
	LC-MS/MS	3.1	9 *N*-nitrosamines 0.1–3.1 ng/L	(Zhao *et al.* 2006)

[a]*N*-nitrosodimethylamine (NDMA).
N/A = not applicable.

The low molecular weights of the *N*-nitrosamines mean that it can be difficult to identify characteristic ions for quantification when analysed by low resolution mass spectrometry (MS) with electron ionization (EI) (Planas *et al.* 2008). Researchers have employed other detection techniques in efforts to attain greater selectivity and sensitivity, particularly, by the use of chemical ionization (CI) as a “softer” ionization source (Charrois *et al.* 2004; Grebel *et al.* 2006; Munch and Bassett, 2004), high resolution mass spectrometry (HRMS) (Ontario MoE, 2004; Planas *et al.* 2008) or tandem mass spectrometry (MS/MS) (Munch and Bassett, 2004; Zhao *et al.* 2006).

Liquid chromatography – tandem mass spectrometry (LC-MS/MS) has only recently been shown to be capable of the analysis of up to nine *N*-nitrosamines in water samples, including the thermally labile *N*-nitrosodiphenylamine (NDPhA), not usually detected by GC-MS (Zhao *et al.* 2006). However, others have measured NDPhA using GC-MS, based on the detection of diphenylamine, assumed to be produced in the GC injection liner from the decomposition of NDPhA (Jurado-Sanchez *et al.* 2007; Planas *et al.* 2008). Furthermore, the availability, relatively low cost and superior separation power of GC-MS instrumentation compared to LC-MS/MS, still makes the former the preferred choice in most laboratories.

6.2.3 Formation and precursors

Several mechanisms have been proposed for *N*-nitrosodimethylamine (NDMA) formation (Table 6.2), however they are not always consistent with experimental observations. Currently dichloramine concentration is considered a key factor in NDMA formation. However, it is likely that a combination of different mechanisms play a part, depending on the reaction conditions and precursors present. Differences in experimental contact times may also lead to different observations. Investigations of specific reaction intermediates and their yields are required to verify the proposed pathways under different treatment scenarios, and in actual water samples. Study of the formation pathways of other *N*-nitrosamines and their corresponding precursors is also needed.

Although most studies on *N*-nitrosamine formation have investigated the role of chlorine or chloramine, other disinfectants have also been shown to

Table 6.2 Mechanisms proposed for NDMA formation.

Mechanism	Notes
Nitrosation	• The nitrosyl cation (NO^+), formed from nitrate acidification, reacts with an amine (e.g. dimethylamine) to produce NDMA (Ayanaba and Alexander, 1974; Mitch and Sedlak, 2002). • Strongly dependent on pH with a maximum rate near pH 3.4. Extremely slow under typical water and wastewater treatment conditions (Mitch and Sedlak, 2002). • A modified reaction mechanism for breakpoint chlorination (i.e. in the presence of hypochlorite) hypothesized that dinitrogen tetroxide is formed as an intermediate (Choi and Valentine, 2003), however NDMA formation under drinking water treatment conditions was not significant. • Nitrosation cannot fully explain *N*-nitrosamine formation in actual water samples.

(*Continued*)

Table 6.2 Mechanisms proposed for NDMA formation (*Continued*).

Mechanism	Notes
Oxidation of UDMH (monochloramine) or Cl-UDMH (dichloramine)	• Unsymmetrical dimethylhydrazine (UDMH) is slowly formed through reaction of monochloramine and dimethylamine, followed by almost instantaneous oxidation of UDMH to a variety of products, including NDMA (Mitch and Sedlak, 2002). • NDMA formation from chloramination of UDMH has been shown to be significantly lower compared to that from chloramination of DMA, suggesting UDMH is not a key intermediate (Schreiber and Mitch, 2006). • An alternative pathway involving dichloramine proposes the formation of a chlorinated UDMH intermediate that is oxidized by dissolved oxygen (Schreiber and Mitch, 2006). • The formation of 5 other *N*-nitrosamines (i.e. NMOR, NEMA, NDEA, NPYR and NPIP) from their corresponding secondary amines were an order of magnitude higher with dichloramine than with monochloramine (Schreiber and Mitch, 2006).
Other Pathways	• Bromide can catalyse the formation of NDMA via both DMA-monochloramine interactions and HOCl-enhanced nitrosation. The reaction is more significant for chloramination than chlorination (Valentine *et al.* 2005). • A pathway of NDMA formation via the oxidation of natural organic matter (NOM) with monochloramine has been hypothesized, although specific precursors and reactions were not determined (Chen and Valentine, 2006). • NDMA formation through ozonation is hypothesized to occur through direct nitrosation (Nawrocki and Andrzejewski, 2011).

produce measurable amounts of NDMA. Chlorine dioxide (ClO_2) and ozone (O_3) can act as oxidants and produce NDMA, but this does not occur in all cases (Andrzejewski *et al.* 2008; Najm and Trussell, 2001; Zhao *et al.* 2008). Inconsistencies between studies suggest that NDMA formation from the action of ClO_2 and O_3 may be limited to very specific conditions and precursors. One study has suggested that O_3 treatment may not produce NDMA in all waters, but may produce NDMA precursors like DMA (Zhao *et al.* 2008). Pre-ozonation before chloramination has also been found to increase *N*-nitrosamine yields from secondary amines, compared to chloramination alone (Bond and Templeton, 2011). Since oxidation of water using ClO_2 or O_3 has been a recommended strategy to reduce *N*-nitrosamines (Zhao *et al.* 2008), more research is needed to clarify possible *N*-nitrosamine formation mechanisms from these oxidants.

The best-known organic nitrogen precursors for *N*-nitrosamines are secondary aliphatic amines (Gerecke and Sedlak, 2003; Mitch *et al.* 2003a; Sacher *et al.* 1997). Dimethylamine, in particular, is widely considered as a model precursor for NDMA (Choi and Valentine, 2003; Fiddler *et al.* 1972; Mitch *et al.* 2003a; Mitch and Sedlak, 2004). Although it has been presumed that DMA exists in natural waters because of animal wastes and microbial respiration, little evidence has been offered to support this because amines are difficult to quantify at the low concentrations expected in natural waters (Gerecke and Sedlak, 2003; Sacher *et al.* 1997). NDMA precursor analysis (Mitch *et al.* 2003a), also known as the NDMA formation potential test, has been employed in many studies to identify *N*-nitrosamine precursors in natural waters and wastewaters (Gerecke and Sedlak, 2003; Mitch *et al.* 2003a; Mitch and Sedlak, 2004). The test involves the reaction of water samples with a large excess of monochloramine over several days, where the aim is to exhaust all NDMA precursors. Many potential precursors have been identified by the formation potential test or similar analysis (Table 6.3).

Table 6.3 Precursors of *N*-nitrosamines identified to date.

Precursors	Notes
Secondary Amines	• Dimethylamine is considered the model NDMA precursor (Choi and Valentine, 2003; Fiddler *et al.* 1972). • Other secondary amines (e.g. diethylamine, methylethylamine and diphenylamine) have also been found to produce analogous *N*-nitrosamines (e.g. NDEA, NEMA and NDPhA) (Yang *et al.* 2009; Zhou *et al.* 2009b). • Aliphatic amines are likely to occur in natural waters as biodegradation products of proteins, amino acids or other nitrogen-containing compounds (Sacher *et al.* 1997).
Natural Organic Matter	• NDMA precursor analysis of drinking water NOM has demonstrated that DMA was not the major NDMA precursor, suggesting NOM contributes NDMA precursors from humic substances or N-containing biopolymers (Chen and Valentine, 2007; Gerecke and Sedlak, 2003).

(*Continued*)

Table 6.3 Precursors of *N*-nitrosamines identified to date (*Continued*).

Precursors	Notes
Anthropogenic Chemicals	• Pharmaceuticals containing DMA functional groups, particularly ranitidine, can form NDMA (Le Roux *et al.* 2011; Shen and Andrews, 2011). The formation cannot be attributed solely to release of DMA (Le Roux *et al.* 2011). Similarly, the personal care products lidocaine and DEET, which contain DEA functional groups, can produce (<1% molar conversion) NDEA (Shen and Andrews, 2011). • NDMA was found to form in similar yields from chloramination of some quarternary ammonium compunds as compared to DMA (Chang *et al.* 2011). • Other molecules that can hydrolyse to form DMA include dithiocarbamates, dithiocarbamate-based fungicides, and substituted phenylurea herbicides (Chen and Young, 2008; Mitch and Sedlak, 2004; Weissmahr and Sedlak, 2000). • *N,N*-Dimethylsulfamide (DMS), a degradation product from the fungicide tolylfluanide, was identified as a major NDMA precursor that contributed to 30–50% of NDMA formation during ozonation of surface waters and groundwater (Schmidt and Brauch, 2008). • A range of other anthropogenic compounds, including common dyes, were found to produce NDMA after ozonation (Nawrocki and Andrzejewski, 2011).
Chemicals and Products used in Water Treatment	• Nitrogen-containing cationic polymers used as coagulation aids (e.g. polydiallyldimethylammonium chloride, poly DADMAC), can form NDMA (Najm and Trussell, 2001; Wilczak *et al.* 2003), reacting with both chlorine (Kohut and Andrews, 2003) and chloramine (Park *et al.* 2009; Wilczak *et al.* 2003) • Ion-exchange resins consisting of quaternary amine functional groups also form NDMA (Gough *et al.* 1977; Najm and Trussell, 2001), postulated to result from precursors present as impurities or from resin degradation (Fiddler *et al.* 1972; Gough *et al.* 1977; Mitch and Sedlak, 2004). NDEA and NDPA can also leach from some resins (Gough *et al.* 1977). • NDMA has been detected in drinking water supplies as a result of contact with rubber components of water supply pipelines (Morran *et al.* 2011), mostly likely due to direct release from the rubber.

The formation potential test has also been used to identify treatment factors that influence *N*-nitrosamine formation. One component of water treatment that has been found to be significant in NDMA formation is the use of organic nitrogen-containing polymers that can contribute significant concentrations of NDMA precursors (Richardson 2005; Wilczak *et al.* 2003). Water supplies derived from wastewater-impacted rivers may also lead to substantial NDMA formation during chloramination, due to the higher concentrations of DMA in wastewater effluent compared to natural waters (Gerecke and Sedlak, 2003). Studies of NDMA formation potentials upon chloramination of waters from five treatment plants in South Australia indicated that plant filtration, polymer dose and sludge supernatant return appeared to influence NDMA formation (Morran *et al.* 2009). However, no clear correlations between any single factor and NDMA formation potentials at the full-scale were identified, suggesting that a combination of many factors must be considered.

6.2.4 Treatment and control

There is extensive literature on the treatment of *N*-nitrosamines in water, and key treatment methods are listed in Table 6.4. *N*-Nitrosamines are very soluble in water, and many common water treatment methods are inefficient for their removal. UV irradiation is the most widely used remediation method for NDMA (Mitch *et al.* 2003b; Stefan and Bolton, 2002) and well-established, full-scale systems are already in place (Jobb *et al.* 1994; Lem, 2005). However UV irradiation for NDMA removal is energy intensive, requiring a UV dosage approximately 10 times higher than that for equivalent virus removal (Mitch *et al.* 2003b).

Table 6.4 *N*-Nitrosamine removal by common and novel water treatment methods.

Treatment	Notes
Granular Activated Carbon	• Not effective (Fleming *et al.* 1996)
Reverse Osmosis	• Low removal (50–75%) of NDMA (Mitch *et al.* 2003b; Van Buynder *et al.* 2009), although removal of *N*-nitrosamines increases with increasing molecular weight (Miyashita *et al.* 2009).
UV Irradiation	• Most widely used removal method for NDMA (Mitch *et al.* 2003b), although dosage required for 90% removal is ~10 times higher than for equivalent virus removal (Mitch *et al.* 2003b). • Rate of removal can be affected by nature of the *N*-nitrosamine side groups, and the presence of organic or particulate matter (Chen *et al.* 2010). *N*-Nitrosamines can reform from degradation products (Liang *et al.* 2003).

(*Continued*)

Table 6.4 *N*-Nitrosamine removal by common and novel water treatment methods (*Continued*).

Treatment	Notes
Advanced Oxidation (UV/O_3, UV/H_2O_2, O_3/H_2O_2)	• Variable removal and relatively high O_3 and H_2O_2 doses required (Lee *et al.* 2007a; Xu *et al.* 2009; Zhao *et al.* 2008). • Relatively slow oxidation rates mean that *N*-nitrosamine removal may not be efficient in waters with high NOM concentrations (Landsman *et al.* 2007). • O_3 may not be practically feasible without UV disinfection due to the potential for bromate formation (Lee *et al.* 2007a). • *N*-Nitrosamine degradation products may also react with hydroxyl radicals, ensuring reformation does not occur (Landsman *et al.* 2007; Xu *et al.* 2009).
Biodegradation	• Demonstrated in laboratory experiments under both aerobic and anaerobic conditions (Gunnison *et al.* 2000; Oliver 1983; Tate and Alexander, 1975), but limited field demonstrations (Drewes *et al.* 2006; Zhou *et al.* 2009a). • Biodegradation rate depends on adaptation of biocommunity (Drewes *et al.* 2006; Patterson *et al.* 2010). • NDMA removal by biodegradation also possible during rapid sand filtration in drinking water and wastewater treatment plants (Hollender *et al.* 2009; Schmidt and Brauch, 2008; Schmidt and Brauch, 2008).
Reduction by Zero-Valent Iron	• Rate of removal increased with nickel addition but catalyst was easily poisoned (Gui *et al.* 2000; Odziemkowski *et al.* 2000).
Adsorption by Cyclodextrin Nanoporous Polymers	• Significant reduction of NDMA (estimated 80%) in bench-scale experiments, but only one study has been performed on water samples and it has not been tested in large-scale systems (Mhlongo *et al.* 2009).

N-Nitrosamine concentrations are often better controlled by removing *N*-nitrosamine precursors, or by controlling treatment parameters implicated in *N*-nitrosamine formation (Table 6.5). In many cases *N*-nitrosamine precursors are more efficiently removed by treatment than the *N*-nitrosamines themselves. For example, while RO removes NDMA inefficiently, it is effective in removing organic NDMA precursors, including colloidal precursors and also dissolved secondary and tertiary amines, which are normally protonated at neutral pH (Mitch *et al.* 2003b; Miyashita *et al.* 2009). However, RO is energy intensive and relatively expensive, and thus controlling formation through pre-treatment may be more cost effective (Charrois and Hrudey, 2007; Chen and Valentine, 2008). Methods to control the levels of nitrogenous substances, such as nitrite, nitrate, ammonia, and hydroxylamine, may be helpful in reducing *N*-nitrosamine concentrations during water treatment (Yang *et al.* 2009).

Table 6.5 *N*-Nitrosamine minimisation by precursor removal or treatment optimisation.

Treatment	Notes
Reverse Osmosis	• Almost complete removal of dissolved NDMA precursors demonstrated in secondary wastewater (Farre *et al.* 2011; Mitch and Sedlak, 2004). • Removal of secondary amines (DMA, MEA, DEA and DPA) better than 99% because they are charged at neutral pH (Miyashita *et al.* 2009)
Pre-oxidation with Ozone	• Up to 95% removal of precursors (Andrzejewski *et al.* 2008) but formation of NDMA has been observed (Zhao *et al.* 2008). • Dose optimisation required to minimise production of O_3-DBPs such as bromate (Lee *et al.* 2007a) and trihalonitromethanes (Krasner *et al.* 2006).
Pre-oxidation with Ferrate	• Good removal (up to 84%) of precursors but requires high doses for slow reacting precursors, including DMA (Lee *et al.* 2008). Less efficient than O_3 or ClO_2 for slow-reacting precursors.
Pre-chlorination	• Can reduce *N*-nitrosamine formation by up to 93% through free chlorination of precursors (Charrois and Hrudey, 2007; Chen and Valentine, 2008) but susceptible to increased formation of halogenated DBPs.
Chloramine Speciation Control	• Dichloramine formation can be minimised by preforming chloramines at $pH > 8.5$ or by adding chlorine before ammonia (Schreiber and Mitch, 2005). • Demonstrated in secondary wastewaters (Mitch *et al.* 2005) but efficacy not known in drinking waters.

6.3 HALONITRILES

6.3.1 Background, toxicity and regulatory status

The halonitriles can be classified in a number of groups: halogenated acetonitriles (HANs) (Table 6.6); halogenated nitriles (HNs) of higher congeners, for example, 3,4-dichlorobutanenitrile, 2,3-dichloro-3-bromopropanenitrile (Table 6.7); halogenated acrylonitriles (e.g. trichloropropenenitrile) and cyanogen halides. Most reports consider the cyanogen halides separately from other HNs and thus cyanogen halides will be discussed in the next section of this chapter. The HANs have been studied most extensively of all HNs. Up to nine HANs can form, and numerous studies have shown that they are nearly ubiquitous in chlorinated drinking water (Krasner *et al.* 1989; Oliver 1983). Although HANs usually form at much lower concentrations than the regulated THMs and HAAs during chlorination, they are relatively toxic and are among the 50 DBPs that were predicted to have the highest carcinogenic potential (Richardson, 2003).

Table 6.6 Halogenated acetonitriles considered, including structure, health values, and structure-activity relationship (SAR) classification (Woo *et al.* 2002), where available. M = likely to be a moderately active multispecies/target carcinogen at relatively high doses or active single species/target carcinogen at low doses; LM = likely to be weakly carcinogenic, or carcinogenic toward a single species/target at relatively high doses.

Compounds	Structure	Health value and source (μg/L)	SAR
Dibromoacetonitrile (DBAN)	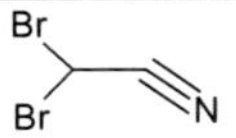	70 (WHO, 2011) 20 (USEPA, 2007)	Not classified
Dichloroacetonitrile (DCAN)	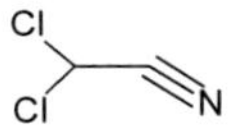	20 (provisional) (WHO, 2011) 6 (USEPA, 2007)	LM
Chloroacetonitrile (CAN)	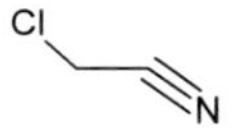	N/A	LM
Trichloroacetonitrile (TCAN)	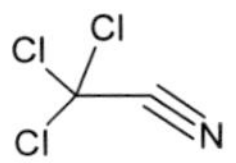	N/A	LM
Bromoacetonitrile (BAN)	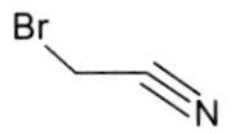	N/A	LM
Bromochloroacetonitrile (BCAN)	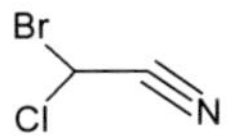	N/A	M
Tribromoacetonitrile (TBAN)	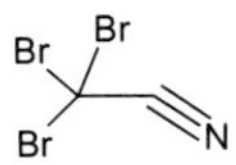	N/A	LM
Bromodichloroacetonitrile (BDCAN)	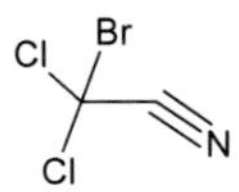	N/A	LM
Dibromochloroacetonitrile (DBCAN)	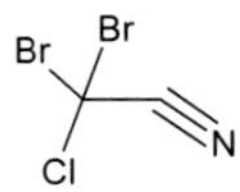	N/A	LM

N/A = Not applicable.

Table 6.7 List of other halogenated nitriles, structure, and structure-activity relationship (SAR) classification (Woo *et al.* 2002), where available. M = likely to be a moderately active multispecies/target carcinogen at relatively high doses or active single species/target carcinogen at low doses. No health value data currently exists for these compounds.

Compounds	Structure	SAR
3,4-Dichlorobutanenitrile	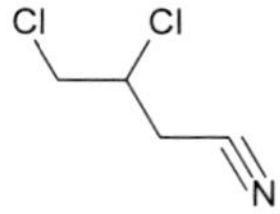	M
cis-2,3,4-Trichloro-2-butenenitrile and *trans*-2,3,4-Trichloro-2-butenenitrile	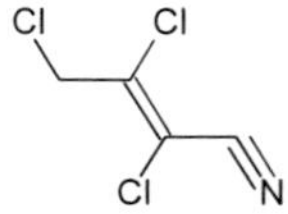 (*trans*)	M
2,3-Dichloro-3-bromopropanenitrile	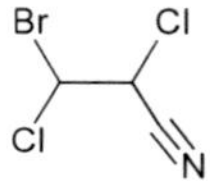	Not classified
3-Bromopropanenitrile	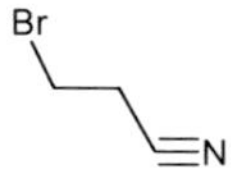	Not classified
Trichloropropenenitrile	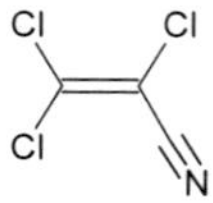	M
2,2-Dichloropropanenitrile (2,2-DCPN)	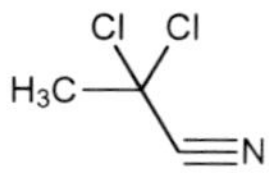	Not classified
2,2-Dibromobutanenitrile (2,2-DBBN)	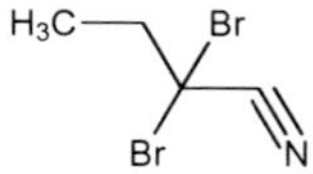	Not classified

HANs are reactive and toxic to gastrointestinal tissues (Lipscomb *et al.* 2009) and have been deemed to be more cytotoxic and genotoxic than the regulated HAAs (Muellner *et al.* 2007). Both DCAN and TCAN may also cause adverse effects on fetal development at high exposure (Smith *et al.* 1989). DCAN, BCAN, CAN and TCAN have been found to be direct acting mutagens (Bull *et al.* 1985; Muller-Pillet *et al.* 2000), and DBAN and BCAN initiated tumours when applied to mouse skin (Bull *et al.* 1985). The brominated HANs generally exhibit higher cytotoxic and genotoxic potencies than the chlorinated analogues (Muellner *et al.*

2007; Muller-Pillet *et al.* 2000). Because of insufficient data, HANs were considered non-classifiable as carcinogenic to humans under the International Agency for Research on Cancer (IARC) (WHO IARC, 1991). In the recent USEPA drinking water standards and health advisory document, both DCAN and DBAN were classified into Group C as possible human carcinogens (USEPA, 2007). While published health values exist for DCAN and DBAN, more toxicological data are required for other HANs to identify appropriate guideline values.

While some longer-chain halonitriles and acrylonitriles have been identified as DBPs (Kristiana *et al.* 2012; Richardson *et al.* 1996; Woo *et al.* 2002), information on their toxicity, occurrence and formation in drinking water appears almost non-existent. However, it has been noted that if 2,2,-dichloropropionitrile was assumed to be of equivalent potency to DBAN, concentrations measured to date would not increase cancer risk significantly, even though it is mutagenic (Bull *et al.* 2011). Like the HANs, the higher order HN were considered non-classifiable as carcinogenic to humans under the IARC (WHO IARC, 1991).

6.3.2 Occurrence in drinking water

HAN concentrations in treated drinking waters are generally below 10 μg/L (Krasner *et al.* 1989; Williams *et al.* 1997), but concentrations as high as 40 μg/L have been reported (McGuire *et al.* 2002; Trehy and Bieber, 1981). Although they are not currently regulated, selected HANs (DCAN, TCAN, BCAN, DBAN) have been measured in several large scale occurrence studies, including surveys of drinking water utilities in U.S. and Canada (Krasner *et al.* 1989; Williams *et al.* 1997), a U. S. nationwide occurrence study of priority unregulated DBPs (Weinberg *et al.* 2002), and the USEPA's Information Collection Rule (McGuire *et al.* 2002), which involved 500 large drinking water plants. Typically DCAN, DBAN and BCAN were most frequently detected, with DCAN being the most prevalent, present in 97% of all samples from the Canadian study (Williams *et al.* 1997). While several disinfectants have been shown to form HANs, plants using chloramines had the highest levels in finished drinking water (McGuire *et al.* 2002) and chloraminated distribution systems also had higher HAN concentrations than chlorinated systems.

Studies in which all 9 HANS have been measured have shown that all can be detected (Kawamoto and Makihata, 2004), although concentrations were generally low (<5 μg/L) and DCAN, BCAN and DBAN remained the predominant species. Ambient bromide levels are known to shift the speciation from the chlorine-containing species to the bromine-containing species (Richardson *et al.* 2003) which is a concern due to the higher carcinogenicity or mutagenicity of brominated DBPs compared to their chlorinated analogues (Muellner *et al.* 2007).

6.3.3 Analysis

Analytical methods for HANs can typically achieve sub-μg/L limits of detection (Table 6.8). While the most commonly used sample preparation method is LLE

(Munch and Hautman, 1995), both closed-loop stripping and SPME have been shown to have wider applicability and sensitivity (Kampioti and Stephanou, 1999; Kristiana *et al.* 2012). To date, there is only one reported method to quantify higher HNs (Kristiana *et al.* 2012).

Table 6.8 Methods for halogenated nitrile analysis and corresponding method detection limits (MDLs).

Extraction method	Instrumentation	Analytes	MDL (µg/L)	Reference
LLE (MTBE)	GC-ECD	4 HAN	<0.006	(Munch and Hautman, 1995)
	GC-MS	9 HAN	0.03–0.26	(Kawamoto and Makihata, 2004)
LLE (Pentane)	GC-ECD	3 HAN	0.05–0.3	(Oliver, 1983)
Closed-loop Stripping	GC-ECD	4 HAN and other DBPs	~0.001	(Kampioti and Stephanou, 1999)
SPME	GC-MS	6 HAN, 2,2-DCPN, 2,2-DBBN	0.001–0.08	(Kristiana *et al.* 2012)

6.3.4 Formation and precursors

While studies to date have provided information on the precursors and chemistry of HN formation Table 6.9, more research is needed to elucidate reaction mechanisms. There is considerable evidence that HNs form with both chlorine (Oliver 1983; Reckhow *et al.* 1990; Trehy *et al.* 1986; Ueno *et al.* 1996) and chloramine (McGuire *et al.* 2002; Yang *et al.* 2010) disinfection, as well as limited evidence that some HNs (DBAN and 3-bromopropanenitrile) can form when ClO_2 is used as a disinfectant (Richardson *et al.* 2003). Unlike other classes of N-DBPs, HAN formation does not appear to be influenced by UV irradiation (Reckhow *et al.* 2010; Shah *et al.* 2011).

Continuing reactions of HANs with residual chlorine can lead to rapid decay of HANs after formation (Reckhow *et al.* 1990). As a result, the concentration of HANs detected in chlorinated waters depends on the balance between formation and degradation rates of each HAN species. Since HAN precursors may be consumed during the initial contact with chlorine, HAN concentrations will often decrease within the distribution system, and this has been demonstrated in some distribution systems (Chen and Weisel, 1998) and formation experiments (Kim *et al.* 2003). Theoretically, monochloramine does not react further with HANs and hence they could accumulate in these systems (Lee *et al.* 2007b).

Table 6.9 Precursors of halogenated nitriles identified to date.

Precursors	Notes
Amino Acids	• DCAN can be produced from the chlorination of amino acids, via the oxidation by chlorine to form nitriles, and subsequent chlorine substitution (Trehy *et al.* 1986). • While certain amino acids (e.g. aspartic acid, asparagine, tryptophan, kynurenine, glutamic acid and histidine) react with chlorine to form HANs, others (e.g. glycine, alanine, serine) do not (Ueno *et al.* 1996; Bond *et al.* 2009). • Some amino acids (e.g. tryptophan, glutamic acid and cysteine) have also been observed to form DCAN during chloramination (Yang *et al.* 2010). • Free amino acid concentrations in natural waters are expected to be too low for amino acids to be the primary HAN precursors (Oliver, 1983; Reckhow *et al.* 2001). Most amino nitrogen in natural waters is present as proteinaceous material or humic structures (Tuschall and Brezonik, 1980).
Natural Organic Matter	• Both algae and nitrogen-containing humic substances have been observed to form HANs during chlorination (Fang *et al.* 2010; Oliver 1983; Reckhow *et al.* 1990) • HAN formation was highest for chlorination of hydrophilic NOM fractions, and lowest for hydrophobic and transphilic fractions (Dotson *et al.* 2009; Hwang *et al.* 2001), correlating with the degree of nitrogen enrichment in each fraction. • Studies using fractionated DOM isolates (Lee *et al.* 2007b) and algal organic matter (Fang *et al.* 2010) suggest that the degree of nitrogen enrichment is an important influence on DCAN yields during chlorination, but not chloramination.
Anthropogenic and Human Sources	• Pharmaceuticals containing DMA functional groups, particularly ranitidine, have been shown to form DCAN (Le Roux *et al.* 2011). • HANs in swimming pools can be derived from nitrogen-containing compounds of human origin, such as urea and hair proteins (Kim *et al.* 2002). Such precursors are unlikely to be present in high quantities in drinking water sources. • Cytosine, one of the four bases found in DNA and RNA, can form DCAN during chloramination (Yang *et al.* 2010).

While occurrence studies generally show higher concentration of HANs in chloraminated waters (McGuire *et al.* 2002), yields of HANs in other studies have been significantly lower after chloramination than chlorination, suggesting the existence of different formation pathways and precursors for each disinfectant

(Dotson *et al.* 2009; Yang *et al.* 2007). Little is known about HAN formation during chloramination, although several studies suggest that monochloramine can produce the nitrile-nitrogen of DCAN (Lee *et al.* 2007b; Yang *et al.* 2007; Yang *et al.* 2010). Pre-formed chloramination has also been shown to form less DCAN than in-line formed chloramination, attributed to HN formation from organic chloramines formed through reaction of free chlorine with organic amines (Hayes-Larson and Mitch, 2010).

6.3.5 Treatment and control

Very limited research has focussed on the removal of HNs in drinking water treatment (Table 6.10). HN formation is controlled by many variables, including temperature, pH, disinfectant concentration and contact time, bromide concentration, and the structural characteristics of organic and inorganic precursors in source water. Hence, HN minimisation in drinking water may be best approached by controlling these variables and by removing HN precursors. This is particularly so, given HANs may serve as precursors for other regulated DBPs (i.e. THMs and HAAs) when they degrade in the presence of free chlorine (Ueno *et al.* 1996).

Table 6.10 Halogenated acetonitrile minimisation by precursor removal or treatment optimisation.

Treatment	Notes
Reverse Osmosis	• Studies of RO treatment for desalination (Agus and Sedlak, 2010) and wastewater reuse (Van Buynder *et al.* 2009) indicate removal efficiencies of 50–75% for dihalogenated acetonitriles, although only a few data points were available for each compound.
pH Control	• Increasing pH enhances base-catalysed decomposition, especially for HANs with a higher degree of halogenation (Chiang *et al.* 2010; Yang *et al.* 2007).
Disinfectant Dose	• While increasing chlorine or monochloramine concentration generally increases HAN formation within short contact times, increased chlorine concentration also enhances HAN degradation (Chiang *et al.* 2010; Yang *et al.* 2007).
UV Photolysis	• Several dihalogenated acetonitriles have been shown to degrade by photolysis (Chen *et al.* 2010), with decay increasing with increasing degree of halogenation and/or replacement of chlorine with bromine.
Precursor Removal	• Removal of the hydrophilic fractions of NOM is expected to reduce HAN formation, however removal of hydrophilic NOM through conventional treatment processes has been found to be highly variable and always poorer than hydrophobic NOM removal (Hwang *et al.* 2001).

(*Continued*)

Table 6.10 Halogenated acetonitrile minimisation by precursor removal or treatment optimisation (*Continued*).

Treatment	Notes
	• Coagulation-dissolved air flotation-filtration was found to reduce algae, low molecular weight DON, and result in lower HAN formation (Chu *et al.* 2011). • Riverbank filtration was found to remove HAN precursors by 30–80% through combined physical, chemical and biological processes, although the main mechanism of removal was not elucidated (Weiss *et al.* 2003). • Pre-ozonation reduced DCAN and BCAN formation in one laboratory study (Chiang *et al.* 2010).

6.4 CYANOGEN HALIDES

6.4.1 Background, toxicity and regulatory status

The cyanogen halides (CNX, Table 6.11) are predominantly chloramination by-products (Cancho *et al.* 2000; Krasner *et al.* 1989) because they undergo chlorine-catalyzed hydrolysis to form cyanate in the presence of free chlorine. The CNXs are toxic because of the cyanide (CN) group (Kononen, 1988) and are rapidly transformed to toxic products by metabolic processes. The acute toxicity of CNXs is similar to the effects of cyanide, and has been demonstrated in mammals (Munro *et al.* 1999). Conversely, the carcinogenic potential of CNXs, assessed by structure-activity relationships, is considered to be low (Woo *et al.* 2002). There is little other data on the oral toxicities of CNXs. Because CNXs are rapidly converted to cyanide within the body, drinking water guidelines have been based on the health guideline for cyanide (NHMRC and NRMMC, 2011; WHO, 2011). The ADWG recommends a guideline of 80 µg/L total cyanogenic compounds (NHMRC and NRMMC, 2011), although it should be noted that WHO have recently removed their previous guideline (70 µg/L total cyanogenic compounds) on the basis that cyanide occurs in drinking water at concentrations well below those of health concern, except in emergency spills (WHO, 2011).

Table 6.11 Cyanogen halides and their structures.

Compound	Structure	Compound	Structure
Cyanogen chloride	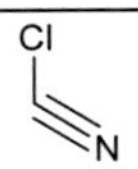	Cyanogen bromide	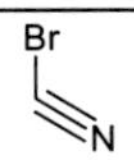

6.4.2 Occurrence in drinking water

Occurrence data for CNXs in drinking water are limited. Cyanogen chloride (CNCl) was first reported in the survey of 35 drinking water utilities in the U.S. (Krasner *et al.* 1989), in which average concentration at utilities that used pre-chlorination followed by ammonia addition (2.2 μg/L) were higher than average concentration for utilities that used only free chlorine (0.4 μg/L). Other studies have also reported CNCl levels that are higher when chloramines are used as the final disinfectant rather than chlorine (Jacangelo *et al.* 1989; Simpson and Hayes, 1998). CNBr has been observed after chloramination or ozonation of water containing bromine and/or bromide (Cancho *et al.* 2000; Diehl *et al.* 2000; Heller-Grossman *et al.* 1999), with maximum concentration of 6.4 μg/L (Cancho *et al.* 2000). CNBr is often the only CNX formed in bromide-containing waters (Cancho *et al.* 2000; Heller-Grossman *et al.* 1999). Ozonation has also been found to favour CNBr formation (Cancho *et al.* 2000).

6.4.3 Analysis

USEPA Method 524.2 has been recommended for CNXs, which involves extraction from water using the purge-and-trap technique followed by GC-MS (Munch 1995; Sclimenti *et al.* 1996). However, the analysis of CNBr using this method is limited by its low vapour pressure and high water solubility (Cancho *et al.* 2000), hence alternative GC-based methods have been proposed, as listed in Table 6.12. More recently, membrane introduction mass spectrometry (MIMS) has been employed, eliminating the need for sample preparation (Yang and Shang, 2005). Analytical detection limits were typically higher than other chromatographic techniques, but still below guideline values, and the method provides linear calibration over a much wider concentration range. The method has the added advantage of potentially allowing on-line monitoring.

Table 6.12 Methods for cyanogen halide analysis and corresponding method detection limits (MDLs).

Extraction method	Instrumentation	MDL (μg/L)		References
		CNCl	CNBr	
Purge-and-trap	GC-MS	0.5	N/A	(Munch, 1995; Sclimenti *et al.* 1996)
Micro LLE	GC-ECD	0.5	N/A	(Sclimenti *et al.* 1996)
Headspace injection	GC-ECD	0.04	0.2	(Xie and Reckhow, 1993)
Headspace SPME	GC-ECD	0.08	0.04	(Cancho *et al.* 2000)
None	MIMS	1.2	3.8	(Yang and Shang, 2005).

N/A = not applicable.

6.4.4 Formation and precursors

Like the HANs, CNX formation differs between chloramination and chlorination, influenced by the susceptibility of CNXs to degrade in the presence of free chlorine (Mitch and Schreiber, 2008; Yang *et al.* 2007). Free chlorine catalyses CNCl hydrolysis to cyanic acid, with a half-life of 30 min under typical chlorine disinfection conditions (Lee *et al.* 2006; Na and Olson, 2006). Thus it is possible that CNCl may decay before it can be observed. On the other hand, monochloramine does not catalyse CNX decay and higher CNCl concentrations were detected after 24 hours when ammonia was added (Na and Olson, 2006). Studies have also demonstrated that monochloramine can contribute to the nitrogen source of CNCl (Mitch and Schreiber, 2008; Yang *et al.* 2007).

Cyanogen bromide also decays in the presence of free chlorine, but to a lesser extent than CNCl (Xie and Reckhow, 1993). The high concentrations of CNBr formed (>30 μg/L) compared to the typically lower concentrations (<1 μg/L) of CNCl in chlorinated systems may be attributed to their different degradation rates in the presence of free chlorine (Heller-Grossman *et al.* 1999). Precursors currently identified for formation of CNXs are listed in Table 6.13.

Table 6.13 Precursors and sources of cyanogen halides.

Precursors	Notes
Amino Acids	• Glycine is an efficient CNX precursor (Shang *et al.* 2000) and one of the most abundant amino acids in natural waters. Threonine, tyrosine and serine also produce minor yields of CNCl during chlorination (Lee *et al.* 2006; Shang *et al.* 2000). • The formation rate of CNCl from glycine is fast (Lee *et al.* 2006). The slow formation observed in some studies (Lee *et al.* 2006; Zheng *et al.* 2004) is attributed to other precursors. • Cyanogen chloride also forms during chloramination of serine, although at serine concentrations higher than expected in natural waters (Zheng *et al.* 2004). Other amino acids such as glycine have not been tested.
Primary Amines	• Chlorination of monomethylamine forms CNCl (Joo and Mitch, 2007), however relatively low concentrations of primary amines in natural waters suggest that other precursors, such as amino acids, are more important.
Natural Organic Matter	• Cyanogen chloride (0–15% yield) has been observed after the reaction of chloramine with humic acid (Ohya and Kanno, 1985), fulvic acid (Xie and Reckhow, 1993), peptides and purines (Yang *et al.* 2010), with aldehydes and nitriles identified as key intermediate products (Yang *et al.* 2010).

(*Continued*)

Table 6.13 Precursors and sources of cyanogen halides (*Continued*).

Precursors	Notes
Aldehydes	• Monochloramine reacts with aldehydes (e.g. formaldehyde) to form intermediates that decay to CNCl (Pedersen *et al.* 1999). An analogous reaction with bromamine or bromochloramine may yield CNBr in the presence of bromide (Lei *et al.* 2006). • Aldehydes can form from the reaction of monochloramine and tertiary alkylamines like trimethylamine (Mitch and Schreiber, 2008), however experimentally observed yields were much lower than model predictions. • The aldehyde reaction pathway cannot account for the increased formation observed with elevated dissolved organic nitrogen concentrations (Joo and Mitch, 2007).
Bromide	• Increasing bromide shifts production of CNCl to CNBr, and can increase the total CNX concentration (Heller-Grossman *et al.* 1999; Xie and Reckhow, 1993). • Fulvic acid has been shown to form CNBr during chloramination when the bromide concentration was >0.3 mg/L (Xie and Reckhow, 1993), presumed to occur when hypobromous acid reacts with nitrogen-containing precursors (Heller-Grossman *et al.* 1999). • CNBr may form after reaction of monobromamine or dibromamine with hydrogen cyanide or cyanide ions (Lei *et al.* 2006).

6.4.5 Treatment and control

Effective removal methods for major CNX precursors like glycine have not been experimentally evaluated. However, the rapid decay of CNCl under typical chlorine disinfection conditions (Lee *et al.* 2006; Na and Olson, 2006) and the sensitivity of CNX formation to pH (Heller-Grossman *et al.* 1999; Yang *et al.* 2007) means that CNX concentrations can be reduced by promoting decomposition and/or reducing formation reaction rates (Table 6.14).

Table 6.14 Summary of treatment methods for cyanogen halides.

Treatment method	Notes
Ozone	• Ozone has been hypothesized to degrade glycine (Lee *et al.* 2006) but this has not been demonstrated in practice.
pH Control	• Cyanogen halide formation decreases with increasing pH (Heller-Grossman *et al.* 1999), potentially from base-catalyzed decomposition (Diehl *et al.* 2000). • Monochloramine and formaldehyde also react approximately eight times slower at pH 6 than at pH 5 (Yang *et al.* 2007).

(*Continued*)

Table 6.14 Summary of treatment methods for cyanogen halides (*Continued*).

Treatment method	Notes
Disinfectant Dose	• Free chlorine catalyses hydrolysis of cyanogen chloride with a half-life of 30 minutes under typical chlorine disinfection conditions (Lee *et al.* 2006; Na and Olson, 2006).
Granular Activated Carbon (GAC)	• Cyanogen chloride and CNBr were completely removed by GAC filtration (Cancho *et al.* 2000), although both formed again after final chlorination, albeit at lower concentrations.

6.5 HALONITROALKANES

6.5.1 Background, toxicity and regulatory status

The halonitromethanes (HNMs) are the main halonitroalkanes to be recognised as DBPs, and this section focuses on the nine possible HNMs (Table 6.15). The only higher alkyl counterpart to have been identified in drinking water is dichloronitroethane, which has been assessed as a moderately relevant carcinogen (Woo *et al.* 2002). Dichloronitropropane has been produced during model studies of chlorination and chloramination of primary amines, although not detected in actual drinking waters (Joo and Mitch, 2007). HNMs are structurally similar to the halomethanes, except that they have a nitro-group bonded to the central carbon atom, instead of a hydrogen atom. Chloropicrin was the first HNM to be identified and is the most commonly detected (Cole *et al.* 2007). Chloropicrin may also be introduced into water systems through agricultural practices, where it is widely used as a soil fumigant. Bromonitromethanes have also been identified in drinking waters (Plewa *et al.* 2004; Richardson *et al.* 1999a), which raises concerns because of their increased toxicity compared to chlorine analogues (Plewa *et al.* 2004; Richardson *et al.* 2003).

Table 6.15 Halonitromethanes considered, including structure.

Compound	Structure	Compound	Structure
Trichloronitromethane (Chloropicrin)	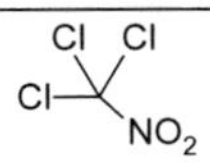	Dichloronitromethane	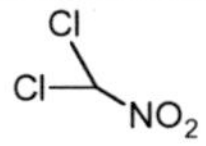
Tribromonitromethane (Bromopicrin)	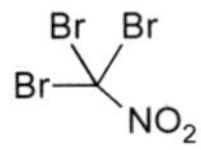	Bromochloronitromethane	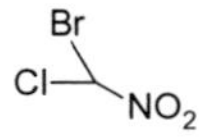
Bromonitromethane	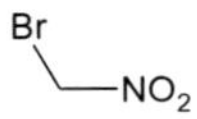	Bromodichloronitromethane	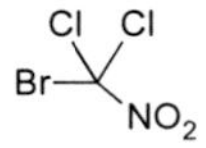

(*Continued*)

Table 6.15 Halonitromethanes considered, including structure (*Continued*).

Compound	Structure	Compound	Structure
Chloronitromethane	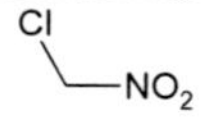	Dibromochloronitromethane	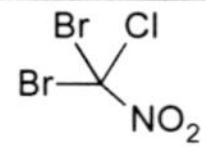
Dibromonitromethane	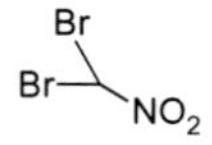		

Bromonitromethanes have been assessed as being among the DBPs with the highest potential toxicities (Woo *et al.* 2002). Bromonitromethanes are extremely cytotoxic and genotoxic to mammalian cells, with the toxicity increasing with bromine substitution (Plewa *et al.* 2004), which is consistent with the generalisation that brominated DBPs are more genotoxic than their chlorinated counterparts (Richardson *et al.* 2007; Woo *et al.* 2002). HNMs are weak mutagens in *Salmonella* TA100 (Kundu *et al.* 2004), but neither chloropicrin nor bromonitromethane have shown mutagenicity in studies using mammalian cells (Liviac *et al.* 2011). *In vivo* toxicity data for the HNMs are extremely limited, and attempts to evaluate the carcinogenic effects of chloropicrin in rats and mice have produced inconclusive results (NIH, 1978). Because of the lack of available information, major health authorities (e.g. WHO, European Union) have not been able to determine guideline values for the presence of HNMs in drinking water and more toxicology data is required to evaluate the health effects of the range of halonitroalkanes for *in vivo* systems.

6.5.2 Occurrence in drinking water

Typical concentrations of halonitroalkanes in finished drinking waters rarely exceed 5 μg/L and are mostly below 1 μg/L (Simpson and Hayes, 1998; Weinberg *et al.* 2002; Williams *et al.* 1997). HNMs have been estimated to typically represent 3–4% of the mass of THMs (Bond *et al.* 2011). Chloropicrin has been the most commonly measured HNM and has been reported in a wide variety of source waters with chlorine or O_3-chlorine treatments, at concentrations up to 6 μg/L (Duguet *et al.* 1985; Hoigné and Bader, 1988; Williams *et al.* 1997). Preceding chlorination with ozonation, in particular, increased chloropicrin concentrations with increasing O_3 dosages (Hoigné and Bader, 1988). Brominated and chlorinated mono-, di- and tri-HNMs have been identified in bench-scale and pilot-scale studies of chlorine and chloramines, with and without ozonation (Krasner *et al.* 2006; Plewa *et al.* 2004; Richardson *et al.* 1999a; Thibaud *et al.* 1987). Pre-ozonation (before chloramination) increased the formation of brominated HNMs in source water with a bromide concentration of 0.14 mg/L, also supporting the finding that ozonation plays an important role in HNM formation (Weinberg *et al.* 2002).

6.5.3 Analysis

Halonitromethanes have been analysed using a number of different sample preparation methods, including LLE and closed-loop stripping, followed by GC separation (Table 6.16). HNM analysis can be challenging. Brominated HNMs are particularly susceptible to degradation, and common quenching agents like sodium thiosulfate and ascorbic acid can contribute to degradation (Merlet *et al.* 1985; Hoigné and Bader 1988; Weinberg *et al.* 2002). The HNMs are also thermally unstable and can decompose in the injection port or GC-MS transfer line, producing unusual mass spectra caused by hydrogen and bromide ion exchanges among some of their fragment ions (Chen *et al.* 2002). Therefore, several methods employ a GC injection temperature of ≤170°C and GC-MS transfer line temperature of ≤225°C in order to avoid decomposition (Chen *et al.* 2002; Richardson *et al.* 2003). Nevertheless, GC-ECD appears to be more sensitive, although clearly less specific, for volatile chlorinated DBPs such as chloropicrin, compared to GC-MS (Nikolaou *et al.* 2002).

Table 6.16 Methods used for halonitromethane analysis and corresponding method detection limits (MDLs).

Extraction method	Instrumentation	Analytes	MDL (μg/L)	References
LLE (MTBE)	GC-ECD	3 HNM	0.001–0.003	(Munch and Hautman, 1995)
Closed-loop Stripping	GC-ECD	Chloropicrin and 13 other DBPs	~0.001	(Kampioti and Stephanou, 1999)
Single drop microextraction	GC-MS	9 HNM	0.06–1.2	(Montesinos *et al.* 2011)

6.5.4 Formation and precursors

Halonitromethanes can be formed from chlorine, chloramine, O_3-chlorine, or O_3-chloramine disinfection (Cole *et al.* 2007; Richardson *et al.* 2007). Precursors that have been identified for HNM formation are listed in Table 6.17. Ozonation prior to chlorination or chloramination increases the formation of HNMs (Chiang *et al.* 2010; Hoigné and Bader 1988; Jacangelo *et al.* 1989; Krasner *et al.* 2006; Plewa *et al.* 2004). Early research (Merlet *et al.* 1985) hypothesized that O_3 was capable of producing nitrated precursors, either by oxidation of amines or production of NO_2^+ which could act as a nitrating agent, forming HNMs upon subsequent chlorination. However, this hypothesis has not been further investigated. Medium pressure UV irradiation has also been found to increase chloropicrin concentrations after chlorination (Reckhow *et al.* 2010), which is attributed to photonitration in the presence of nitrate and NOM. Low pressure UV, which emits no radiation in the nitrate-absorbing region, did not have the same effect. In another study of UV irradiation, HNMs did not increase in the absence of nitrate (Shah *et al.* 2011).

Table 6.17 Precursors of halonitromethanes identified to date.

Precursors	Notes
Nitrite	• Nitrite oxidation can produce nitrated organic compounds like nitromethane (Merlet *et al.* 1985), nitrophenol (Thibaud *et al.* 1987) and nitrated forms of humic acid (Choi and Richardson, 2004), which form chloropicrin after chlorination. • Typically low concentrations of nitrite limit this pathway.
Primary Amines	• Degradation of dichlorinated primary amines can form halonitroalkanes, after oxidation and subsequent chlorine addition (Joo and Mitch, 2007). • Monomethylamine and *n*-propylamine produced chloropicrin and dichloronitropropane, respectively, with higher yields during chloramination than chlorination (Joo and Mitch, 2007). Trimethylamine has been observed to form chloropicrin during chlorination (Mitch and Schreiber, 2008). • Yields of HNMs from primary amines are orders of magnitude lower than yields of other DBPs (Joo and Mitch, 2007).
Ion Exchange Resins	• Anion-exchange resins containing alkyl amine groups (e.g. trimethylamine, tributylamine) produced up to 10 µg/L chloropicrin following chlorination or chloramination of synthetic groundwater (Kemper *et al.* 2009). Resins with dialkylethanolamine moieties (e.g. dimethylethanolamine) were less susceptible to chloropicrin formation.
Natural Organic Matter	• HNM formation was significantly higher from hydrophilic NOM fractions than hydrophobic and transphilic fractions (Hu *et al.* 2010). Low molecular weight hydrophilic organic matter can persist during conventional water treatment processes. • Chloropicrin can be generated from chloramination of some amino acids (e.g. glycine, tryptophan), dipeptides (e.g. glycineglycine), purines (e.g. adenine) and pyrimidines (e.g. cytosine) (Yang *et al.* 2010).

6.5.5 Treatment and control

Chloropicrin can persist in aqueous environments for up to 10 days in the absence of light (Wilhelm *et al.* 1996). Precursor removal or treatment (Table 6.18) is therefore necessary prior to distribution (Mezyk *et al.* 2006).

Table 6.18 Halonitromethane minimisation by precursor removal or treatment.

Treatment	Notes
Activated Carbon	• Activated carbon partially removes chloropicrin (Duguet *et al.* 1985; Hoigné and Bader, 1988) and has been recommended for inclusion after ozonation/chlorination (Hoigné and Bader, 1988). The extent of removal of other halonitroalkanes is not known.
Biological Treatment	• Riverbank filtration has been shown to remove 70–100% of chloropicrin precursors (Weiss *et al.* 2003). • Chloropicrin can be biodehalogenated to nitromethane by the soil microbe *Pseudomonas putida* (Castro *et al.* 2002).
Zero Valent Iron	• Degradation of chloropicrin by zero-valent iron produces nitromethane, chloronitromethane, dichloronitromethane, and methylamine (Pearson *et al.* 2005). • Degradation is rapid at high initial concentrations of chloropicrin (~mg/L), but the efficacy of this treatment at typical drinking water concentrations (low μg/L) has not been verified.
Advanced Oxidation and UV Irradiation	• Degradation of chloropicrin by UV photolysis has been demonstrated in theoretical laboratory studies (Chen *et al.* 2010; Wilhelm *et al.* 1996), but not yet in water treatment. The rate of photolytic degradation has been predicted to be related to the degree of halogenation and/or bromination (Chen *et al.* 2010). • AOP degradation of 8 halonitromethanes in water has been modelled, but not yet tested (Mezyk *et al.* 2006). Bromide, chloride and nitrite are anticipated to be degradation products.
Temperature	• Chloropicrin concentrations in a New Jersey water distribution system during the cold season were on average less than half that measured during the warm season (Chen and Weisel, 1998), a trend observed for other DBPs as well.

6.6 HALOACETAMIDES

6.6.1 Background, toxicity and regulatory status

Haloacetamides (HAAms) have only recently been identified as drinking water DBPs (Krasner *et al.* 2006). They are considered to be produced almost exclusively from the degradation of the corresponding HANs (Glezer *et al.* 1999; Reckhow *et al.* 2001), although recent research has suggested that alternative pathways that do not include HAN degradation do exist (Chu *et al.* 2010). Limited research has been conducted on the toxicity of HAAms (Table 6.19). All of the HAAms, except for DCAAm, have been found to be genotoxic, more toxic than the regulated HAAs, but comparable to other N-DBPs, including the HANs and HNMs (Plewa *et al.* 2008). Like other halogenated DBPs, the combined cytotoxicites and genotoxicities of HAAms generally increase with the degree of halogenation, and in the order of iodo- > bromo- > chloro- groups. No carcinogenicity studies have been reported for this class, but TCAAm and DCAAm were considered to be of low concern for carcinogenic potential when analysed by structure-activity relationships (Woo *et al.* 2002). As HAAms are still a relatively new family of DBPs, drinking water guidelines have not been established by any major health authority.

Table 6.19 Haloacetamides considered and their structures.

Compound	Structure	Compound	Structure
Chloroacetamide (CAAm)	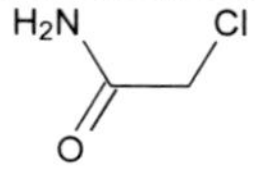	Trichloroacetamide (TCAAm)	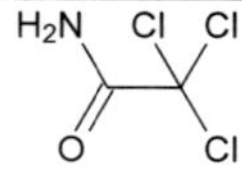
Bromoacetamide (BAAm)	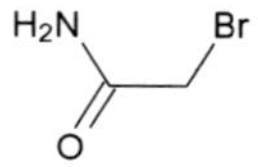	Tribromoacetamide (TBAAm)	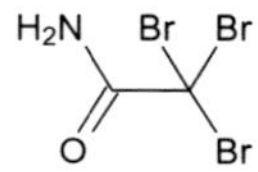
Dichloroacetamide (DCAAm)	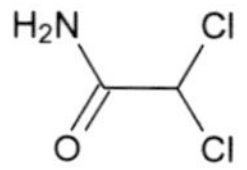	Bromoiodoacetamide (BIAAm)	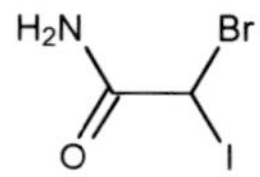
Dibromoacetamide (DBAAm)	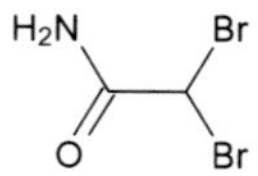		

6.6.2 Occurrence, formation and removal in drinking water

HAAms are typically present in drinking water at concentrations approximately one order of magnitude less than the HAAs, and have been found in drinking waters treated with chlorine, chloramines and/or ClO_2 (Krasner *et al.* 2006; Weinberg *et al.* 2002). Preliminary findings suggest that HAAms may form higher concentrations with chloramination compared to chlorination (Richardson *et al.* 2007). DCAAm is typically the major haloacetamide species (Krasner *et al.* 2006), which corresponds to the predominance of DCAN. CAAm, BAAm, DCAAm, DBAAm, and TCAAm have all been detected in finished drinking waters up to a total concentration of 14 µg/L (Krasner *et al.* 2006). Recently, BCAAm, BDCAAm and DBCAAm were identified, but not quantified, in chlorinated RO concentrates (Pressman *et al.* 2010), and were assumed to form from hydrolysis of halonitriles present at high concentrations. An iodinated HAAm, BIAAm, was also recently identified, but not quantified, in finished drinking waters from source waters that contained relatively high bromide and iodide levels (Plewa *et al.* 2008). Like HANs, HAAms can form from hydrophilic organic matter, particularly amino acid precursors (Chu *et al.* 2010), and removal of these precursors is influenced by the removal of low molecular weight DON in water treatment processes (Chu *et al.* 2011).

6.6.3 Analysis

The analysis of HAAms is usually carried out by LLE followed by GC-ECD or GC-MS (Plewa *et al.* 2008; Weinberg *et al.* 2002). One comparative study found that LLE with MtBE or ethyl acetate, followed by GC-ECD, was better than SPE followed by LC-MS, acid-catalyzed hydrolysis to convert HAAms to the corresponding HAAs, or silylation (Weinberg *et al.* 2002). Practical quantification limits of 0.1 μg/L were achieved for all five HAAms with the LLE GC-ECD method.

6.7 HALOAMINES

Haloamines include chloramines (monochloramine, dichloramine and trichloramine) and bromamines (e.g. monobromamine, dibromamine and bromochloramine), and are formed by the reaction of free chlorine or bromine with ammonia (Hofmann and Andrews, 2001; Lei *et al.* 2004). Haloamines are recognized as a potentially important oxidant species when they form during water disinfection (Valentine, 1986). Organic chloramines may also be formed from the reaction of chlorine or inorganic chloramine with amines and amino acids (Westerhoff and Mash, 2002) and organic nitrogen from algae (Fang *et al.* 2010). Organic chloramines can exist as short-lived intermediates in the formation of other N-DBPs (Fang *et al.* 2010; Hayes-Larson and Mitch, 2010).

The WHO health guideline for monochloramine is 3 mg/L (WHO, 2011), but toxicity data for dichloramine and trichloramine are not sufficient for derivation of health-based guidelines. Based on the existing studies, monochloramine appears to have some mutagenic potential *in vitro*, but no significant mutagenic properties *in vivo*, and the IARC has classified the chloramines in Group 3 (not classifiable as to its carcinogenicity to humans). No studies have been reported on the toxicities of the bromamines. In the structure activity relationships analysis (Woo *et al.* 2002), haloamines were predicted to pose a low concern for carcinogenicity, because of their structural analogy to chloramine, which has negative cancer bioassay data (Selkirk and Soward, 1993). Toxicological studies conducted with *N*-chloropiperidine and *N*-chlorodiethylamine as model compounds suggest organic chloramines may be toxic (Bempong *et al.* 1985; Scully and Bempong, 1982), however determination of their environmental significance requires identification of the specific organic chloramines actually formed upon chlorination of natural waters. Biomedical studies of organic chloramines have also indicated toxicological activity, but do not provide sufficient information to assess risks related to their occurrence in drinking water (Bull *et al.* 2011).

Typical monochloramine residuals in finished water range between 1–3 mg/L, when monochloramine is used as a primary disinfectant (Wolfe *et al.* 1984). Because organic chloramines are less effective as disinfectants than inorganic chloramines (Gagnon *et al.* 2008), their presence in water treatment systems can lead to overestimation of disinfection capability. The production of organic chloramines has been found to be faster, and in greater concentrations, with chlorination than with chloramination (Lee and Westerhoff, 2009), resulting in an overestimation of disinfection capability.

While bromamine chemistry has been studied for several decades (Galal-Gorchev and Morris, 1965; Valentine, 1986), the understanding of these DBPs is far from adequate. The significance of bromamine formation during drinking water treatment remains uncertain, since there are currently no available data on concentrations in actual treated drinking waters. Bromamines may form during chlorination, chloramination and ozonation of bromide-containing waters (Galal-Gorchev and Morris, 1965; Hofmann and Andrews, 2001). Bromamines are unstable and decay under the conditions encountered in water treatment and distribution systems (Lei *et al.* 2004). While bromamine formation is favoured for inhibiting bromate formation during ozonation (Hofmann and Andrews, 2001), bromamines may lead to the formation of other DBPs of health concern such as cyanogen bromide (Diehl *et al.* 2000; Luong *et al.* 2002). Bromamine formation may also result in a rapid loss of chlorine or chloramine residual after disinfection, because of reactions of bromamine with HOCl and NH_3.

Common analytical methods for chloramines include UV or colorimetric detection (Table 6.20), however these methods are unable to distinguish between organic and inorganic chloramines (Lee *et al.* 2007c; Shang *et al.* 2000). Membrane introduction mass spectrometry is the only analytical method developed to date that can distinguish between inorganic and organic chloramines (Shang *et al.* 2000), but the method is limited by instrument availability.

Table 6.20 Methods used for haloamine analysis.

Analytes	Method	Notes	Reference
Chloramines	*N,N*-diethyl-*p*-phenylenediamine (DPD) colorimetric method	Unable to distinguish between dichloramine and organic chloramines	(Clesceri *et al.* 1998)
	MIMS	Identifies all chloramine species, method limited by instrument availability	(Shang *et al.* 2000)
	MonochlorF method	Measures monochloramine only	(Lee *et al.* 2007c)

(Continued)

Table 6.20 Methods used for haloamine analysis (*Continued*).

Analytes	Method	Notes	Reference
Chloramines (*Continued*)	UV Spectrophotometry	Monochloramine in the presence of dichloramine can be calculated by measuring at 245 nm and 295 nm, and solving simultaneous equations. An unidentified product of monochloramine decomposition interferes in this method	(Valentine *et al.* 1986)
Bromamines	UV Spectrophotometry	λ_{max} of 278 nm and 232 nm for monobromamine and dibromamine respectively	(Lei *et al.* 2004).

6.8 NON-HALOGENATED NITRILES

Six non-halogenated nitriles (Table 6.21) have been identified, but not quantified, in drinking water samples (Richardson *et al.* 1996; Woo *et al.* 2002) and in disinfection experiments of NOM-containing water (Richardson *et al.* 1999b; Pressman *et al.* 2010). Very little is known about this class of DBPs. While no toxicity information exists, like the haloacetamides and haloamines, this class of compounds are predicted to be of low concern for carcinogenic potential (Woo *et al.* 2002). There are few indicators as to possible formation mechanisms, although benzeneacetonitrile was detected in waters that were chlorinated following TiO_2/UV treatment (Richardson *et al.* 1996). Benzeneacetonitrile and heptane nitrile can both form from ozonation and without further chlorine-based disinfection (Richardson *et al.* 1999b). A laboratory study of primary amine precursors also resulted in the detection of propionitrile from the chlorination of propylamine (Joo and Mitch, 2007), presumably formed by elimination of two hydrochloric acid molecules from the dichloramine.

Table 6.21 Structure of non-halogenated nitriles identified to date in drinking water.

Compound	Structure	Compound	Structure
Benzonitrile	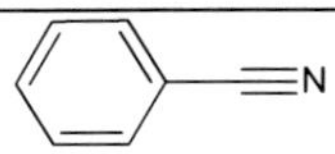	Heptane nitrile	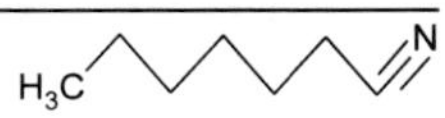

(*Continued*)

Table 6.21 Structure of non-halogenated nitriles identified to date in drinking water (*Continued*).

Compound	Structure	Compound	Structure
Benzeneacetonitrile	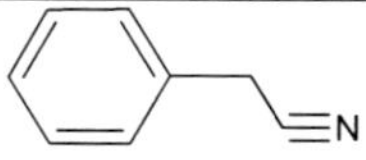	3-Methylbutane nitrile	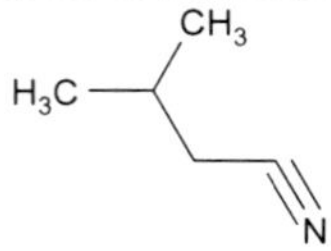
2-Methylbenzonitrile	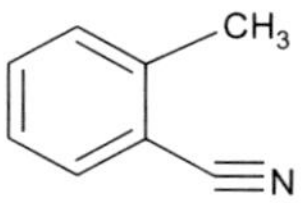	Propionitrile	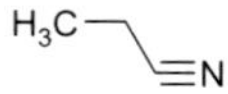

6.9 SUMMARY AND CONCLUSIONS

- There is a growing base of knowledge on NDMA formation in drinking water, however comparatively little research literature discusses the occurrence and control of other N-DBPs. Many N-DBPs (e.g. HANs and HNMs) are considered more genotoxic and cytotoxic than currently regulated DBPs, with brominated N-DBPs often more potent than their chlorinated analogues. There are still significant knowledge gaps in occurrence and toxicity data for many N-DBPs, and these must be addressed to assess properly the health impacts of these compounds.
- A wide variety of mechanisms have been implicated in N-DBP formation. Drinking water treatment plays a key role in N-DBP formation because of its influence on precursor availability and chosen disinfection strategy. Organic compounds that have been identified as precursors for N-DBPs are largely related to the dissolved organic nitrogen content in natural waters, including some amino acids, humic substances and algae suspensions. However, nitrogen-containing polymers and ion-exchange resins, often used in water treatment, have also been shown to act as N-DBP precursors, particularly for *N*-nitrosamines and HNMs.
- Some N-DBPs have been reported to occur at higher concentrations after chloramination than chlorination, particularly NDMA, which continues to form in the distribution system. Chloramine may serve as a nitrogen source for some DBPs, including HANs and cyanogen halides. Conversely, the presence of free chlorine can cause degradation after initial formation for many N-DBPs (e.g. HANs, HAAms and cyanogen halides), potentially resulting in decreasing concentrations within distribution systems. The presence of bromide in source waters can also shift the speciation of N-DBPs from chlorinated to brominated, which is of concern due to the higher toxicities of many brominated N-DBPs.

- While numerous biological, chemical and physical treatment processes have been suggested for N-DBP minimisation, few have been explored at full-scale. Different methods are effective for removal of different N-DBPs. However, precursor removal and disinfection optimisation are both suggested as important avenues of N-DBP minimisation.

REFERENCES

Agus E. and Sedlak D. L. (2010). Formation and fate of chlorination by-products in reverse osmosis desalination systems. *Water Res.*, **44**, 1616–1626.

Andrzejewski P., Kasprzyk-Hordern B. and Nawrocki J. (2008). *N*-nitrosodimethylamine (NDMA) formation during ozonation of dimethylamine-containing waters. *Water Res.*, **42**, 863–870.

Ayanaba A. and Alexander M. (1974). Transformations of methylamines and formation of a hazardous product, dimethylnitrosamine, in samples of treated sewage and lake water. *J. Environ. Qual.*, **3**, 83–89.

Bempong M. A., Montgomery C. and Scully F. E. (1985). Mutangenicity and clastogenicity of *N*-chloropiperidine. *J. Environ. Pathol. Tox. Oncol.*, **6**, 241–251.

Bond T., Henriet O., Goslan E. H., Parsons S. A. and Jefferson B. (2009). Disinfection byproduct formation and fractionation behavior of natural organic matter surrogates. *Environ. Sci. Technol.*, **43**, 5982–5989.

Bond T., Huang J., Templeton M. R. and Graham N. (2011). Occurrence and control of nitrogenous disinfection by-products in drinking water – a review. *Water Res.*, **45**, 4341–4354.

Bond T. and Templeton M. R. (2011). Nitrosamine formation from the oxidation of secondary amines. *Water Sci. Technol. Water Supply*, **11**, 256–265.

Bull R. J., Meier J. R., Robinson M., Ringhand H. P., Laurie R. D. and Stober J. A. (1985). Evaluation of mutagenic and carcinogenic properties of brominated and chlorinated acetonitriles – byproducts of chlorination. *Fundam. Appl. Toxicol.*, **5**, 1065–1074.

Bull R. J., Reckhow D. A., Li X. F., Humpage A. R., Joll C. and Hrudey S. E. (2011). Potential carcinogenic hazards of non-regulated disinfection by-products: haloquinones, halo-cyclopentene and cyclohexene derivatives, *N*-halamines, halonitriles, and heterocyclic amines. *Toxicology*, **286**, 1–19.

Cancho B., Ventura F. and Galceran M. T. (2000). Simultaneous determination of cyanogen chloride and cyanogen bromide in treated water at sub-μg/L levels by a new solid-phase microextraction-gas chromatographic-electron-capture detection method. *J. Chromatogr. A*, **897**, 307–315.

Castro C. E., Wade R. S. and Belser N. O. (2002). Biodehalogenation. The metabolism of chloropicrin by Pseudomonas sp. *J. Agric. Food. Chem.*, **31**, 1184–1187.

Chang H. H., Chen C. Y. and Wang G. S. (2011). Identification of potential nitrogenous organic precursors for C-, *N*-DBPs and characterization of their DBPs formation. *Water Res.*, **45**, 3753–3764.

Charrois J. W. A., Arend M. W., Froese K. L. and Hrudey S. E. (2004). Detecting *N*-nitrosamines in drinking water at nanogram per liter levels using ammonia positive chemical ionization. *Environ. Sci. Technol.*, **38**, 4835–4841.

Charrois J. W. A. and Hrudey S. E. (2007). Breakpoint chlorination and free-chlorine contact time: Implications for drinking water *N*-nitrosodimethylamine concentrations. *Water Res.*, **41**, 674–682.

Chen B., Lee W., Westerhoff P. K., Krasner S. W. and Herckes P. (2010). Solar photolysis kinetics of disinfection byproducts. *Water Res.*, **44**, 3401–3409.

Chen P. H., Richardson S. D., Krasner S. W., Majetich G. and Glish G. L. (2002). Hydrogen abstraction and decomposition of bromopicrin and other trihalogenated disinfection byproducts by GC/MS. *Environ. Sci. Technol.*, **36**, 3362–3371.

Chen Z. and Valentine R. L. (2006). Modeling the formation of *N*-nitrosodimethylamine (NDMA) from the reaction of natural organic matter (NOM) with monochloramine. *Environ. Sci. Technol.*, **40**, 7290–7297.

Chen Z. and Valentine R. L. (2007). Formation of *N*-nitrosodimethylamine (NDMA) from humic substances in natural water. *Environ. Sci. Technol.*, **41**, 6059–6065.

Chen Z. and Valentine R. L. (2008). The influence of the pre-oxidation of natural organic matter on the formation of *N*-nitrosodimethylamine (NDMA). *Environ Sci. Technol.*, **42**, 5062–5067.

Chen W. J. and Weisel C. P. (1998). Halogenated DBP concentrations in a distribution system. *J. Am. Water Works Assoc.*, **90**, 151–163.

Chen W. H. and Young T. M. (2008). NDMA formation during chlorination and chloramination of aqueous diuron solutions. *Environ. Sci. Technol.*, **42**, 1072–1077.

Cheng R. C., Hwang C. J., Andrews-Tate C., Guo Y., Carr S. and Suffet I. H. (2006). Alternative methods for the analysis of NDMA and other nitrosamines in water. *J. Am. Water Works Assoc.*, **98**, 82–96.

Chiang P. C., Chang E. E., Chuang C. C., Liang C. H. and Huang C. P. (2010). Evaluating and elucidating the formation of nitrogen-contained disinfection by-products during pre-ozonation and chlorination. *Chemosphere*, **80**, 327–333.

Choi J. and Richardson S. D. (2004). Formation of halonitromethanes in drinking water. Proceedings AWWA Water Quality Technology Conference, AWWA, San Antonio, TX.

Choi J. and Valentine R. L. (2003). *N*-Nitrosodimethylamine formation by free-chlorine-enhanced nitrosation of dimethylamine. *Environ. Sci. Technol.*, **37**, 4871–4876.

Chu W. H., Gao N. Y. and Deng Y. (2010). Formation of haloacetamides during chlorination of dissolved organic nitrogen aspartic acid. *J. Hazard. Mater.*, **173**, 82–86.

Chu W. H., Gao N. Y., Templeton M. R. and Yin D. Q. (2011). Comparison of inclined plate sedimentation and dissolved air flotation for the minimisation of subsequent nitrogenous disinfection by-product formation. *Chemosphere*, **83**, 647–651.

Clesceri L. S., Greenberg A. E. and Eaton A. D. (1998). Standard Methods for the Examination of Water and Wastewater, 20th edn. APHA, AWWA, WEF, Washington, DC.

Cole S. K., Cooper W. J., Fox R. V., Gardinali P. R., Mezyk S. P., Mincher B. J. and O'Shea K. E. (2007). Free radical chemistry of disinfection byproducts. 2. Rate constants and degradation mechanisms of trichloronitromethane (chloropicrin). *Environ. Sci. Technol.*, **41**, 863–869.

Diehl A. C., Speitel G. E., Symons J. M., Krasner S. W., Hwang S. J. and Barrett S. E. (2000). DBP formation during chloramination. *J. Am. Water Works Assoc.*, **92**, 76–90.

Dotson A., Westerhoff P. and Krasner S. W. (2009). Nitrogen enriched dissolved organic matter (DOM) isolates and their affinity to form emerging disinfection by-products. *Water Sci. Technol.*, **60**, 135–143.

Drewes J. E., Hoppe C. and Jennings T. (2006). Fate and transport of *N*-nitrosamines under conditions simulating full-scale groundwater recharge operations. *Water Environ. Res.*, **78**, 2466–2473.

Duguet J. P., Tsutsumi Y., Bruchet A. and Mallevialle J. (1985). Chloropicrin in potable water: conditions of formation and evolution along various treatment processes. In: Water Chlorination: Chemistry, Environmental Impact and Health Effects, R. L. Jolley, R. J. Bull, W. P. Davis, S. Katz, M. H. Roberts and V. A. Jacobs (eds), Lewis Publishers, Chelsea, MI, 1201–1213.

Fang J., Yang X., Ma J., Shang C. and Zhao Q. (2010). Characterization of algal organic matter and formation of DBPs from chlor(am)ination. *Water Res.*, **44**, 5897–5906.

Farre M. J., Keller J., Holling N., Poussade Y. and Gernjak W. (2011). Occurrence of *N*-nitrosodimethylamine precursors in wastewater treatment plant effluent and their fate during ultrafiltration-reverse osmosis membrane treatment. *Water Sci. Technol.*, **63**, 605–612.

Fiddler W., Pensabene J. W., Doerr R. C. and Wasserman A. E. (1972). Formation of *N*-nitrosodimethylamine from naturally occurring quaternary ammonium-compounds and tertiary-amines. *Nature*, **236**, 307.

Fleming E. C., Pennington J. C., Wachob B. G., Howe R. A. and Hill D. O. (1996). Removal of *N*-nitrosodimethylamine from waters using physical-chemical techniques. *J. Hazard. Mater.*, **51**, 151–164.

Gagnon G. A., Baribeau H., Rutledge S. O., Dumancic R., Oehmen A., Chauret C. and Andrews S. (2008). Disinfectant efficacy in distribution systems: A pilot-scale assessment. *J. Water Supply Res. Technol. AQUA*, **57**, 507–518.

Galal-Gorchev H. and Morris J. C. (1965). Formation and stability of bromamide, bromimide, and nitrogen tribromide in aqueous solution. *Inorg. Chem.*, **4**, 899–905.

Gerecke A. C. and Sedlak D. L. (2003). Precursors of *N*-mitrosodimethylamine in natural waters. *Environ. Sci. Technol.*, **37**, 1331–1336.

Glezer V., Harris B., Tal N., Iosefzon B. and Lev O. (1999). Hydrolysis of haloacetonitriles: linear free energy relationship, kinetics and products. *Water Res.*, **33**, 1938–1948.

Gough T. A., Webb K. S. and McPhail M. F. (1977). Volatile nitrosamines from ion-exchange resins. *Food Cosmet. Toxicol.*, **15**, 437–440.

Grebel J. E., Young C. C. and Suffet I. H. M. (2006). Solid-phase microextraction of *N*-nitrosamines. *J. Chromatogr. A*, **1117**, 11–18.

Gui L., Gillham R. W. and Odziemkowski M. S. (2000). Reduction of *N*-nitrosodimethylamine with granular iron and nickel enhanced iron. 1. Pathways and kinetics. *Environ. Sci. Technol.*, **34**, 3489–3494.

Gunnison D., Zappi M. E., Teeter C., Pennington J. C. and Bajpai R. (2000). Attenuation mechanisms of *N*-nitrosodimethylamine at an operating intercept and treat groundwater remediation system. *J. Hazard. Mater.*, **73**, 179–197.

Hayes-Larson E. L. and Mitch W. A. (2010). Influence of the method of reagent addition on dichloroacetonitrile formation during chloramination. *Environ. Sci. Technol.*, **44**, 700–706.

Heller-Grossman L., Idin A., Limoni-Relis B. and Rebhun M. (1999). Formation of cyanogen bromide and other volatile DBPs in the disinfection of bromide-rich lake water. *Environ. Sci. Technol.*, **33**, 932–937.

Hofmann R. and Andrews R. C. (2001). Ammoniacal bromamines: a review of their influence on bromate formation during ozonation. *Water Res.*, **35**, 599–604.

Hoigné J. and Bader H. (1988). The formation of trichloronitromethane (chloropicrin) and chloroform in a combined ozonation/chlorination treatment of drinking water. *Water Res.*, **22**, 313–319.

Hollender J., Zimmermann S. G., Koepke S., Krauss M., McArdell C. S., Ort C., Singer H., von Gunten U. and Siegrist H. (2009). Elimination of organic micropollutants in a municipal wastewater treatment plant upgraded with a full-scale post-ozonation followed by sand filtration. *Environ. Sci. Technol.*, **43**, 7862–7869.

Hu J., Song H., Addison J. W. and Karanfil T. (2010). Halonitromethane formation potentials in drinking waters. *Water Res.*, **44**, 105–114.

Hwang C. J., Krasner S. W., Sclimenti M. J., Amy G. L., Dickenson E., Bruchet A., Prompsy C., Gisele F., Croue J. P., Violleau D. and Leenheer J. (2001). Polar NOM: Characterization DBPs, Treatment. AWWARF, Denver, CO.

Jacangelo J. G., Patania N. L., Reagan K. M., Aieta E. M., Krasner S. W. and McGuire M. J. (1989). Ozonation – assessing its role in the formation and control of disinfection by-products. *J. Am. Water Works Assoc.*, **81**, 74–84.

Jobb D. B., Hunsinger R. B., Meresz O. and Taguchi V. (1994). Removal of *N*-Nitrosodimethylamine from the Ohsweken Water Supply, Final Report, Ontario Ministry of Environment. Toronto, Ontario, Canada.

Joo S. H. and Mitch W. A. (2007). Nitrile, aldehyde, and halonitroalkane formation during chlorination/chloramination of primary amines. *Environ. Sci. Technol.*, **41**, 1288–1296.

Jurado-Sanchez B., Ballesteros E. and Gallego M. (2007). Comparison of the sensitivities of seven *N*-nitrosamines in pre-screened waters using an automated preconcentration system and gas chromatography with different detectors. *J. Chromatogr. A*, **1154**, 66–73.

Kampioti A. A. and Stephanou E. G. (1999). Simultaneous determination of halogenated neutral and acidic disinfection by-products in drinking water by closed-loop stripping extraction and capillary gas chromatography. *J. Chromatogr. A*, **857**, 217–229.

Kawamoto T. and Makihata N. (2004). Distribution of bromine/chlorine-containing disinfection by-products in tap water from different water sources in the hyogo prefecture. *J. Health Sci.*, **50**, 235–247.

Kemper J. M., Westerhoff P., Dotson A. and Mitch W. A. (2009). Nitrosamine, dimethylnitramine, and chloropicrin formation during strong base anion-exchange treatment. *Environ. Sci. Technol.*, **43**, 466–472.

Kim J., Chung Y., Shin D., Kim M., Lee Y., Lim Y. and Lee D. (2003). Chlorination by-products in surface water treatment process. *Desalination*, **151**, 1–9.

Kim H., Shim J. and Lee S. (2002). Formation of disinfection by-products in chlorinated swimming pool water. *Chemosphere*, **46**, 123–130.

Kohut K. D. and Andrews S. A. (2003). Polyelectrolyte age and *N*-nitrosodimethylamine formation in drinking water treatment. *Water Qual. Res. J. Can.*, **38**, 719–735.

Kononen D. W. (1988). Acute toxicity of cyanogen chloride to daphnia-magna. *Bull. Environ. Conam. Toxicol.*, **41**, 371–377.

Krasner S. W., McGuire M. J., Jacangelo J. G., Patania N. L., Reagan K. M. and Aieta E. M. (1989). The occurrence of disinfection by-products in United-States drinking-water. *J. Am. Water Works Assoc.*, **81**, 41–53.

Krasner S. W., Weinberg H. S., Richardson S. D., Pastor S. J., Chinn R., Sclimenti M. J., Onstad G. D. and Thruston A. D. (2006). Occurrence of a new generation of disinfection byproducts. *Environ. Sci. Technol.*, **40**, 7175–7185.

Kristiana I., Joll C. and Heitz A. (2012). Analysis of halonitriles in drinking water using solid-phase microextraction and gas chromatography-mass spectrometry. *J. Chromatogr. A.*, **1225**, 45–54.

Kundu B., Richardson S. D., Granville C. A., Shaughnessy D. T., Hanley N. M., Swartz P. D., Richard A. M. and DeMarini D. M. (2004). Comparative mutagenicity of halomethanes and halonitromethanes in Salmonella TA100: structure-activity analysis and mutation spectra. *Mutat. Res.-Fund. Mol. Mech.*, **554**, 335–350.

Landsman N. A., Swancutt K. L., Bradford C. N., Cox C. R., Kiddle J. J. and Mezyk S. P. (2007). Free radical chemistry of advanced oxidation process removal of nitrosamines in water. *Environ. Sci. Technol.*, **41**, 5818–5823.

Le Roux J., Gallard H. and Croue J. P. (2011). Chloramination of nitrogenous contaminants (pharmaceuticals and pesticides): NDMA and halogenated DBPs formation. *Water Res.*, **45**, 3164–3174.

Lee C., Lee Y., Schmidt C., Yoon J. and Von Gunten U. (2008). Oxidation of suspected *N*-nitrosodimethylamine (NDMA) precursors by ferrate (VI): kinetics and effect on the NDMA formation potential of natural waters. *Water Res.*, **42**, 433–441.

Lee J. H., Na C. Z., Ramirez R. L. and Olson T. M. (2006). Cyanogen chloride precursor analysis in chlorinated river water. *Environ. Sci. Technol.*, **40**, 1478–1484.

Lee W. and Westerhoff P. (2009). Formation of organic chloramines during water disinfection – chlorination versus chloramination. *Water Res.*, **43**, 2233–2239.

Lee W., Westerhoff P. and Croue J.-P. (2007b). Dissolved organic nitrogen as a precursor for chloroform, dichloroacetonitrile, *N*-nitrosodimethylamine, and trichloronitromethane. *Environ. Sci. Technol.*, **41**, 5485–5490.

Lee W., Westerhoff P., Yang X. and Shang C. (2007c). Comparison of colorimetric and membrane introduction mass spectrometry techniques for chloramine analysis. *Water Res.*, **41**, 3097–3102.

Lee C., Yoon J. and Von Gunten U. (2007a). Oxidative degradation of *N*-nitrosodimethylamine by conventional ozonation and the advanced oxidation process ozone/hydrogen peroxide. *Water Res.*, **41**, 581–590.

Lei H. X., Marinas B. J. and Minear R. A. (2004). Bromamine decomposition kinetics in aqueous solutions. *Environ. Sci. Technol.*, **38**, 2111–2119.

Lei H. X., Minear R. A. and Marinas B. J. (2006). Cyanogen bromide formation from the reactions monobromamine and dibromamine with cyanide ion. *Environ. Sci. Technol.*, **40**, 2559–2564.

Lem W. (2005). Case Study: Application of A Medium Pressure UV System for the Treatment of *N*-Nitrosodimethylamine (NDMA). Calgon Carbon Corporation, Pittsburgh, PA.

Liang S., Min J. H., Davis M. K., Green J. F. and Remer D. S. (2003). Use of pulsed-UV processes to destroy NDMA. *J. Am. Water Works Assoc.*, **95**, 121–131.

Lipscomb J. C., El-Demerdash E. and Ahmed E. A. (2009). Haloacetonitriles: metabolism and toxicity. *Rev. Environ. Contam. Tox.*, **198**, 169–200.

Liviac D., Creus A. and Marcos R. (2011). Mutagenic analysis of six disinfection by-products in the Tk gene of mouse lymphoma cells. *J. Hazard. Mater.*, **190**, 1045–1052.

Luong T. V., Peters C. J. and Perry R. (2002). Influence of bromide and ammonia upon the formation of trihalomethanes under water-treatment conditions. *Environ. Sci. Technol.*, **16**, 473–479.

McGuire M. J., McLain J. L. and Oblensky A. (2002). Information Collection Rule Data Analysis. AWWARF, Denver, CO.

Merlet N., Thibaud H. and Dore M. (1985). Chloropicrin formation during oxidative treatments in the preparation of drinking-water. *Sci. Total Environ.*, **47**, 223–228.

Mezyk S. P., Helgeson T., Cole S. K., Cooper W. J., Fox R. V., Gardinali P. R. and Mincher B. J. (2006). Free radical chemistry of disinfection-byproducts. 1. Kinetics of hydrated electron and hydroxyl radical reactions with halonitromethanes in water. *J. Phys. Chem. A*, **110**, 2176–2180.

Mhlongo S. H., Mamba B. B. and Krause R. W. (2009). Monitoring the prevalence of nitrosamines in South African waters and their removal using cyclodextrin polyurethanes. *Phys. Chem. Earth*, **34**, 819–824.

Mitch W. A., Gerecke A. C. and Sedlak D. L. (2003a). A *N*-nitrosodimethylamine (NDMA) precursor analysis for chlorination of water and wastewater. *Water Res.*, **37**, 3733–3741.

Mitch W. A., Oelker G. L., Hawley E. L., Deeb R. A. and Sedlak D. L. (2005). Minimization of NDMA formation during chlorine disinfection of municipal wastewater by application of pre-formed chloramines. *Environ. Eng. Sci.*, **22**, 882–890.

Mitch W. A. and Schreiber I. M. (2008). Degradation of tertiary alkylamines during chlorination/chloramination: implications for formation of aldehydes, nitriles, halonitroalkanes, and nitrosamines. *Environ. Sci. Technol.*, **42**, 4811–4817.

Mitch W. A. and Sedlak D. L. (2002). Formation of *N*-nitrosodimethylamine (NDMA) from dimethylamine during chlorination. *Environ. Sci. Technol.*, **36**, 588–595.

Mitch W. A. and Sedlak D. L. (2004). Characterization and fate of *N*-nitrosodimethylamine precursors in municipal wastewater treatment plants. *Environ. Sci. Technol.*, **38**, 1445–1454.

Mitch W. A., Sharp J. O., Trussell R. R., Valentine R. L., Alvarez-Cohen L. and Sedlak D. L. (2003b). *N*-nitrosodimethylamine (NDMA) as a drinking water contaminant: a review. *Environ. Eng. Sci.*, **20**, 389–404.

Miyashita Y., Park S. H., Hyung H., Huang C. H. and Kim J. H. (2009). Removal of *N*-nitrosamines and their precursors by nanofiltration and reverse osmosis membranes. *J. Environ. Eng.-ASCE*, **135**, 788–795.

Montesinos I., Cardador M. J. and Gallego M. (2011). Determination of halonitromethanes in treated water. *J. Chromatogr. A*, **1218**, 2497–2504.

Morran J., Slyman N. and Newcombe G. (2009). Factors affecting NDMA formation in treated drinking water. *Water*, **36**, 57–61.

Morran J., Whittle M., Fabris R. B., Harris M., Leach J. S., Newcombe G. and Drikas M. (2011). Nitrosamines from pipeline materials in drinking water distribution systems. *J. Am. Water Works Assoc.*, **103**, 76–83.

Muellner M. G., Wagner E. D., McCalla K., Richardson S. D., Woo Y. T. and Plewa M. J. (2007). Haloacetonitriles vs. regulated haloacetic acids: are nitrogen-containing DBPs more toxic? *Environ. Sci. Technol.*, **41**, 645–651.

Muller-Pillet V., Joyeux M., Ambroise D. and Hartemann P. (2000). Genotoxic activity of five haloacetonitriles: comparative investigations in the single cell gel electrophoresis (Comet) assay and the Ames-fluctuation test. *Environ. Mol. Mutagen.*, **36**, 52–58.

Munch J. W. (1995). Method 524.2, Measurement of Purgeable Organic Compounds in Water by Capillary Column Gas Chromatography/Mass Spectrometry. USEPA, Cincinnati, OH.

Munch J. W. and Bassett M. V. (2004). Method 521, Determination of Nitrosamines in Drinking Water by Solid Phase Extraction and Capillary Column Gas Chromatography with Large Volume Injection and Chemical Ionization Tandem Mass Spectrometry (MS/MS). USEPA, Cincinnati, OH.

Munch D. J. and Hautman D. P. (1995). Method 551.1. Determination of Chlorinated Disinfection Byproducts, Chlorinated Solvents, and Halogenated Pesticides/Herbicides in Drinking Water by Liquid/Liquid Extraction and Gas Chromatography with Electron Capture Detection. USEPA, Cincinnati, OH.

Munro N. B., Talmage S. S., Griffin G. D., Waters L. C., Watson A. P., King J. F. and Hauschild V. (1999). The sources, fate, and toxicity of chemical warfare agent degradation products. *Environ. Health Perspect.*, **107**, 933–974.

Na C. Z. and Olson T. M. (2006). Mechanism and kinetics of cyanogen chloride formation from the chlorination of glycine. *Environ. Sci. Technol.*, **40**, 1469–1477.

Najm I. and Trussell R. R. (2001). NDMA formation in water and wastewater. *J. Am. Water Works Assoc.*, **93**, 92–99.

Nawrocki J. and Andrzejewski P. (2011). Nitrosamines and water. *J. Hazard. Mater.*, **189**, 1–18.

NHMRC and NRMMC (2011). Australian Drinking Water Guidelines Paper 6 National Water Quality Management Strategy. NHMRC, NRMMC, Commonwealth of Australia, Canberra.

NIH (1978). Bioassay of Chloropicrin for Possible Carcinogenicity. 78–1315, National Cancer Institute, DHEW and NIH, Bethesda, MD.

Nikolaou A. D., Lekkas T. D., Golfinopoulos S. K. and Kostopoulou M. N. (2002). Application of different analytical methods for determination of volatile chlorination by-products in drinking water. *Talanta*, **56**, 717–726.

Odziemkowski M. S., Gui L. and Gillham R. W. (2000). Reduction of *N*-nitrosodimethylamine with granular iron and nickel-enhanced iron. 2. Mechanistic studies. *Environ. Sci. Technol.*, **34**, 3495–3500.

Ohya T. and Kanno S. (1985). Formation of cyanide ion of cyanogen chloride through the cleavage of aromatic rings by nitrous-acid or chlorine. VIII. On the reaction of humic-acid with hypochlorous acid in the presence of ammonium ion. *Chemosphere*, **14**, 1717–1722.

Oliver B. G. (1983). Dihaloacetonitriles in drinking-water – algae and fulvic-acid as precursors. *Environ. Sci. Technol.*, **17**, 80–83.

Ontario MoE (2004). The Determination of *N*-nitrosamines in Water by Gas Chromatography-High Resolution Mass Spectrometry (GC-HRMS). NITROSO-E3388, Toronto, Canada.

Park S. H., Wei S., Mizaikoff B., Taylor A. E., Favero C. and Huang C. H. (2009). Degradation of amine-based water treatment polymers during chloramination as *N*-nitrosodimethylamine (NDMA) Precursors. *Environ. Sci. Technol.*, **43**, 1360–1366.

Patterson B. M., Shackleton M., Furness A. J., Pearce J., Descourvieres C., Linge K. L., Busetti F. and Spadek T. (2010). Fate of nine recycled water trace organic contaminants and metal(loid)s during managed aquifer recharge into a anaerobic aquifer: column studies. *Water Res.*, **44**, 1471–1481.

Pearson C. R., Hozalski R. M. and Arnold W. A. (2005). Degradation of chloropicrin in the presence of zero-valent iron. *Environ. Toxicol. Chem.*, **24**, 3037–3042.

Pedersen E. J., Urbansky E. T., Marinas B. J. and Margerum D. W. (1999). Formation of cyanogen chloride from the reaction of monochloramine with formaldehyde. *Environ. Sci. Technol.*, **33**, 4239–4249.

Planas C., Palacios Ó., Ventura F., Rivera J. and Caixach J. (2008). Analysis of nitrosamines in water by automated SPE and isotope dilution GC/HRMS: occurrence in the different steps of a drinking water treatment plant, and in chlorinated samples from a reservoir and a sewage treatment plant effluent. *Talanta*, **76**, 906–913.

Plewa M. J., Muellner M. G., Richardson S. D., Fasano F., Buettner K. M., Woo Y. T., McKague A. B. and Wagner E. D. (2008). Occurrence, synthesis, and mammalian cell cytotoxicity and genotoxicity of haloacetamides: an emerging class of nitrogenous drinking water disinfection byproducts. *Environ. Sci. Technol.*, **42**, 955–961.

Plewa M. J., Wagner E. D., Jazwierska P., Richardson S. D., Chen P. H. and McKague A. B. (2004). Halonitromethane drinking water disinfection byproducts: chemical characterization and mammalian cell cytotoxicity and genotoxicity. *Environ. Sci. Technol.*, **38**, 62–68.

Pressman J. G., Richardson S. D., Speth T. F., Miltner R. J., Narotsky M. G., Hunter E. S. III, Rice G. E., Teuschler L. K., McDonald A., Parvez S., Krasner S. W., Weinberg H. S., McKague A. B., Parrett C. J., Bodin N., Chinn R., Lee C. T. and Simmons J. E. (2010). Concentration, chlorination, and chemical analysis of drinking water for disinfection byproduct mixtures health effects research: US EPA's Four Lab Study. *Environ. Sci. Technol.*, **44**, 7184–7192.

Raksit A. and Johri S. (2001). Determination of *N*-nitrosodimethylamine in environmental aqueous samples by isotope-dilution GC/MS-SIM. *J. AOAC Int.*, **84**, 1413–1419.

Reckhow D. A., Linden K. G., Kim J., Shemer H. and Makdissy G. (2010). Effect of UV treatment on DBP formation. *J. Am. Water Works Assoc.*, **102**, 100–113.

Reckhow D. A., Platt T. L., MacNeill A. L. and McClellan J. N. (2001). Formation and degradation of dichloroacetonitrile in drinking waters. *J. Water Supply Res. Technol. AQUA*, **50**, 1–13.

Reckhow D. A., Singer P. C. and Malcolm R. L. (1990). Chlorination of humic materials – by-product formation and chemical interpretations. *Environ. Sci. Technol.*, **24**, 1655–1664.

Richardson S. D. (2003). Disinfection by-products and other emerging contaminants in drinking water. *TrAC, Trends Anal. Chem.*, **22**, 666–684.

Richardson S. D. (2005). New disinfection by-product issues: Emerging DBPs and alternative routes of exposure. *Global NEST J.*, **7**, 43–60.

Richardson S. D., Plewa M. J., Wagner E. D., Schoeny R. and DeMarini D. M. (2007). Occurrence, genotoxicity, and carcinogenicity of regulated and emerging disinfection by-products in drinking water: a review and roadmap for research. *Mutat. Res.-Rev. Mutat.*, **636**, 178–242.

Richardson S. D., Thruston A. D., Caughran T. V., Chen P. H., Collette T. W., Floyd T. L., Schenck K. M., Lykins B. W., Sun G.-R. and Majetich G. (1999a). Identification of new drinking water disinfection byproducts formed in the presence of bromide. *Environ. Sci. Technol.*, **33**, 3378–3383.

Richardson S. D., Thruston A. D., Caughran T. V., Chen P. H., Collette T. W., Floyd T. L., Schenck K. M., Lykins B. W., Sun G. R. and Majetich G. (1999b). Identification of new ozone disinfection byproducts in drinking water. *Environ. Sci. Technol.*, **33**, 3368–3377.

Richardson S. D., Thruston A. D., Collette T. W., Patterson K. S., Lykins B. W. and Ireland J. C. (1996). Identification of TiO_2/UV disinfection byproducts in drinking water. *Environ. Sci. Technol.*, **30**, 3327–3334.

Richardson S. D., Thruston A. D., Rav-Acha C., Groisman L., Popilevsky I., Juraev O., Glezer V., McKague A. B., Plewa M. J. and Wagner E. D. (2003). Tribromopyrrole, brominated acids, and other disinfection byproducts produced by disinfection of drinking water rich in bromide. *Environ. Sci. Technol.*, **37**, 3782–3793.

Sacher F., Lenz S. and Brauch H. J. (1997). Analysis of primary and secondary aliphatic amines in waste water and surface water by gas chromatography mass spectrometry after derivatization with 2,4-dinitrofluorobenzene or benzenesulfonyl chloride. *J. Chromatogr. A*, **764**, 85–93.

Schmidt C. K. and Brauch H. J. (2008). *N*, *N*-dimethosulfamide as precursor for *N*-nitrosodimethylamine (NDMA) formation upon ozonation and its fate during drinking water treatment. *Environ. Sci. Technol.*, **42**, 6340–6346.

Schreiber I. M. and Mitch W. A. (2005). Influence of the order of reagent addition on NDMA formation during chloramination. *Environ. Sci. Technol.*, **39**, 3811–3818.

Schreiber I. M. and Mitch W. A. (2006). Nitrosamine formation pathway revisited: the importance of chloramine speciation and dissolved oxygen. *Environ. Sci. Technol.*, **40**, 6007–6014.

Sclimenti M. J., Hwang C. J. and Krasner S. W. (1996). A comparison of analytical techniques for determining cyanogen chloride in chloraminated drinking water. In: Water Disinfection and Natural Organic Matter: Characterization and Control, R. A. Minear and G. L. Amy (eds.), ACS Symposium Series, **649**, American Chemical Society, Washington DC, pp. 126–136.

Scully F. E. Jr. and Bempong M. A. (1982). Organic *N*-chloramines: chemistry and toxicology. *Environ. Health Perspect.*, **46**, 111–116.

Selkirk J. K. and Soward S. M. (1993). Compendum of abstracts from long-term cancer studies reported by the National Toxicology Program from 1976 to 1992. *Environ. Health Perspect.*, **101**, 1–295.

Shah A. D., Dotson A. D., Linden K. G. and Mitch W. A. (2011). Impact of UV disinfection combined with chlorination/chloramination on the formation of halonitromethanes and haloacetonitriles in drinking water. *Environ. Sci. Technol.*, **45**, 3657–3664.

Shang C., Gong W. L. and Blatchley E. R. (2000). Breakpoint chemistry and volatile byproduct formation resulting from chlorination of model organic-*N* compounds. *Environ. Sci. Technol.*, **34**, 1721–1728.

Shen R. and Andrews S. A. (2011). Demonstration of 20 pharmaceuticals and personal care products (PPCPs) as nitrosamine precursors during chloramine disinfection. *Water Res.*, **45**, 944–952.

Simpson K. L. and Hayes K. P. (1998). Drinking water disinfection by-products: an Australian perspective. *Water Res.*, **32**, 1522–1528.

Smith M. K., Randall J. L., Stober J. A. and Read E. J. (1989). Developmental toxicity of dichloroacetonitrile – a by-product of drinking-water disinfection. *Fundam. Appl. Toxicol.*, **12**, 765–772.

Stefan M. I. and Bolton J. R. (2002). UV direct photolysis of *N*-nitrosodimethylamine (NDMA): kinetic and product study. *Helvetica Chimica Acta.*, **85**, 1416–1426.

Tate R. L. and Alexander M. (1975). Stability of nitrosamines in samples of lake water, soil, and sewage. *J. Natl. Cancer I.*, **54**, 327–330.

Thibaud H., Delaat J., Merlet N. and Dore M. (1987). Chloropicrin formation in aqueous-solution – effect of nitrites on precursors formation during the oxidation of organic-compounds. *Water Res.*, **21**, 813–821.

Trehy M. L. and Bieber T. L. (1981). Detection, identification and quantitiative analysis of dihaloacetonitriles in chlorinated natural waters. In: Advances in the Identification and Analysis of Organic Pollutants in Water, L. H. Keith (ed.), Ann Arbor Science, Ann Arbor, MI, **2**, 941–975.

Trehy M. L., Yost R. A. and Miles C. J. (1986). Chlorination byproducts of amino acids in natural waters. *Environ. Sci. Technol.*, **20**, 1117–1122.

Tuschall J. R. Jr. and Brezonik P. L. (1980). Characterization of organic nitrogen in natural waters: its molecular size, protein content, and interactions with heavy metals. *Limnol. Oceanogr.*, **25**, 495–504.

Ueno H., Moto T., Sayato Y. and Nakamuro K. (1996). Disinfection by-products in the chlorination of organic nitrogen compounds: by-products from kynurenine. *Chemosphere*, **33**, 1425–1433.

USEPA (2007). Drinking Water Standards and Health Advisories Table. http://www.epa.gov/region09/water/drinking/files/dwsha_0607.pdf (accessed 7 February 2012).

Valentine R. L. (1986). Bromochloramine oxidation of *N*, *N*-diethyl-para-phenylenediamine in the presence of monochloramine. *Environ. Sci. Technol.*, **20**, 166–170.

Valentine R. L., Brandt K. I. and Jafvert C. T. (1986). A spectrophotometric study of the formation of an unidentified monochloramine decomposition product. *Water Res.*, **20**, 1067–1074.

Valentine R. L., Choi J., Chen Z., Barrett E. and Hwang C. (2005). Factors Affecting the Formation of NDMA in Water and Occurrence. AWWARF, Denver, CO.

Van Buynder P., Lugg R., Rodriguez C., Bromley M., Filmer J., Blair P., Handyside M., Higginson S., Turner N., Lord O., Taylor P., Courtney K., Newby C., Heitz A., Linge K., Blythe J., Busetti F. and Toze S. (2009). Premier's Collaborative Research Program (2005–2008). Characterising Treated Wastewater For Drinking Purposes Following Reverse Osmosis Treatment. Technical Report. DoHWA, Perth, Western Australia.

Weinberg H. S., Krasner S. W., Richardson S. D. and Thruston A. D. Jr. (2002). The Occurrence of Disinfection By-Products (DBPs) of Health Concern in Drinking Water: Results of a Nationwide DBP Occurrence Study. EPA/600/R-02/068, USEPA, Athens, GA.

Weiss W. J., Bouwer E. J., Ball W. P., O'Melia C. R., Lechevallier M. W., Arora H. and Speth T. F. (2003). Riverbank filtration – fate of DBP precursors and selected microorganisms. *J. Am. Water Works Assoc.*, **95**, 68–81.

Weissmahr K. W. and Sedlak D. L. (2000). Effect of metal complexation on the degradation of dithiocarbamate fungicides. *Environ. Toxicol. Chem.*, **19**, 820–826.

Westerhoff P. and Mash H. (2002). Dissolved organic nitrogen in drinking water supplies: a review. *J. Water Supply Res. Technol. AQUA*, **51**, 415–448.

WHO (2011). Guidelines for Drinking Water Quality. 4th edn, World Health Organization, Geneva. http://www.who.int/water_sanitation_health/publications/2011/dwq_guidelines/en/index.html (accessed 07 February 2012).

WHO IARC (1991). Halogenated acetonitriles. *IARC Monographs on the Evaluation of Carcinogenic Risks to Humans*, **52**, 269–293.

Wilczak A., Assadi-Rad A., Lai H. H., Hoover L. L., Smith J. F., Berger R., Rodigari F., Beland J. W., Lazzelle L. J., Kinicannon E. G., Baker H. and Heaney C. T. (2003). Formation of NDMA in chloraminated water coagulated with DADMAC cationic polymer. *J. Am. Water Works Assoc.*, **95**, 94–106.

Wilhelm S. N., Shepler K., Lawrence L. J. and Lee H. (1996). Environmental fate of chloropicrin. In: Fumigants: Environmental Fates, Analysis, Exposure, J. N. Seiber, J. A. Knuteson, J. E. Woodrow, N. L. Wolfe, M. V. Yates and S. R. Yates (eds), ACS Symposium Series, **652**, American Chemical Society, Washington DC, pp. 80–93.

Williams D. T., LeBel G. L. and Benoit F. M. (1997). Disinfection by-products in Canadian drinking water. *Chemosphere*, **34**, 299–316.

Wolfe R. L., Ward N. R. and Olson B. H. (1984). Inorganic chloramines as drinking-water disinfectants – a review. *J. Am. Water Works Assoc.*, **76**, 74–88.

Woo Y. T., Lai D., McLain J. L., Manibusan M. K. and Dellarco V. (2002). Use of mechanism-based structure-activity relationships analysis in carcinogenic potential ranking for drinking water disinfection by-products. *Environ. Health Perspect.*, **110**, 75–87.

Xie Y. and Reckhow D. A. (1993). A rapid and simple analytical method for cyanogen chloride and cyanogen bromide in drinking water. *Water Res.*, **27**, 507–511.

Xu B. B., Chen Z. G., Qi F. and Xu Z. Z. (2009). Degradation of *N*-nitrosodimethylamine (NDMA) in water by UV/O-3. *J. Water Supply Res. Technol. AQUA*, **58**, 135–145.

Yang L., Chen Z. L., Shen J. M., Xu Z. Z., Liang H., Tian J. Y., Ben Y., Zhai X., Shi W. X. and Li G. B. (2009). Reinvestigation of the nitrosamine-formation mechanism during ozonation. *Environ. Sci. Technol.*, **43**, 5481–5487.

Yang X., Fan C., Shang C. and Zhao Q. (2010). Nitrogenous disinfection byproducts formation and nitrogen origin exploration during chloramination of nitrogenous organic compounds. *Water Res.*, **44**, 2691–2702.

Yang X. and Shang C. (2005). Quantification of aqueous cyanogen chloride and cyanogen bromide in environmental samples by MIMS. *Water Res.*, **39**, 1709–1718.

Yang X., Shang C., Lee W., Westerhoff P. and Fan C. H. (2008). Correlations between organic matter properties and DBP formation during chloramination. *Water Res.*, **42**, 2329–2339.

Yang X., Shang C. and Westerhoff P. (2007). Factors affecting formation of haloacetonitriles, haloketones, chloropicrin and cyanogen halides during chloramination. *Water Res.*, **41**, 1193–1200.

Zhao Y. Y., Boyd J., Hrudey S. E. and Li X. F. (2006). Characterization of new nitrosamines in drinking water using liquid chromatography tandem mass spectrometry. *Environ. Sci. Technol.*, **40**, 7636–7641.

Zhao Y. Y., Boyd J. M., Woodbeck M., Andrews R. C., Qin F., Hrudey S. E. and Li X. F. (2008). Formation of *N*-nitrosamines from eleven disinfection treatments of seven different surface waters. *Environ. Sci. Technol.*, **42**, 4857–4862.

Zheng A. P., Dzombak D. A. and Luthy R. G. (2004). Formation of free cyanide and cyanogen chloride from chloramination of publicly owned treatment works secondary effluent: Laboratory study with model compounds. *Water Environ. Res.*, **76**, 113–120.

Zhou Q. L., McCraven S., Garcia J., Gasca M., Johnson T. A. and Motzer W. E. (2009a). Field evidence of biodegradation of *N*-nitrosodimethylamine (NDMA) in groundwater with incidental and active recycled water recharge. *Water Res.*, **43**, 793–805.

Zhou W. J., Boyd J. M., Qin F., Hrudey S. E. and Li X. F. (2009b). Formation of *N*-nitrosodiphenylamine and two new *N*-containing disinfection byproducts from chloramination of water containing diphenylamine. *Environ. Sci. Technol.*, **43**, 8443–8448.

Chapter 7

NDMA and other *N*-nitrosamines – health risk assessment and management

J. M. Boyd, J. W. A. Charrois, R. Hofmann and S. E. Hrudey

7.1 *N*-NITROSAMINES IN DRINKING WATER

N-Nitrosamines are a class of probable human carcinogens (IARC 1978) that were first identified as drinking water disinfection byproducts (DBPs) in 1989. That discovery of N-nitrosodimethylamine (NDMA) in Ohsweken, Ontario, Canada (Taguchi *et al.* 1994) was followed by the detection of NDMA in California drinking water as a result of ground water contamination with rocket fuel, which contained NDMA precursors. A subsequent state-wide survey detected NDMA in non-rocket fuel contaminated wells. Since then, *N*-nitrosamines have been identified in several other locations in North America (Barrett *et al.* 2003; Boyd *et al.* 2010; Charrois *et al.* 2004; Charrois *et al.* 2007; USEPA, 2011c; Zhao *et al.* 2006), Japan (Asami *et al.* 2009), China (Wang *et al.* 2010), and the UK (Templeton and Chen, 2010). A total of 7 *N*-nitrosamines have been identified as DBPs: NDMA, N-nitrosopyrrolidine (NPyr), N-nitrosomorpholine (NMor), N-nitrosopiperidine (NPip), N-nitrosodibutylamine (NDBA), N-nitrosodiethylamine (NDEA) and N-nitrosodiphenylamine (NDPhA) (Table 7.1). NDMA remains the most commonly detected *N*-nitrosamine.

N-Nitrosamines have been part of the U.S. Environmental Protection Agency's (USEPA) future regulatory planning as part of the third Candidate Contaminant List (CCL-3) and the Unregulated Monitoring Rule - 2 (UCMR-2) (USEPA, 2005). The CCL-3 is a list of water contaminants considered for future regulation and contains 5 *N*-nitrosamines: NDMA, NDEA, NDPA, NPyr and NDPhA (USEPA, 2009). Between 2007–2010 U.S. drinking water treatment plants monitored for 6 *N*-nitrosamines (NDMA, NDEA, NDPA, NPyr, *N*-nitrosomethylethylamine (NMEA), and NDBA) as part of the UCMR-2. Initial results for the UCMR-2 show that a number of very large and extra large size drinking water systems have reported NDMA concentrations above the California Notification Limit of 10 ng/L (USEPA, 2011c).

Table 7.1 Structures, lifetime cancer risk estimates and target organs of the 6 *N*-nitrosamine DBPs identified in drinking water.

***N*-Nitrosamine**	**Abbrev.**	**Structure**	**10^{-6} upperbound lifetime cancer risk from drinking water consumption (USEPA, 2011a)**	**Target organ (Reference)**
N-nitrosodimethylamine	NDMA	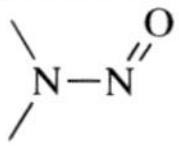	0.7 ng/L	Liver (Peto *et al.* 1991)
N-nitrosopyrrolidine	NPyr	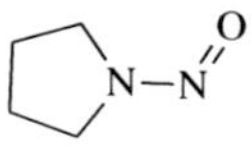	20 ng/L	Liver (Gray *et al.* 1991)
N-nitrosopiperidine	NPip	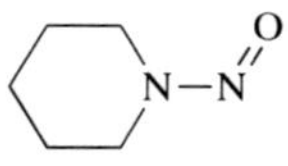	NA	Liver, esophagus, jaw (Gray *et al.* 1991)
N-nitrosodiphenylamine	NDPhA	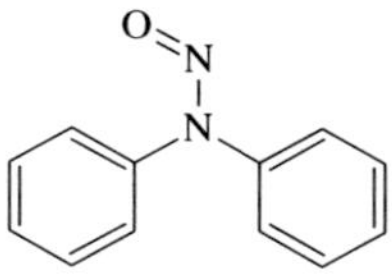	7000 ng/L	Bladder (Cardy *et al.* 1979)
N-nitrosomorpholine	NMor	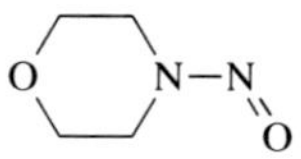	NA	Liver (Lijinsky *et al.* 1988)
N-nitrosodibutylamine	NDBA	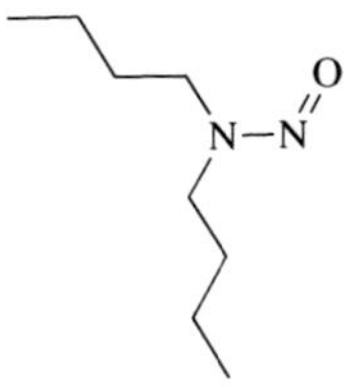	6 ng/L	Bladder, esophagus (Okada and Ishidate, 1977)

NA = Not available.

In most locations concentrations are below 10 ng/L, however NDMA concentrations up to 630 ng/L (USEPA, 2011c) have been reported. This extreme value, which overtook the previous reported high of 180 ng/L (Charrois *et al.* 2004) may yet be corrected or revised as one of five values reported above

100 ng/L out of 18,040 samples analysed for NDMA (and five other *N*-nitrosamines, NDEA, NDBA, NDPA, NMEA and NPYR) under the UCMR-2 and reported in preliminary form in July 2011. A summary of this largest international database on *N*-nitrosamines in drinking water is provided in Table 7.2 (USEPA, 2011c).

Table 7.2 Preliminary *N*-nitrosamine monitoring data results for UCMR-2 (USEPA, 2011).

N-Nitrosamine	MRL ng/L	Number of analyses	Number of detections (% of analyses)	Number of PWS with analyses	Number of PWS with detects (% of PWS)	Maximum detection concentration ng/L
NDEA	5	18,038	46 (0.26%)	1198	26 (2.2%)	100
NDMA	2	18,040	1,841 (10.2%)	1198	324 (27%)	630
NDBA	4	18,043	9 (0.05%)	1198	5 (0.4%)	21
NDPA	7	18,049	0 (0%)	1198	0 (0%)	
NMEA	3	18,043	3 (0.02%)	1198	3 (0.3%)	5
NPYR	2	18,043	41 (0.23%)	1198	21 (1.8%)	24

7.2 GUIDELINES AND REGULATION OF *N*-NITROSAMINES IN DRINKING WATER

N-Nitrosamine guidelines have been adopted by many jurisdictions around the world (Table 7.3). As the next step in the CCL-3 process, the USEPA administrator has announced that the agency will consider regulating *N*-nitrosamines as a group. The Ontario and California guidelines were put in place as a result of the detection of NDMA in their drinking water. Most other guidelines are only recently approved or are still under consideration. In addition, most guidelines are for NDMA only or NDMA and a few other *N*-nitrosamines.

Table 7.3 *N*-Nitrosamine regulations and guidelines currently in effect or under consideration.

Jurisdiction	*N*-Nitrosamine	Concentration	Reference
Ontario	NDMA	9 ng/L	GovOnt (2002)
Canada	NDMA	40 ng/L	Health Canada (2010)

(*Continued*)

Table 7.3 *N*-Nitrosamine regulations and guidelines currently in effect or under consideration (*Continued*).

Jurisdiction	*N*-Nitrosamine	Concentration	Reference
California	NDMA	10 ng/L	
	NDEA	10 ng/L	CalDPH (2007)
	NDPA	10 ng/L	
Massachusetts	NDMA	10 ng/L	MassDEP (2004)
Australia	NDMA	100 ng/L	NHMRC/NRMMC (2011)
World Health Organization	NDMA	100 ng/L	WHO (2008)

7.3 TOXICITY AND HUMAN HEALTH EFFECTS

7.3.1 Mode of action

Once *N*-nitrosamines enter the body, they are metabolized by enzymes to a free radical ion that is responsible for toxic effects. Specifically, this radical can add methyl groups onto certain molecules, like DNA. If left unrepaired, a methylated DNA base will result in a DNA mutation that may lead to carcinogenesis. Figure 7.1 shows the toxic mode of action for NDMA, which is similar for the other *N*-nitrosamines except for NDPhA.

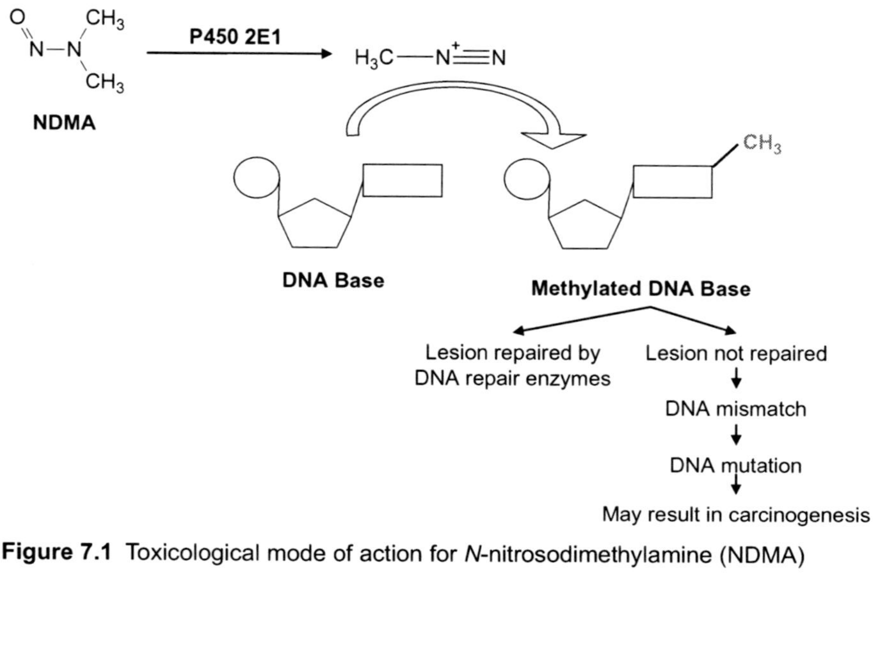

Figure 7.1 Toxicological mode of action for *N*-nitrosodimethylamine (NDMA)

7.3.2 Human toxicity

7.3.2.1 Occupational exposure

Occupational exposure to *N*-nitrosamines may occur in several industries including foundries (Ducos *et al.* 1988), leather tanneries (Rounbehler *et al.* 1979) and manufacturing of rubber products and tires (Fajen *et al.* 1979). Straif *et al.* (2000) found an increased occurrence of cancers of the esophagus, oral cavity and pharynx in rubber plant workers (Straif *et al.* 2000) indicating the potential hazard of occupational exposure to these chemicals. Reh *et al.* (2000) identified *N*-nitrosamine-related DNA adducts in the blood of rubber plant workers (Reh *et al.* 2000) indicating that occupational exposure to *N*-nitrosamines can result in the formation of DNA adducts, which is the first step in carcinogenesis (Figure 7.1).

7.3.2.2 Endogenous formation

A second major exposure source for *N*-nitrosamines is endogenous formation. This refers to formation of *N*-nitrosamines within the body as opposed to exogenous exposure from *N*-nitrosamines already formed in the environment (like DBPs). A major site of endogenous formation is in the stomach where amines and nitrates/nitrites from our food undergo nitrosation to form *N*-nitrosamines (Tannenbaum, 1980; Vermeer *et al.* 1998; Krul *et al.* 2004). Several studies have investigated the human health effects of exposure to endogenously formed *N*-nitrosamines, however there are still major uncertainties involved.

7.3.2.3 Drinking water exposure

At this time there have been no epidemiology studies investigating human health effects from exposure to *N*-nitrosamines in drinking water.

7.3.3 Rodent toxicity

As data on the effect of *N*-nitrosamines on human health are limited, rodent toxicity data are used for risk analysis. *N*-Nitrosamine toxicity has been well studied in rodents. In general, *N*-nitrosamines have been able to cause cancer in most rodent species they have been tested on. In addition, *N*-nitrosamines display tissue selectivity (Table 7.1). The rat bioassay conducted by Peto *et al.* (1991) is considered the best data set for chronic NDMA effects and has been used for the majority of risk assessments currently used for guideline development. The study tested NDMA on over 2000 rats and observed a dose-response relationship between NDMA dose and liver cancer (Peto *et al.* 1991). Other rodent studies have been performed using the other *N*-nitrosamines identified as DBPs. Lifetime cancer risk estimates and target organs for each of those *N*-nitrosamines are shown in Table 7.1.

7.4 NDMA EXPOSURE

Drinking water is one of many possible sources of *N*-nitrosamine exposure. *N*-Nitrosamines have been known to be contained in many food items particularly smoked meats, cheeses and beer. They have also been detected in cosmetics and are present in tobacco smoke. *N*-Nitrosamines can also form endogenously in our body.

7.4.1 Sources

Exposure to NDMA and other *N*-nitrosamines can come from several sources. Drinking water may contain *N*-nitrosamines formed from disinfection. Humans are exposed to DBPs via ingestion from drinking, and inhalation and absorbtion through the skin from showering. Recent studies have suggested that inhalation and dermal routes of exposure may be more important for DBPs like trihalomethanes than ingestion (Nuckols *et al.* 2005). While the majority of the identified *N*-nitrosamine DBPs are considered to be semi-volatile, they are not nearly as volatile as the trihalomethanes. For example, the volatility Henry's Law constant for NDMA is $0.248\ m^3 \cdot Pa/mol$ (Haruta *et al.* 2011) compared to $122\ m^3 \cdot Pa/mol$ (Gossett, 1987) for chloroform, about a 500-fold difference, meaning that much more chloroform will be volatilized compared to NDMA. Uptake of NDMA through the skin is estimated at 10^{-4} cm/h (Schäfer *et al.* 2010). Penetration of chloroform through human skin is estimated at 0.14 cm/h (Nakai *et al.* 1999). The lower permeability of NDMA indicates that both inhalation and dermal routes of exposure are more important for trihalomethanes than for *N*-nitrosamines. Typically drinking water concentrations are below 10 ng/L as indicated by Table 7.2.

Exposure to pre-formed *N*-nitrosamines in food occurs when secondary amines in foods react with nitrate to form *N*-nitrosamines. Foods with historically high levels of NDMA include smoked meats, beer, and dairy products.

Endogenous formation of *N*-nitrosamines also occurs from the reaction of amines and nitrate or nitrite in the body. However the amount of NDMA formed in the body can be quite variable. The relative contributions of these exposures can be quite uncertain. This subject is elaborated in Section 7.4.3.

Exposure to *N*-nitrosamines may also come from several exogenous sources including cigarette smoke, rubber products and pesticides (WHO, 2008).

7.4.2 Daily NDMA exposure estimates

In 2002 the World Health Organization estimated a worst-case average daily intake of NDMA from contaminated food, air and water of 5–16 μg/kg/day for individuals aged 20–59 (Liteplo *et al.* 2002). The contribution to this from drinking water was estimated to be 0.3–1 μg/kg/day assuming an average NDMA concentration in drinking water of 12 ng/L and a maximum concentration of 40 ng/L. This is estimated to be a worst case scenario and total exogenous (external)

N-nitrosamine exposure has likely decreased in the past two decades because of efforts to remove *N*-nitrosamines from foods. However, even so this suggests that *N*-nitrosamine exposure from drinking water is likely low (considering a person drinking 10 ng/L and drinking 2 L per day).

7.4.3 Contributions to total NDMA exposure

Fristachi and Rice attempted to quantify the amount of NDMA contributed by endogenous formation, food, and water (Fristachi and Rice 2007). They estimated that drinking water contributed less than 0.02% of NDMA exposure assuming average NDMA concentrations in drinking water of 2.1 ng/L. Foods and beverages contributed approximately 0.5%. Endogenous formation was estimated at about 99.5%. This suggests that NDMA exposure from drinking water is not significant compared to endogenous formation.

There are several factors to keep in mind when considering these results. Data regarding the presence of NDMA in foods is from the early 1990's. These concentrations have likely decreased because of efforts to prevent *N*-nitrosamine formation. In addition, estimating the amount of NDMA formed from endogenous nitrosation is very difficult and a lot of uncertainty is involved. The authors estimated NDMA concentrations in drinking water based on a survey by Valentine *et al.* and used NDMA concentrations of 2.1 ng/L in their calculations. This was a combination of chlorinating and chloraminating systems, even though chloraminating systems tend to have higher concentrations of NDMA and other *N*-nitrosamines. No distinction was made between the chlorinating or chloramination systems, although it is clear from the data that the chloraminating systems had fewer non-detects than the chlorinating systems. In addition, the six systems with elevated NDMA concentrations (10–24 ng/L) were removed from the model as outliers. Considering that many guidelines under consideration (Table 7.3) would consider all of these values as acceptable, their removal may be an underestimation of the contribution from drinking water. In addition, the UCMR-2 occurrence results indicate that only 7% of the ~400 samples with NDMA concentrations >10 ng/L were found in systems serving <50,000 population, although the UCMR-2 data base proportionally under-represents small systems.

7.5 ABILITY OF DRINKING WATER NDMA EXPOSURES TO EXPLAIN BLADDER CANCER

7.5.1 *N*-Nitrosamine-specific bladder cancer epidemiology

Zeegers *et al.* (2006) performed a large prospective cohort study on 120,852 individuals in the Netherlands, which after 9.3 years of follow-up yielded 889 cases of bladder cancer. These are substantial numbers to work with. This study found no association between nitrate intake in food, in water, in total nitrate intake with bladder cancer. Nitrate has been used as a surrogate for endogenous *N*-nitrosamine formation potential.

Larsson *et al.* (2009) performed a prospective cohort study on 82,202 individuals in Sweden, which after 9.4 years of follow-up yielded 485 cases of bladder cancer. With these substantial numbers they explored associations of bladder cancer with total meat intake, red meat intake, processed meat intake, chicken/poultry and fried meats/fish. No category of meat consumption was associated with bladder cancer. The case control studies on gastric cancer have relied upon food questionnaires that assessed *N*-nitrosamines and/or NDMA based upon the contributions largely from processed or smoked meats, so the lack of any association of bladder cancer with any meat category suggests that no association would be found with NDMA or *N*-nitrosamines if similar methods had been applied to estimate associations with NDMA or *N*-nitrosamines in the Larsson *et al.* (2009) study.

The most recent study on bladder cancer did estimate NDMA exposure explicitly (Jakszyn *et al.* 2011). This was also a large cohort study with a total of 481,419 participants across 10 European countries. After 8.7 years of follow-up, 1,001 participants were diagnosed with bladder cancer. No association between bladder cancer and estimated NDMA intake was found over a range from the lowest quartile of 0–0.05 μg/d NDMA (0–0.0008 μg/kg-d assuming 60 kg body weight) to a highest quartile of 0.19–21 μg/d NDMA (0.003–0.35 μg/kg-d).

7.5.2 *N*-Nitrosamine risk assessment for drinking water

Urinary bladder cancer is the cancer outcome most consistently observed in epidemiology studies of long-term exposures to chlorinated drinking water (Chapter 11). Estimates of new cases of bladder cancer for 2011 are: 7200 and 69250 for Canada and the United States respectively (ACS, 2011; CCS, 2011).

Epidemiology based population-attributable risk (PAR) estimates for bladder cancer, resulting from long-term exposures to chlorinated DBPs, was estimated to be 14%–16% in one study (King and Marrett, 1996). The USEPA used a PAR range of 2%–17%, based on five epidemiology studies (including King and Marrett, 1996), for economic cost-benefit calculations for the Stage 2 DBP Rule (Odom *et al.* 1999). Based on the widest range of PAR estimates (2%–17%), if a causal link was actually established, approximately 150 to 1200 new Canadian and 1400 to 12,000 new American bladder cancer cases would be attributed to DBPs. Alternatively, if exposure was eliminated, the estimates would reflect numbers of cases that would be avoided. It is important to recognize that given the high degree of uncertainty, the U.S. Environmental Protection Agency acknowledged that the lower boundary PAR estimates could be as low as zero cases (USEPA, 2003).

Bladder cancer risks from PAR estimates are derived from chlorinated DBP exposures, typically indexed by THM and HAA concentrations. However, mechanistic evidence for bladder cancer based on drinking water THM and HAA exposures is lacking (see Chapters 10 and 11). *N*-Nitrosamines are more potent carcinogens by 2 to 4 orders of magnitude, compared to THMs based on oral slope factors ($[(mg/kg)/d]^{-1}$) and drinking water unit risks $(\mu g/L)^{-1}$ (derived from USEPA Integrated Risk Information System (IRIS) estimates) (Table 7.4). As used here, "oral cancer slope factor (CSF)" is defined as an upper-bound estimate of the lifetime human cancer risk per mg of agent/kg body weight/day (USEPA, 2011b). "Unit risk" is calculated from the oral CSF, which is an estimate of 1 in a million upper bound, lifetime cancer risk per μg/L drinking water (USEPA, 2011b).

Table 7.4 *N*-Nitrosamine and THM unit risk and CSF comparison – USEPA Integrated Risk Information System (IRIS) data base estimates.

DBP	Oral CSF $[(mg/kg)/d]^{-1}$	Drinking water unit risk (10^{-6} per μg/L)
N-Nitrosamines		
NDMA	51	1400
NMEA	22	630
NDEA	150	4300
NDPA	7	200
NPYR	2.1	61
NDBA	5.4	160
THMs		
Chloroform	NA	NA
BDCM	0.062	1.8
DBCM	0.084	2.4
Bromoform	0.0079	0.23

NA = not applicable.

Because the CSF and unit risk are based on a policy default linear extrapolation model, the cancer potency estimated by the model among different DBPs can be compared in direct proportion of the CSF or the unit risk of the respective DBPs. Hence the conclusion about *N*-nitrosamines showing 100 to 10,000 fold higher potency than THMs follows from the comparisons in Table 7.4. Of course, chloroform has no cancer slope factor at all (Chapter 1).

NDMA concentrations from food ingestion based on estimates from Liteplo *et al.* (2002) can be used with the USEPA slope factors from Table 7.4 to estimate annual bladder cancer risk estimates from food ingestion for Canada and the United States using a 70 year life expectancy and population estimates of 35 million for Canada and 313 million for the United States. For drinking water exposure to NDMA, the massive UCMR-2 data base (USEPA 2011c) shows that most water utilities have NDMA concentrations below 10 ng/L. Using 10 ng/L as an upper bound lifetime average daily exposure for an entire nation, the drinking water unit risk data for NDMA in Table 7.4, can be used to estimate an expected upper bound for cases of cancer predicted for drinking water exposure to NDMA. These estimates are summarized in Table 7.5.

Table 7.5 Comparison of annual cancer cases using NDMA CSF estimates for any cancer vs. PAR epidemiology estimates for bladder cancer.

	Canada			U.S.A.		
	Food[a]	Water[b]	PAR est.[c]	Food[a]	Water[b]	PAR est.[c]
High est.	250	<7	1200	2200	<60	12,000
Low est.	100	<7	150	900	<60	1400

[a]food estimate based high and low dietary NDMA intake according to Liteplo *et al.* 2002 and application of CSF for NDMA.
[b]water estimate based on lifetime nationwide average NDMA concentration assumed to be <10 ng/L and unit risk for NDMA (equivalent to CSF).
[c]PAR estimate based on using high PAR estimate of 17% and low estimate of 2% applied to all national bladder cancer cases per year.

Comparing the PAR estimates with the theoretical NDMA annual cancer risk estimates two things become clear. Firstly, the upper limit PAR estimates for bladder cancer cases far exceed the upper-bound annual cancer case estimates for NDMA for any cancers. Secondly when exposure from food is considered, drinking water estimates account for a small proportion of the theoretical number of cases compared to ingestion of food. Based on the limited drinking water DBP occurrences of NDMA, the risk of bladder cancer that could be attributed to NDMA as a DBP in drinking water appears to be too low for NDMA to qualify as the yet unidentified DBP needed to explain epidemiologic estimates of bladder cancer risk (Chapters 10 and 11).

7.6 RISK MANAGEMENT TO REDUCE NDMA DURING DRINKING WATER TREATMENT

7.6.1 NDMA removal by treatment

NDMA is a small, water-soluble molecule (Log $K_{OW} = -0.57$; OEHHA, 2006), and is difficult to remove by conventional treatment. Coagulation using alum, ferric chloride, or ferric sulphate has been shown to have no measurable impact on NDMA concentrations (Chung *et al.* 2009). Air stripping is impractical due to a low volatility (Henry's Law constant of $0.248\ m^3 \cdot Pa/mol$ (Haruta *et al.* 2011)). Membrane treatment has proven ineffective, with nanofiltration and reverse osmosis demonstrating 50% or less removal in several studies (Mitch *et al.* 2003; Plumlee *et al.* 2008). NDMA reacts very slowly with ozone (reaction rate coefficientof $0.052\ M^{-1}s^{-1}$; Lee *et al.* 2007b), making ozonation impractical.

NDMA is only weakly adsorbed to activated carbon or other common adsorbents. Fleming *et al.* (1996) reported little measurable removal by silica, XAD-7 resin, and zeolite-Y (Fleming *et al.* 1996). The same study determined that adsorption using coconut-based and bituminous coal-based activated carbons (Calgon CSC and F300, respectively), as well as Ambersorb 563 and 572, could lead to greater than 99% adsorption, but only under very favourable conditions of high adsorbent loadings and long contact times, with the isotherm parameters suggesting that adsorbent usage rates would be at least an order of magnitude higher than when adsorbing other contaminants commonly treated by activated carbon. Dobbs and Cohen (1980) performed batch adsorption isotherms using Calgon F300 for 143 different contaminants, and NDMA was among the most poorly adsorbed, with the Freundlich K parameter several orders of magnitude lower than compounds that are traditionally poorly adsorbed, such as chloroform and carbon tetrachloride. Similar results are reported by Kommineni *et al.* (2003), who conducted batch isotherm tests on Calgon BCP MOD1 (bituminous coal-based activated carbon), Calgon PCB (coconut-based activated carbon), Ambersorb 600 and 572, and Alltech 5776 (Kommineni *et al.* 2003). In every case, the NDMA isotherm adsorption parameters were many times lower than what is commonly considered acceptable in the water treatment industry from a practical and economic perspective.

NDMA is biodegradable, with studies showing considerable on-site decomposition over time in contaminated ground. Whether drinking water treatment can be specifically designed to remove NDMA through a biological process, such as biologically active filtration is currently unknown. Chung *et al.* (2009) reported that when NDMA was combined with 10 g/L of activated sludge from a wastewater treatment plant in a batch experiment, less than 20% biodegradation of NDMA was observed over 24 hours. In contrast, Schmidt *et al.* (2008) measured NDMA across three drinking water treatment plants that used ozonation followed by either sand or activated carbon filtration (Schmidt and Brauch, 2008). Under these conditions, the removal of 5 to 10 ng/L of NDMA (up to 89% of the NDMA initially present) was observed across the filters, with the authors attributing the removal to biodegradation. This mechanism was not confirmed, however. More work is needed in this area.

7.6.1.1 UV photolysis

One of the first communities to implement NDMA control in the treatment of its drinking water was Ohsweken, Ontario, Canada (Jobb *et al.* 1994). In the late

1980s and early 1990s, NDMA concentrations in treated water ranged up to 115 ng/L, which exceeded the Ontario limit of 9 ng/L. The NDMA was formed within the treatment plant, apparently through reaction between the chlorine that was applied prior to coagulation and unidentified precursors in the source water. A 30 kW medium pressure UV reactor was installed to treat 20 m^3/h, with NDMA consistently reduced to below 5 ng/L.

Subsequent experience has shown that UV photolysis is an effective and practical method to destroy already-formed NDMA, and it is by far the most common form of NDMA treatment (Lee *et al.* 2005a; Mitch *et al.* 2003). Common UV doses for disinfection are in the order of 40 mJ/cm^2, while the dose required for 90% reduction of NDMA is in the order of 1000 mJ/cm^2 (Sharpless and Linden, 2003). This translates in practice to EE/O values (electrical energy required per order of reduction) of roughly 0.2 to 0.3 kWh per 1000 gallons per log (90%) of NDMA removal, but can range from 0.05 (in the case of waters with UV transmissivity of more than 98% per cm) to over 0.7 kWh/1000 gallons/order (Calgon, 2011; Trojan, 2011). Low pressure high output (LPHO) UV systems tend to offer better energy efficiency than medium pressure (MP) systems, especially if nitrate is present which absorbs some of the low wavelength light from MP systems; however, practical issues such as plant footprint and the need to treat other contaminants often make the preferred selection of LPHO versus MP UV systems site-specific. Both LPHO and MP systems are commonly used for NDMA photolysis.

Studies have reported that the rate of NDMA photolysis is pH dependent (Lee *et al.* 2005b; Stefan and Bolton, 2002; Xu *et al.* 2009). In general, the rate is relatively consistent in the pH range of 6 to 8, but is roughly twice as fast at pH 3 compared to pH 7. At pH values greater than 8, the rate of photolysis dramatically decreases such that by pH 9.4, the rate of photolysis was reported to be seven times slower than at pH 7 (Xu *et al.* 2009). UV photolysis of NDMA should therefore be undertaken at pH values below 8 to optimize performance.

One of the main by-products of NDMA photolysis is dimethylamine (DMA). Since DMA is a precursor to NDMA formation by reacting with chlorine or chloramines, it has been suggested that NDMA photolysis at a treatment plant may lead to subsequent NDMA reformation in the chlorinated or chloraminated distribution system. Experience at Ohsweken demonstrated up to 30 ng/L of NDMA reformation in the distribution system (Jobb *et al.* 1994). Mitch *et al.* (2003) argue, however, that DMA formation from NDMA photolysis would be minor and would contribute only negligibly to subsequent NDMA formation (Mitch *et al.* 2003). The production of NDMA in the distribution system at Ohsweken was likely due to the reaction between chlorine and other NDMA precursors that had not been removed during treatment. In general, no other significant by-products of NDMA photolysis have been identified to date, with the possible exception of nitrite. Medium pressure lamps emit light at low wavelengths (below 240 nm), which can cause nitrite formation from the photolysis of nitrate. At common nitrate concentrations (less than 10 mg/L as nitrogen), the expected yield of nitrite during NDMA treatment would be below common regulatory limits (1 mg/L as nitrogen).

7.6.1.2 Advanced oxidation

The reaction rate coefficient between NDMA and hydroxyl radicals formed during advanced oxidation is $4.5 \times 10^8\ M^{-1}s^{-1}$(Lee *et al.* 2007b). This is a relatively slow reaction. Sharpless and Linden (2003) observed under laboratory conditions that for UV-hydrogen peroxide advanced oxidation, there was little to no advantage to be gained by applying hydrogen peroxide over the direct UV photolysis of NDMA, when using either low pressure or medium pressure UV lamps (Sharpless and Linden, 2003).

7.6.1.3 Point of use/point of entry treatment systems (POU/POE)

There are no obvious solutions for NDMA removal by the consumer. At a municipal scale, UV photolysis is practical, but household UV devices are designed to deliver doses of approximately 40 mJ/cm^2 compared to the ~1000 mJ/cm^2 dose needed for effective NDMA control. Other common POU/POE systems include activated carbon, membrane treatment (including reverse osmosis), and ion exchange. While some NDMA removal may occur through these devices, they are not specifically designed for NDMA control, and their performance will generally be poor and inconsistent or unpredictable.

7.6.2 Preventing NDMA formation

7.6.2.1 Precursor removal and oxidation

NDMA precursors can include a wide range of compounds that are difficult to identify in any given water source. In general, these precursors can include organic tertiary and quaternary amines (Kemper *et al.* 2010; Mitch and Schreiber, 2008; Mitch *et al.* 2003), and both humic and non-humic fractions of natural organic matter (Chen and Valentine, 2007; Dotson *et al.* 2007). Designing treatment technologies that specifically target NDMA precursor removal is difficult. Research shows poor correlations between bulk parameters such as total organic carbon, UV absorbance at 254 nm, and NDMA formation potential (Chen and Westerhoff, 2010), so it is difficult to predict whether treatment processes such as enhanced coagulation that target bulk organic removal would be effective for NDMA minimization. In contrast, studies tend to show consistent success in preoxidization prior to chlorination/chloramination to minimize NDMA formation. The application of chlorine as little as one minute prior to ammonia has been shown to reduce subsequent NDMA formation, with longer free-chlorine contact times and doses generally yielding greater reductions in NDMA formation upon chloramination (Charrois and Hrudey, 2007; Chen and Valentine, 2007). Ozone and chlorine dioxide oxidation of natural waters prior to chloramination have also been shown to reduce subsequent NDMA formation by 32–94% (Lee *et al.* 2007a).

7.6.2.2 Chloramine chemistry

NDMA formation is typically higher in distribution systems that use chloramines compared to chlorine (Barrett *et al.* 2003). Utilities normally use chloramines when chlorine is not practical, so it is likely unreasonable to assume that reverting from chloramines to chlorine would commonly be possible to solve an NDMA problem. Instead, solutions should target optimizing chloramination chemistry to minimize NDMA.

A study by Schreiber and Mitch (2006) theorizes that dichloramines may be instrumental in the formation pathway of NDMA, and that steps should be taken to minimize the presence of dichloramines (Schreiber and Mitch, 2006, this book Chapter 5). This includes maintaining high pH (e.g. >8.5), especially at the point of chloramine formation, and a molar chlorine to nitrogen ratio of below 1. This latter point suggests that when generating chloramines in the main process stream, it is preferable to apply ammonia downstream of already-blended chlorine, since the chlorine to ammonia-nitrogen ratio will be less than 1 in the local vicinity of the ammonia injector. Conversely, if chlorine is added to already-blended ammonia, there will be an excess of chlorine in the local region of the chlorine injector, possibly leading to dichloramine and therefore NDMA formation, especially if the pH is below 8. An alternative solution is to generate chloramines offline in a system that promotes exclusively monochloramine formation (i.e. $pH > 8.5$, Cl:N molar ratio less than 1), prior to adding the chloramines to the main flow.

7.6.2.3 Polymers and ion exchange resins

In the past, certain amine-based polymers used in water treatment were shown to react with chlorine or chloramines to form NDMA (Kohut and Andrews, 2003; Wilczak *et al.* 2003). Likewise, ion exchange resins containing quaternary ammonium ions were implicated as a source of NDMA, either when chlorinated water was passed through the resin, or by leaching precursors into the water to subsequently form NDMA upon chlorination or chloramination (Siddiqui and Atasi, 2004). Chemical suppliers are increasingly aware of these problems and have worked to find solutions in recent years. Utilities concerned about this potential problem should discuss the issue with their chemical suppliers.

7.7 CONCLUSIONS

N-Nitrosamines have, not surprisingly, attracted a lot of attention as DBPs.

- *N*-Nitrosamines, NDMA in particular have emerged as a class of DBPs attracting major attention. Several jurisdictions have adopted drinking water guidelines for NDMA and the USEPA is considering regulating *N*-nitrosamines as a group.

- *N*-Nitrosamines are widely accepted as cancer-causing in rodents by a genotoxic mechanism which generates the regulatory interest.
- Despite causing cancer in rodents with much higher potency than any of the regulated DBPs, the low concentrations and occurrence of *N*-nitrosamines in drinking water are not able to explain epidemiologic bladder cancer predictions.
- Drinking water provides a very small fraction of population exposure to *N*-nitrosamines making reduction of *N*-nitrosamines in drinking water an ineffective way of controlling human exposure to *N*-nitrosamines.
- Risk management options within drinking water treatment for *N*-nitrosamines provide few practical options for removing *N*-nitrosamines (with a possible exception of high dose (~1000 mJ/cm^2) UV photolysis) but there are viable options for preventing *N*-nitrosamine formation.

REFERENCES

ACS (2011). *Cancer Facts and Figures 2011* [online]. American Cancer Society Available at: http://www.cancer.org/Research/CancerFactsFigures/CancerFactsFigures/cancer-facts-figures-2011 (accessed 10 January 2012).

Asami M., Oya M. and Kosaka K. (2009). A nationwide survey of NDMA in raw and drinking water in Japan. *Sci. Total Environ.*, **407**(11), 3540–3545.

Barrett S., Hwang C., Guo Y., Andrews S. A. and Valentine R. L. (2003). Occurrence of NDMA in drinking water: a North American Survey, 2001–2002. Proceedings of the AWWA Annual Conference, Anaheim, CA.

Boyd J. M., Zhao Y. Y., Wagner M., Qin F., Levallois P., Legay C., Charrois J. W. A., Richardson S. D., Hrudey S. E. and Li X. F. (2010). Occurrence of *N*-nitrosodiphenylamine and *N*-nitrosodimethylamine in 38 drinking water systems in Canada and the USA. Proceedings of the IWA World Water Congress and Exhibition, Montreal, Quebec.

Bradley P. M., Carr S. A., Baird R. B. and Chapelle F. H. (2005). Biodegradation of *N*-nitrosodimethylamine in soil from a water reclamation facility. *Biorem. J.*, **9**(2), 115–120.

Calgon (2011). Personal communication with Keith Bircher, Calgon Carbon Corporation.

CalDPH (2011). NDMA and other *N*-nitrosamines – drinking water issues [online]. California Department of Public Health Drinking Water Program Available at: http://www.cdph.ca.gov/certlic/drinkingwater/Pages/NDMA.aspx (accessed 10 January 2012).

CCS (2011). *Canadian Cancer Statistics 2011* [online]. Canadian Cancer Society Available at http://www.cancer.ca/statistics (accessed 10 January 2012).

Cardy R. H., Lijinsky W. and Hildebrandt P. K. (1979). Neoplastic and nonneoplastic urinary bladder lesions induced in Fischer 344 rats and B6C3F1 hybrid mice by *N*-nitrosodiphenylamine. *Ecotoxicol. Environ. Safe*, **3**(1), 29–35.

Charrois J. W. A., Arend M. W., Froese K. L. and Hrudey S. E. (2004). Detecting *N*-nitrosamines in drinking water at nanogram per liter levels using ammonia positive chemical ionization. *Environ. Sci. Technol.*, **38**(18), 4835–4841.

Charrois J. W. A., Boyd J. M., Froese K. L. and Hrudey S. E. (2007). Occurrence of *N*-nitrosamines in Alberta public drinking-water distribution systems. *J. Environ. Eng. Sci.*, **6**, 103–114.

Charrois J. W. A. and Hrudey S. E. (2007). Breakpoint chlorination and free-chlorine contact time: implications for drinking water *N*-nitrosodimethylamine concentrations. *Water Res.*, **41**(3), 674–682.

Chen Z. and Valentine R. L. (2007). Formation of *N*-nitrosodimethylamine (NDMA) from humic substances in natural water. *Environ. Sci. Technol.*, **41**(17), 6059–6065.

Chen B. and Westerhoff P. (2010). Predicting disinfection by-product formation potential in water. *Water Res.*, **44**(13), 3755–3762.

Chung J., Yoon Y., Kim M., Lee S. B., Kim H. J. and Choi C. K. (2009). Removal of radio *N*-nitrosodimethylamine (NDMA) from drinking water by coagulation and powdered activated carbon (PAC) adsorption. *Drink. Water Eng. Sci.*, **2**, 49–55.

Dobbs R. A. and Cohen J. M. (1980). Carbon Adsorption Isotherms for Toxic Organics. United States Environmental Protection Agency, EPA-600/8-80–023. Washington, D. C., USA.

Dotson A., Westerhoff P. and Krasner S. W. (2007). Nitrosamine formation from natural organic matter isolates and sunlight photolysis of *N*-nitrosamines. In: Proceedings of the American Water Works Association Annual Conference and Exposition, Toronto, ON, Canada.

Ducos P., Gaudin R., Maire C., Mavelle T., Bouchikhi B. and Debry G. (1988). Occupational exposure to volatile *N*-nitrosamines in foundries using the ashland core-making process. *Environ. Res.*, **47**(1), 72–78.

Fajen J. M., Carson G. A., Rounbehler D. P., Fan T. Y., Vita R., Goff U. E., Wolf M. H., Edwards G. S., Fine D. H., Reinhold V. and Biemann K. (1979). *N*-Nitrosamines in the Rubber and Tire Industry. *Science*, **205**(4412), 1262–1264.

Fleming E. C., Pennington J. C., Wachob B. G., Howe R. A. and Hill D. O. (1996). Removal of *N*-nitrosodimethylamine from waters using physical-chemical techniques. *J. Hazard. Mater.*, **51**(1–3), 151–164.

Fristachi A. and Rice G. (2007). Estimation of the total daily oral intake of NDMA attributable to drinking water. *J. Water Health.*, **5**(3), 341–355.

Gossett J. M. (1987). Measurement of Henry's Law constants for C1 and C2 chlorinated hydrocarbons. *Environ. Sci Technol.*, **21**(2), 202–208. [conversion to SI units according to Sander: http://www.rolf-sander.net/henry/henry-conv.html] (accessed 3 January 2012).

GovOnt (2002). Safe Drinking Water Act. Ontario Regulation 169/03, Schedule 2. Government of Ontario.

Gray R., Peto R., Brantom P. and Grasso P. (1991). Chronic nitrosamine ingestion in 1040 rodents: the effect of the choice of nitrosamine, the species studied and the age of starting exposure. *Cancer Res.*, **51**(23), 6470–6491.

Haruta S., Jiao W., Chen W., Chang A. C. and Gan J. (2011). Evaluating Henry's law constant of *N*-Nitrosodimethylamine (NDMA). *Water Sci. Technol.*, **64**(8), 1636–1641. [conversion to SI units according to Sander: http://www.rolf-sander.net/henry/henry-conv.html]

Health Canada. (2010). *N-nitrosodimethylamine (NDMA) in drinking water* [online]. Available at: http://www.hc-sc.gc.ca/ewh-semt/consult/_2010/ndma/draft-ebauche-eng.php (accessed 10 January 2012).

IARC (1978). *IARC Monographs on the Evaluation of the Carcinogenic Risk of Chemicals to Humans: Some N-Nitroso Compounds.* International Agency for Research on Cancer, Lyon, France, p. 365.

Jakszyn P., Gonzalez C. A., Lujan-Barroso L., Ros M. M., Bueno-de-Mesquita H. B., Roswall N., Tjønneland A. M., Buchner F. L., Egevad L., Overvad K., Raaschou-Nielsen O., Clavel-Chapelon F., Boutron-Ruault M.-C., Touillaud M. S., Chang-Claude J., Allen N. E., Kiemeney L. A., Key T. J., Kaaks R., Boeing H., Weikert S., Trichopoulou A., Oikonomou E., Zylis D., Palli D., Berrino F., Vineis P., Tumino R., Mattiello A., Peeters P. H. M., Parr C. L., Gram I. T., Skeie G., Sanchez M.-J., Larrañaga N., Ardanaz E., Navarro C., Rodríguez L., Ulmert D., Ehrnstrom R., Hallmans G., Ljungberg B., Roddam A. W., Bingham S. A., Khaw K.-T., Slimani N., Boffetta P. A., Jenab M., Mouw T., Michaud D. S. and Riboli E. (2011). Red meat, dietary *N*-nitrosamines, and heme iron and risk of bladder cancer in the European Prospetive Investigation into Cancer and Nutrition (EPIC). *Cancer Epidemiology Biomarkers & Prevention*, **20**(3), 555–558.

Jobb D. B., Hunsinger R. B., Meresz O. and Taguchi V. (1994). Removal of *N*-nitrosodimethylaine from the Ohsweken (Six Nations) Water Supply, Final Report, Ontario Ministry of Environment, Toronto, Canada. Available at: http://www.archive.org/stream/removalofnnitros00ontauoft/removalofnnitros00ontauoft_djvu.txt (accessed March 7, 2012).

Kemper J. M., Walse S. S. and Mitch W. A. (2010). Quaternary amines as nitrosamine precursors: a role for consumer products? *Environ. Sci. Technol.*, **44**(4), 1224–1231.

King W. D. and Marrett L. D. (1996). Case-control study of bladder cancer and chlorination by products in treated water (Ontario, Canada). *Cancer Cause. Control*, **7**, 596–604.

Kohut K. D. and Andrews S. A. (2003). Polyelectrolytes as amine precursors for N-nitrosodimethylamine in drinking water. In the Proceedings of the AWWA Annual Conference, Anaheim, CA.

Kommineni S., Ela W. P., Arnold R. G., Huling S. G., Hester B. J. and Betterton E. A. (2003). NDMA treatment by sequential GAC adsorption and Fenton-driven destruction. *Environ. Eng. Sci.*, **20**(4), 361–373.

Krul C. A. M., Zeilmaker M. J., Schothorst R. C. and Havenaar R. (2004). Intragastric formation and modulation of *N*-nitrosodimethylamine in a dynamic in vitro gastrointestinal model under human physiological conditions. *Food Chem. Toxicol.*, **42**, 51–63.

Larsson S. C., Johansson J.-E., Andersson S.-O. and Wolk A. (2009). Meat intake and bladder cancer risk in a Swedish prospective cohort. *Cancer Causes Control*, **20**, 35–40.

Lee C., Choi W., Kim Y. G. and Yoon J. (2005a). UV photolytic mechanism of *N*-nitrosodimethylamine in water: dual pathways to methylamine versus dimethylamine. *Environ. Sci. Technol.*, **39**(7), 2101–2106.

Lee C., Choi W. and Yoon J. (2005b). UV photolytic mechanism of *N*-nitrosodimethylamine in water: roles of dissolved oxygen and solution pH. *Environ. Sci. Technol.*, **39**(24), 9702–9709.

Lee C., Schmidt C., Yoon J. and von Gunten U. (2007a). Oxidation of *N*-nitrosodimethylamine (NDMA) precursors with ozone and chlorine dioxide: kinetics and effect on NDMA formation potential. *Environ. Sci. Technol.*, **41**(6), 2056–2063.

Lee C., Yoon J. and von Gunten U. (2007b). Oxidative degradation of *N*-nitrosodimethylamine by conventional ozonation and the advanced oxidation process ozone/hydrogen peroxide. *Water Res.*, **41**(3), 581–590.

Lijinsky W., Kovatch R. M., Riggs C. W. and Walters P. T. (1988). Dose response study with *N*-nitrosomorpholine in drinking water of F-344 rats. *Cancer Res.*, **48**, 2089–2095.

Liteplo R. G., Meek M. E. and Windle W. (2002). *N*-nitrosodimethylamine; Concise International Chemcial Assessment Document (CICAD) 38. Wissenschaftliche Verlagsgesellschaft, Stuttgart, Germany.

MassDEP (2004). Current regulatory limit: *N*-nitrosodimethylamine (NDMA) [online]. Massachusetts Department of Environmental Protection Available at: http://www.mass.gov/dep/water/drinking/standards/ndma.htm (accessed 11 January 2012).

Mitch W. A. and Schreiber I. M. (2008). Degradation of tertiary alkylamines during chlorination/chloramination: implications for formation of aldehydes, nitriles, halonitroalkanes, and *N*-nitrosamines. *Environ. Sci. Technol.*, **42**(13), 4811–4817.

Mitch W. A., Sharp J. O., Trussell R. R., Valentine R. L., Alvarez-Cohen L. and Sedlak D. L. (2003). *N*-nitrosodimethylamine (NDMA) as a drinking water contaminant: a review. *Environ. Eng. Sci.*, **20**(5), 389–404.

Nakai J. S., Stathopulos P. B., Campbell G. L., Chu I., Li-Muller A. and Aucoin R. (1999). Penetration of chloroform, trichloroethylene, and tetrachloroethylene through human skin. *J. Toxicol. Environ. Health A*, **58**(3), 157–170.

NHMRC/NRMMC (2011). Australian Drinking Water Guidelines. National Health and Medical Research Council/Natural Resources Management Ministerial Council. Canberra.

Nuckols J. R., Ashley D. L., Lyu C., Gordon S. M., Hinckley A. F. and Singer P. (2005). Influence of tap water quality and household water use activities on indoor air and internal dose levels of trihalomethanes. *Environ. Health Perspect.*, **113**(7), 863–870.

Odom R., Regli S., Messner M., Cromwell J. and Javdan M. (1999). Benefit-cost analysis of the Stage 1 D/DBP Rule. *J. Am. Water Works Assoc.*, **91**(4), 137–147.

OEHHA (2006). Public Health Goal for *N*-nitrosodimethylamine in drinking water [online]. Office of Environmental Health Hazard Assessment; California Environmental Protection Agency Available at: http://oehha.ca.gov/water/phg/pdf/122206NDMAphg.pdf (accessed 12 January 2012)

Okada M. and Ishidate M. (1977). Metabolic fate of *N*-*n*-butyl-*N*-(4-hydroxybutyl)-nitrosamine and its analogues: selective induction of urinvary bladder tumors in the rat. *Xenobiotica*, **7**(1–2), 11–24.

Peto R., Gray R., Brantom P. and Grasso P. (1991). Effects on 4080 rats of chronic ingestion of *N*-nitrosodiethylamine or *N*-nitrosodimethylamine – a detailed dose-response study. *Cancer Res.*, **51**(23), 6415–6451.

Plumlee M. H., Lopez-Mesas M., Heidlberger A., Ishida K. P. and Reinhard M. (2008). *N*-nitrosodimethylamine (NDMA) removal by reverse osmosis and UV treatment and analysis via LC-MS/MS. *Water Res.*, **42**(1–2), 347–355.

Reh B. D., DeBord D. G., Butler M. A., Reid T. M., Mueller C. and Fajen J. M. (2000). O6-methylguanine DNA adducts associated with occupational nitrosamine exposure. *Carcinogenesis*, **21**(1), 29–33.

Rounbehler D. P., Krull I. S., Goff E. U., Mills K. M., Morrison J., Edwards G. S., Fine D. H., Fajen J. M., Carson G. A. and Rheinhold V. (1979). Exposure to *N*-Nitrosodimethylamine in a Leather Tannery. *Food Cosmet. Toxicol.*, **17**(5), 487–491.

Schäfer A. I., Mitch W., Walewijk S., Munoz A., Teuten E., Reinhard M., Isabel C. E. and Andrea I. S. (2010). Micropollutants in water recycling: a case study of *N*-Nitrosodimethylamine (NDMA) exposure from water versus food. In: Escobar I. C. and Schafer A. I. (eds) *Sustainability Science and Engineering: Sustainable Water for the Future – Water Recycling vs. Desalination*. Elsevier, Vol. 2, Amsterdam, The Netherlands, pp. 203–228.

Schmidt C. K. and Brauch H.-J. (2008). *N,N*-dimethosulfamide as precursor for *N*-nitrosodimethylamine (NDMA) formation upon ozonation and its fate during drinking water treatment. *Environ. Sci. Technol.*, **42**(17), 6340–6346.

Schreiber I. M. and Mitch W. A. (2006). Nitrosamine formation pathway revisited: the importance of chloramine speciation and dissolved oxygen. *Environ. Sci. Technol.*, **40**(19), 6007–6014.

Sharpless C. M. and Linden K. G. (2003). Experimental and model comparisons of low- and medium-pressure Hg lamps for the direct and H_2O_2 assisted UV photodegradation of *N*-nitrosodimethylamine in simulated drinking water. *Environ. Sci. Technol.*, **37**(9), 1933–1940.

Siddiqui M. and Atasi K. Z. (2004). *N*-nitrosodimethylamine: a disinfectant byproduct and its occurrence in wastewater. *Water Environ. Res.*, **76**(4), 316–326.

Stefan M. I. and Bolton J. R. (2002). UV direct photolysis of *N*-nitrosodimethylamine (NDMA): kinetic and product study. *Helv. Chim. Acta*, **85**(5), 1416–1426.

Straif K., Keil U., Taeger D., Holthenrich D., Sun Y., Bungers M. and Weiland S. K. (2000). Exposure to *N*-nitrosamines, carbon black, asbestos, and talc and mortality from stomach, lung, and laryngeal cancer in a cohort of rubber workers. *Am. J. Epidemiol.*, **152**(4), 297–306.

Taguchi V., Jenkins S. D. W., Wang D. T., Palmentier J. P. F. P. and Reiner E. J. (1994). Determination of *N*-nitrosodimethylamine by isotope-dilution, high-resolution mass-spectrometry. *Can. J. Appl. Spectrosc.*, **39**(3), 87–93.

Tannenbaum S. R. (1980). A model of estimation of human exposure to endogenous *N*-nitrosodimethylamine. *Oncology*, **17**, 232–235.

Templeton M. R. and Chen Z. (2010). NDMA and seven other *N*-nitrosamines in selected UK drinking water supply systems. *J. Water Supply Res. Technol. – AQUA*, **59**(4), 277–283.

Trojan (2011). Personal communication with Alan Royce, Trojan Technologies.

USEPA (2003). National primary drinking water regulations: disinfectants and disinfection by-products; national primary and secondary drinking water regulations: approval of analytical methods for chemical contaminants, proposed rule. U.S. Environmental Protection Agency. *Fed. Regist.*, **68**(159), 49,548–49,681.

USEPA (2005). Unregulated contaminant monitoring regulation (UCMR) for public water systems revisions. U.S. Environmental Protection Agency. *Fed. Regist.*, **70**(161), 49,094–49,138.

USEPA (2009). Contaminant candidate list 3 (CCL-3) – final. U.S. Environmental Protection Agency. *Fed. Regist.*, **74**, 51,850–51,862.

USEPA (2011a). Integrated Risk Information System (IRIS) [online]. U.S. Environmental Protection Agency Available from: http://www.epa.gov/iris/ (accessed 10 January 2012).

USEPA (2011b). Integrated Risk Information System (IRIS). Glossary of IRIS terms [online]. U.S. Environmental Protection Agency Available from: http://www.epa.gov/iris/help_gloss.htm (accessed 10 January 2012).

USEPA (2011c). Unregulated Contaminant Monitoring Rule 2 (UCMR-2) (2008–2010) Occurrence Data [online]. U.S. Environmental Protection Agency Available at: http://water.epa.gov/lawsregs/rulesregs/sdwa/ucmr/data.cfm (accessed 13 June 2011).

Vermeer I. T. M., Pachen D. M. F. A., Dallinga J. W., Kleinjans J. C. S. and van Maanen J. M. S. (1998). Volatile *N*-nitrosamine formation after intake of nitrate at the ADI level in combination with an amine-rich diet. *Environ. Health Perspect.*, **106**, 459–463.

Wang W. F., Hu J. Y., Yu J. W. and Yang M. (2010). Determination of *N*-nitrosodimethylamine in drinking water by UPLC-MS/MS. *J. Environ. Sci. China*, **22**(10), 1508–1512.

Wilczak A., Assadi-Rad A., Lai H. H., Hoover L. L., Smith J. F., Berger R., Rodigari F., Beland J. W., Lazzelle L. J., Kinicannon E. G., Baker H. and Heaney C. T. (2003). Formation of NDMA in chloraminated water coagulated with DADMAC cationic polymer. *J. AWWA*, **95**(9), 94–106.

WHO (2008). *N-nitrosodimethylamine in Drinking-Water*. World Health Organization, p. 29.

Xu B., Chen Z., Qi F., Shen J. and Wu F. (2009). Factors influencing the photodegradation of *N*-nitrosodimethylamine in drinking water. *Front. Environ. Sci. Eng. China*, **3**(1), 91–97.

Zeegers M. P. Selen R. F. M., Kleinjans J. C. S., Goldbohm R. A. and van den Brandt P. A. (2006). Nitrate intake does not influence bladder cancer risk: the Netherlands Cohort Study. *Environ. Health Persp.*, **114**, 1527–1531.

Zhao Y. Y., Boyd J., Hrudey S. E. and Li X. F. (2006). Characterization of new *N*-nitrosamines in drinking water using liquid chromatography tandem mass spectrometry. *Environ. Sci. Technol.*, **40**(24), 7636–7641.

Chapter 8

Recent novel DBPs halogenated benzoquinones

Y. Zhao, W. Wang, S. E. Hrudey and X.-F. Li

8.1 INTRODUCTION

A number of epidemiologic studies have indicated associations between consumption of chlorinated drinking water and increased risk of bladder cancer, but causative agents have not been identified (Chapters 10 and 11). Halobenzoquinones (HBQs) have been predicted to be a class of DBPs that might be formed in drinking water and be of potential toxicological relevance (Bull *et al.* 2006). For example, 2,6-dichloro-1,4-benzoquinone (DCBQ), 2,6-dichloro-3-methyl-1,4-benzoquinone (DCMBQ), and 2,3,6-trichloro-1,4-benzoquinone (TCBQ) were estimated to have very low toxic thresholds (lowest observed adverse effect level – LOAEL) of 0.049, 0.079, and 0.033 mg/kg day^{-1}, based on quantitative structure toxicity relationship (QSTR) analysis (Bull *et al.* 2006). Water Research Foundation Project 4089 was funded, in part, to develop analytical methods for the determination of HBQs in drinking water (Li *et al.* 2010). This chapter conveys what was involved in documenting a new class of toxicologically relevant DBPs by describing the necessary analytical methods development of HBQs, including water sample collection, storage, extraction, and LC-MS/MS analysis. For the purposes of this book, this elaboration illustrates how much effort is required to identify DBPs which may be more toxicologically relevant than those currently regulated or many other identified DBPs.

A preliminary survey was conducted to determine the occurrence of HBQs in drinking water treatment plants. Four HBQs with differing substitution species (chlorine, bromide, and methyl group), numbers, and positions were selected as target HBQs: DCBQ, DCMBQ, TCBQ, and 2,6-dibromo-1,4-benzoquinone (DBBQ) (Figure 8.1).

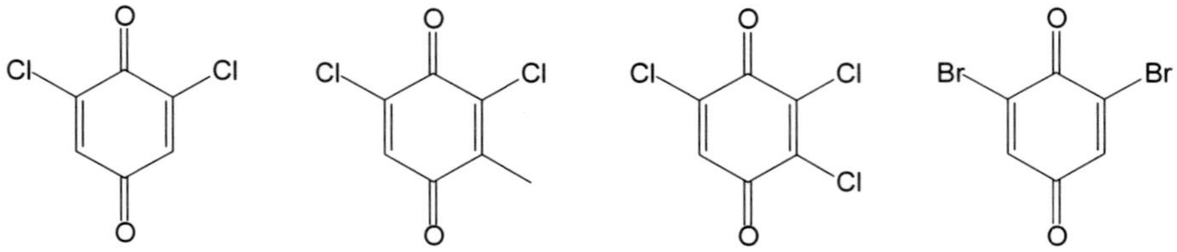

Figure 8.1 Chemical structure of four target HBQs

8.2 SAMPLING PROCEDURES

8.2.1 Water sample collection and storage

Water samples were collected using 4-L amber glass bottles with PTFE-lined screw caps. All the bottles were pre-cleaned with methanol followed by water, and oven baked overnight at 180°C. The bottles were rinsed with water samples prior to sampling. From each sampling site, two water samples were collected by grab sampling. One of the water samples had 0.25% (vol/vol) formic acid (FA) added to quench residual free chlorine (West and Rollefson, 1936; Hoq *et al.* 1991) and stabilize HBQs (8.2.2), and the other one was collected without any agents added. Immediately after collection, the samples were transported back to the laboratory overnight in coolers with ice packs. For each set of samples, one field blank sample of water (Optima LC/MS grade) with FA (0.25%, vol/vol) added, was included and underwent the sample collection and transportation procedure together with the water samples. Once transported back to the laboratory, the water samples were stored at 4°C until analysis.

8.2.2 Sample de-chlorination and preservation

8.2.2.1 Effects of pH on HBQs

Degradation has been a major problem hampering the investigation of HBQs in drinking water. We investigated the stability of the DCBQ standard in aqueous solutions at acidic, neutral, and basic conditions. Aqueous samples with DCBQ at 100 ng/mL were prepared in different buffer solutions: acidic conditions with 0.1 mol/L of ammonium acetate at pH 4.5, neutral conditions with 0.1 mol/L of sodium phosphate at pH 7.0, and basic conditions with 0.1 mol/L of sodium bicarbonate at pH 9.5. Each solution was analyzed by LC-MS/MS at different storage times. The stability of DCBQ at different pH conditions was measured by the relative peak areas versus storage time.

Figure 8.2 shows the stability of aqueous DCBQ in acidic and neutral conditions. In the acidic solution (pH 4.5), the average of the relative peak areas of DCBQ to the first injection was 88 ± 7% over 21 h (Figure 8.2a). In the neutral pH solution (pH 7), the relative peak areas of DCBQ clearly decreased with the increase of storage time, and the half-life (peak area decreased by 50%) of the analytes was 6–7 h (Figure 8.2b). In the basic solution (pH 9.5), no peak of DCBQ was detected even when it was injected immediately after preparation. The results indicated that HBQs are stable in acidic solution, but are not stable in neutral and basic solutions. This can be explained by the chemical property of HBQs: they are subject to hydroxylation via substitution reaction, in which halogen is displaced by a hydroxyl group. In basic conditions, the hydroxylation reaction is enhanced by the higher concentration of hydroxyl ion (OH^-); in acidic conditions, however, the reaction is restrained by the presence of highly abundant protons (H^+).

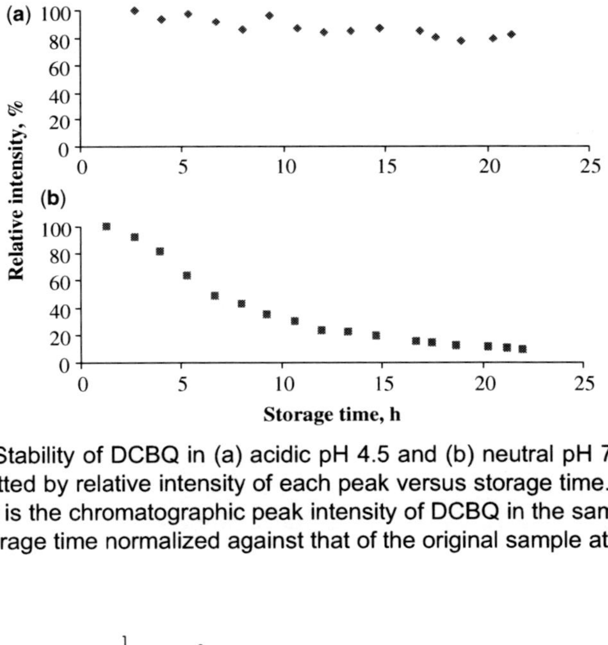

Figure 8.2 Stability of DCBQ in (a) acidic pH 4.5 and (b) neutral pH 7.0 solutions. Data are plotted by relative intensity of each peak versus storage time. The relative intensity (%) is the chromatographic peak intensity of DCBQ in the sample after the specified storage time normalized against that of the original sample at time zero

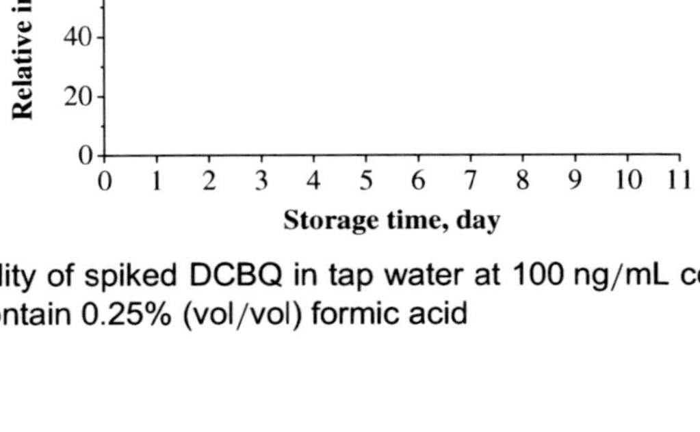

Figure 8.3 Stability of spiked DCBQ in tap water at 100 ng/mL concentration after acidification to contain 0.25% (vol/vol) formic acid

8.2.2.2 Stability of DCBQ in acidified water samples

We examined the stability of DCBQ in acidified water samples. Water samples were prepared from tap water spiked with DCBQ at 100 ng/mL to which 0.25% (vol/vol) FA was added. After different storage times of 0, 2, 5, 7 and 10 days, water samples were analyzed for the DCBQ. Figure 8.3 shows the concentration of DCBQ with storage time. The results show that DCBQ is stable during the 10-day storage time in acidified water samples.

8.3 SOLID PHASE EXTRACTION

Solid phase extraction was developed for the purification and concentration of HBQs from water samples. A VISIPREP SPE manifold (Supelco, Bellefonte, PA) with flow control liners was used. Based on the chemical molecular structures of HBQs, these chemicals are weakly hydrophobic. Two types of SPE cartridges, Oasis HLB (6 mL, 200 mg packing per cartridge; Waters, Milford, MA) and HyperSep/Hypercarb (6 mL, 1 g bed weight per cartridge; Thermo, US), were examined for the extraction of HBQs from water. Both cartridges were suitable, but the Oasis HLB required less elution solvent and had higher recoveries for the HBQs compared to the HyperSep/Hypercarb.

The Waters Oasis HLB cartridge was used for further optimization. A series of optimization experiments were compared for spiked water samples: 500-mL tap water sample with added 0.25% FA, containing each of the four HBQs at 30 ng/L. We used mixture solvents of water (FA 0.25%) and methanol (FA 0.25%) as washing solvents, and different compositions were compared for their efficiencies to remove interference and the recoveries of the HBQs. The results showed that the composition of washing solvents was a very important factor to efficiently retain the HBQs on the cartridge while removing sample matrices. Figure 8.4 illustrates the chromatograms of the extracts when the washing step was performed with varying solvent compositions: (a) no wash; (b) 30% methanol in 0.25% aqueous FA; (c) 50% methanol in 0.25% aqueous FA. Without washing, Figure 8.4a shows that the HBQs in the extract are barely detected due to severe matrix interference. Using 30% methanol in 0.25% aqueous FA to wash the SPE cartridge dramatically improved the separation and detection of the HBQs in the extract (Figure 8.4b), but the background interference was still severe. Increasing methanol to 50% in 0.25% aqueous FA as the washing solvent dramatically reduced the background peaks and enhanced the detection of the HBQs (Figure 8.4c). Further increasing the methanol content in the washing solvent also resulted in the loss of the HBQs. The results showed that the minimum interference and the maximum recovery of the HBQs were achieved with methanol/water (v/v 50/50 with 0.25% FA).

The optimized SPE procedures were: (1) prior to the SPE, the cartridge was preconditioned with 6 mL of methanol followed by 6 mL of water and 6 mL of aqueous FA (0.25%, v/v); (2) water sample (500 mL) was forced through the cartridge at a flow rate of approximately 8 mL/min, and twelve samples were processed in parallel; (3) after sample loading, the cartridges were washed with 6 mL of aqueous FA (0.25%, v/v) and 6 mL of methanol/aqueous FA (50/50, v/v), and then dried for 10 min under vacuum; (4) the analytes were then eluted from the cartridges with 6 mL methanol containing 0.25% FA. After SPE, the extract was condensed to 100 μL (avoiding complete dryness), and then reconstituted to 500 μL with aqueous FA (0.25%, v/v).

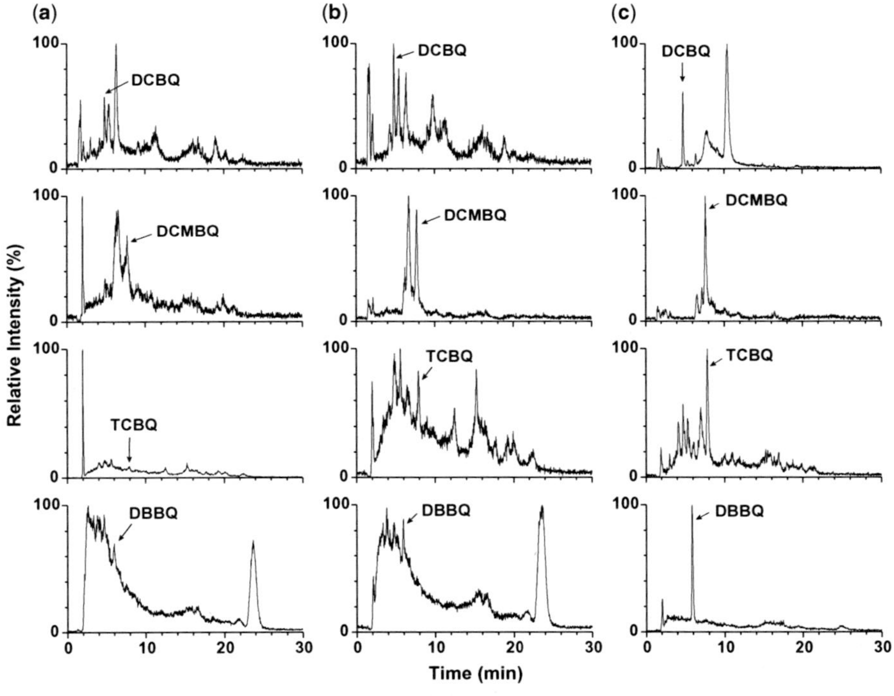

Figure 8.4 MRM chromatograms of DCBQ, DCMBQ, TCBQ and DBBQ obtained from the extracts when the SPE was performed with varying washing solvents: (a) without washing, (b) washing with 30% methanol in 0.25% aqueous FA, (c) washing with 50% methanol in 0.25% aqueous FA

8.4 ANALYTICAL METHOD

8.4.1 Liquid chromatography

An Agilent 1100 series LC system consisting of a binary pump and an autosampler (Agilent, Waldbronn, Germany) was used for the HPLC separation of the four

HBQs at room temperature (22°C). LC separation of the four HBQs was compared for four columns: C18, C8, activated graphic carbon, and phenyl. The C18 column provided better separation of these compounds than the other columns. A Luna C18 (2) column (100 × 2.0 mm i.d., 3 μm; Phenomenex, Torrance, CA, USA) was used for the following experiment.

Mobile phase composition of the organic mobile phase was compared for methanol with 0.25% FA and acetonitrile with 0.25% FA, using water containing 0.25% FA (vol/vol) as the aqueous mobile phase. Results show that using methanol with 0.25% FA as organic phase resulted in lower background and better separation than using acetonitrile with 0.25% FA as the organic phase. After a thorough optimization, the optimal gradient began with 20% methanol with 0.25% FA (B); linearly increased B to 90% in 20 min, and kept steady for 5 min; decreased B to 20% at 25 min, and held for 15 min. A 10-port 2-position switching valve (Rheodyne, Rohnert Park, CA, USA) was used between the HPLC column outlet and the inlet of the MS for directing the LC effluent to the MS detector: the LC effluent was diverted to waste for the initial 10 min, then directly sprayed to the MS during the elution time of 10 to 25 min, and finally diverted to waste during the elution time of 25–40 min. The flow rate of the mobile phase was 150 μL/min, and the injection volume was 20 μL.

Figure 8.5 shows the typical chromatograms of the four HBQs obtained under the optimized gradient LC separation and MS detection conditions (MRM). All four peaks are symmetrical. The elution order of DCBQ, DBBQ, DCMBQ, and TCBQ is according to the strength of hydrophobic interaction of the analytes with the C18 stationary phase.

8.4.2 Mass spectrometry

8.4.2.1 ESI-MS

A QTRAP 5500 mass spectrometer (AB Sciex, Concord, ON, Canada) was used for MS experiments of HBQs in infusion experiments (Syringe Pump, Harvard Apparatus, Holliston, MA, USA) using individual standard solutions having concentrations between 1 and 5 μg/mL. Both ESI and atmospheric pressure chemical ionization with both positive and negative modes were examined for the 4 HBQs. Monitoring of the traditional protonated molecular $[M+1]^+$ ions in positive mode and deprotonated molecular $[M-1]^-$ ions in negative mode showed low sensitivity and were not reproducible. However, the mass spectra generated in negative ESI mode gave the most abundant and stable protonated molecule $[M+H]^-$ ions for all the HBQs: 177 for DCBQ, 191 for DCMBQ, 211 for TCBQ, and 267 for DBBQ.

To confirm the identification of the $[M+H]^-$ as precursor ion (Q1) under negative ESI, we examined their accurate mass using a hybrid quadrupole time-of-flight mass spectrometer (API Q-STAR Pulsar I; AB Sciex, Concord, ON, Canada) equipped with a regular ion-spray source. The MS conditions used in this study were negative TOF-MS mode, ion source voltage −4500 V, declustering potential (DP1) −55 V, focusing potential −300 V, declustering potential (DP2) −15 V, ion source gas (N_2) 50 arbitrary unit, curtain gas (N_2) 25 arbitrary unit, accumulation time 1 s, and scan range (m/z) 50–300 amu. Figure 8.6 shows their accurate MS spectra, with the isotopic information of 35/37Cl and 79/81Br, in comparison with the theoretical values. The accurate mass measurements of $[M+H]^-$ and isotopic patterns were consistent with their theoretical values. These results confirmed that the precursor $[M+H]^-$ ions instead of the traditional $[M-H]^-$ ions are produced in negative ESI.

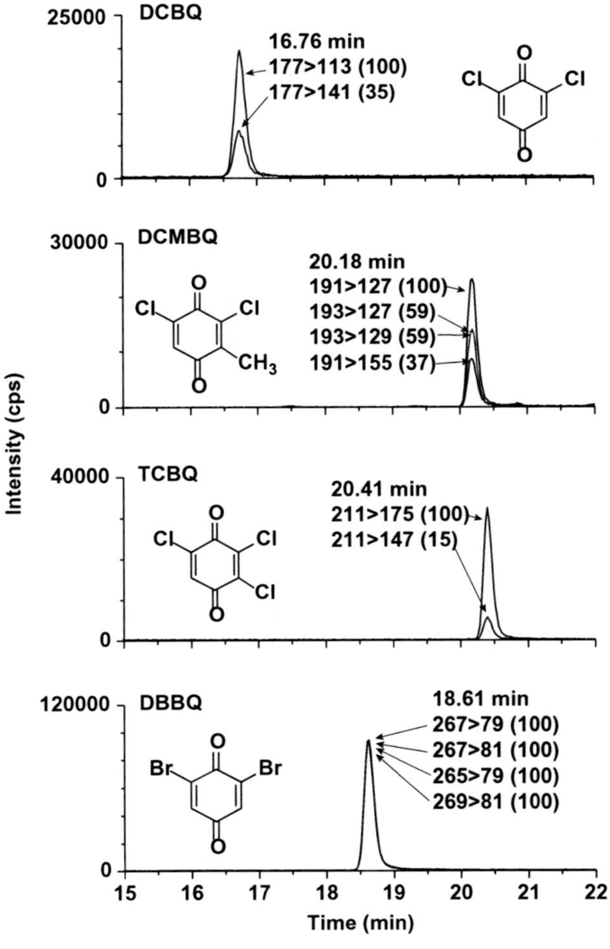

Figure 8.5 MRM chromatograms of the standard solution (5 ng/mL) of DCBQ, DCMBQ, TCBQ, and DBBQ under the optimal conditions

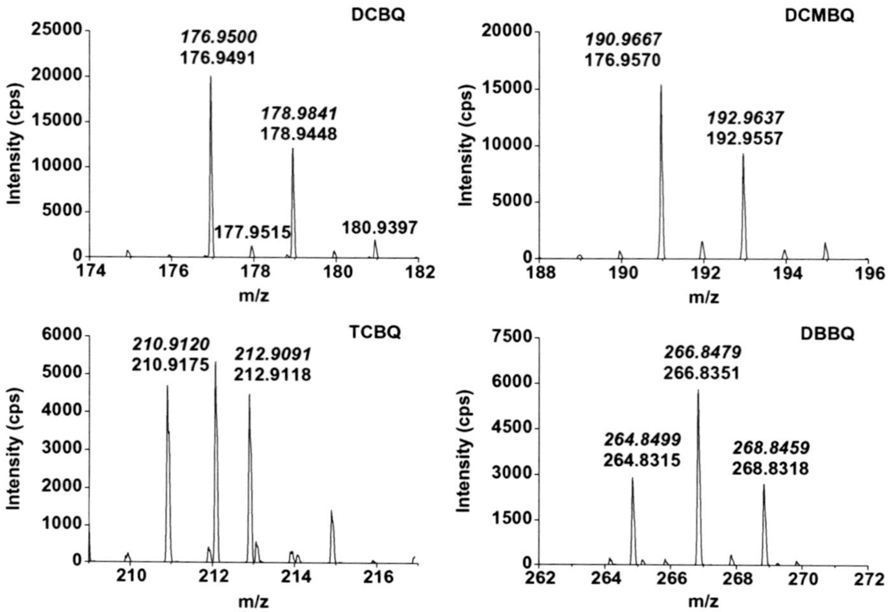

Figure 8.6 Accurate mass of the precursor ions of the $[M+H]^-$ of DCBQ, DCMBQ, TCBQ, and DBBQ from TOF SCAN measurement under ESI in comparison with their theoretical values (in italics)

To understand the formation of $[M+H]^-$ produced as the major parent ions, we estimated the pKa values of the four HBQs based on measurement of the pH of aqueous haloquinone at different concentrations. Table 8.1 shows the pKa values we estimated to be 6.3, 5.9, 5.8, and 6.0 (at $25.0 \pm 0.1°C$) for DCBQ, DCMBQ, TCBQ, and DBBQ, respectively. In haloquinone solutions acidified with 0.25% FA to pH 2.6–2.8, the predominant species (99.9%) at equilibrium would be neutral molecular species, and <0.1% would be present as the dissociated anionic $[M-H]^-$ species in solution. Under the negative ESI conditions, the neutral haloquinone molecules could be reduced to dihydrohaloquinones, following an electrochemistry process documented previously: quinone $+ 2H^+ + 2e =$ dihydroquinone (0.70V). The pathway of $[M+H]^-$ formation under negative ESI is proposed as Figure 8.7. Two steps might be involved: (1) reduction of C=O groups in HBQs to C—OH forming the $[M+2H]$ intermediates; (2) rapid de-protonation of the $[M+2H]$ to produce $[M+H]^-$. This explanation is consistent with the observation of $[M+H]^-$ ions as the most abundant ions from negative ESI of the four HBQs in acidic solutions (pH 2.6–2.8 with 0.25% FA). This explained a previous observation that $[M+H]^-$ ions are the most abundant ions from negative ESI of the four HBQs in acidic solutions.

Table 8.1 Estimated pK_a values of DCBQ, DCMBQ, TCBQ, and DBBQ along with those of formic acid and acetic acid for validation of the estimation.

Conc. (mol/L)	Measured pH [a](25.0 ± 0.1°C)	Estimated pK_a [b]	Mean pK_a [c]
DCBQ			
1.0×10^{-3}	4.58	6.15	6.3
5.0×10^{-4}	4.85	6.39	
2.5×10^{-4}	4.96	6.30	
DCMBQ			
1.0×10^{-3}	4.45	5.88	5.9
5.0×10^{-4}	4.56	5.79	
2.5×10^{-4}	4.76	59.8	
TCBQ			
1.0×10^{-3}	4.33	5.64	5.7
5.0×10^{-4}	4.58	5.84	
2.5×10^{-4}	4.67	5.74	
DBBQ			
1.0×10^{-3}	4.58	6.15	6.0
3.3×10^{-4}	4.69	5.88	
Formic acid			
1.0×10^{-4}	3.94	3.88	3.9 [d]
1.0×10^{-5}	4.44	3.88	
Acetic acid			
1.0×10^{-4}	4.43	4.86	4.9 [e]
1.0×10^{-5}	4.95	4.90	

[a]pH values were measured using Φ250 pH / Temp / mV Meter (Beckman Coulter, Inc., Fullerton, CA).
[b]pK_a values were estimated from $K_a = [H^+]^2 / (C\text{-}[H^+])$, C stands for the total concentration of the HBQs (Conc., mol/L).
[c]pK_a values were estimated for every weak acids at each of the conc. levels, and the means were given. Here, we did not estimate the pK_a values from plotting lg$C$ against pH (lg$C$ = −2 pH − lgK_a, pK_a = interception), because the HBQ properties of low solubility and weak acidity make it difficult to obtain aqueous HBQs with wide pH ranges.
[d]The empirical constant pK_a of formic acid is 3.8.
[e]The empirical constant pK_a of acetic acid is 4.8.

8.4.2.2 Multiple reaction monitoring (MRM)

To obtain selective determination for the HBQs in real water samples, the multiple reaction monitoring (MRM) method was established. Firstly, product ions of the precursor ion $[M+H]^-$ for each HBQs were examined in MS/MS by collision induced dissociation (CID). MS/MS mass spectra of the precursor ion $[M+H]^-$

gave abundant and representative product ions of $[M + H - HX]^-$, $[M + H - HX - CO]^-$, $[M + H - CO]^-$, and X^-. Specific mass transitions, from precursor ion to product ion, for each HBQ are presented in Table 8.2. These mass transitions were chosen for quantification and confirmation purposes, respectively, in the MRM detection mode.

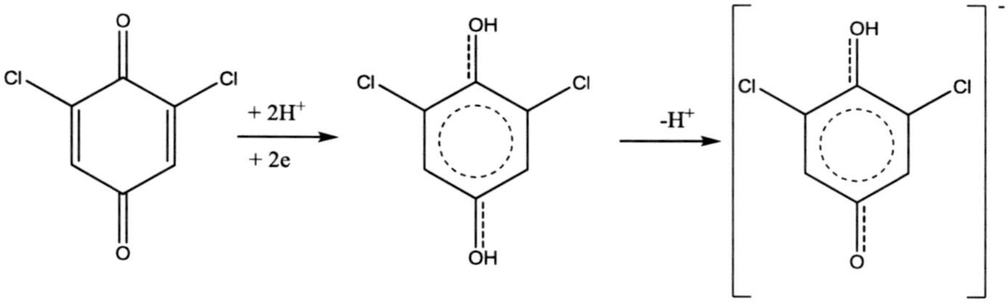

Figure 8.7 A proposed mechanism for production of chloro-quinone molecular ions (precursor ion) $[M + H]^-$

Table 8.2 The precursor ion (Q1) and product ions (Q3) of four HBQs.

	Q1	Q3	
DCBQ	177	141, 113, 35	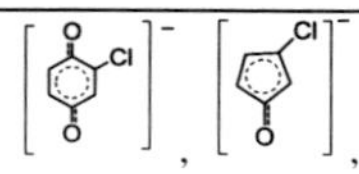 , Cl^-
DCMBQ	191/193	163, 155, 127, 35	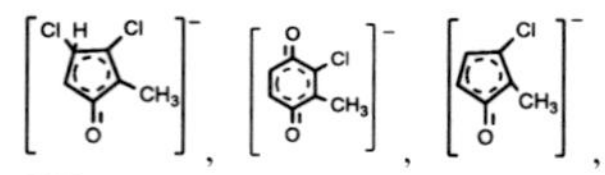 , Cl^-
TCBQ	211	175, 147, 35	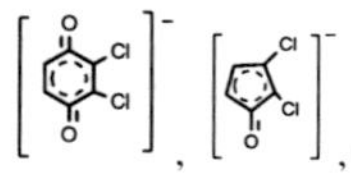 , Cl^-
DBBQ	267/269	157/159, 79/81	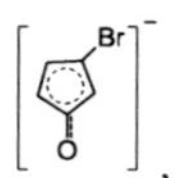 , Br^-

To enhance the sensitivity of MRM analysis, MS parameters were optimized for each precursor ion and product ion. Optimization experiments were carried out at the analytical flow rate by infusing the prepared solution (10–100 ng/mL in water, 0.25% FA) directly into the mobile phase via flow injection analysis (FIA). Optimal MS parameters were: ion spray voltage, −4500 V; source temperature, 475°C; gas I, 50 arbitrary unit; gas II, 60 arbitrary unit; curtain gas, 30 arbitrary unit. Optimal compound-dependent parameters, including declustering potential (DP), entrance potential (EP), collision energy (CE), and collision cell exit potential (CXP) for each mass transition (Q1 → Q3, m/z) are presented in Table 8.3.

Table 8.3 MRM parameters used for the four HBQs (negative ESI).

Analytes	Mass transitions (m/z) (Q1 → Q3)	DP	CE	CXP	EP
DCBQ	177 → 113[a], 141	−100	−24	−13	−10
DCMBQ	191 → 127[a], 155	−85	−24	−11	−10
	193 → 127/129	−85	−24	−11	−10
TCBQ	211 → 175[a], 147	−80	−18	−11	−10
DBBQ	267 → 157/159,	−100	−20	−7	−10
	267 → 79/ 81[a]	−100	−50	−10	−10
	265 → 79	−100	−50	−10	−10
	269 → 81	−100	−50	−10	−10

[a]The more abundant product ion used for quantitative analysis.

8.4.3 LC-ESI-MS/MS method performance

8.4.3.1 HBQs confirmation

The four HBQs in treated water were identified on the basis of matching (1) retention time, (2) MRM mass transitions, and (3) isotopic (^{35,37}Cl or ^{79,81}Br) patterns of MRM mass transition ions with those of the standards. Figure 8.8 show an example of confirmation of the identification of DCBQ, DCMBQ, TCBQ, and DBBQ in an authentic water sample by comparing the selected MRM mass transitions and the isotopic patterns of each HBQ between the water sample and the standards.

8.4.3.2 Quantification

The standard addition method was used for the calibration of analytes because isotope labelled internal standards are not available for the HBQs. From the 500 µL reconstituted SPE extract, an aliquot of the extract (120 µL) was directly analyzed with LC–MS/MS to estimate the concentration levels of the HBQs; then appropriate levels of the standards were added to the following samples. Four 90-µL aliquots were used to prepare a set of solutions to establish a calibration curve by adding 10 µL of mix standards at various concentrations depending on their estimated levels. DCBQ was spiked at 10, 50, and 100 ng/L, and DCMBQ, TCBQ, and DBBQ were spiked at 1, 5, and 10 ng/L. The peak areas of the spiked samples determined with LC–MS/MS (MRM) were calibrated against the spiked concentrations. The four-point calibration curves offered linear coefficients of 0.993–0.999, and these were used to determine concentrations of the HBQs in water samples. By using the standard addition method, the matrix effects on quantification were reduced.

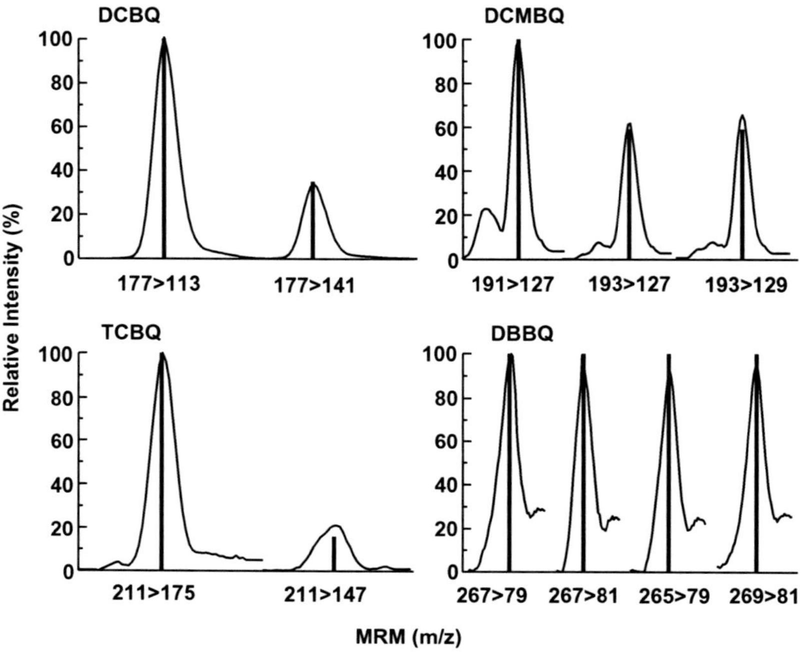

Figure 8.8 Relative intensities of the MRM ion transitions (m/z) of DCBQ, DCMBQ, TCBQ, and DBBQ detected in a water sample (peaks) compared with those of the standards (solid lines)

Of the selected MRM mass transitions, the most sensitive one was used to quantify the concentration of HBQ in water samples: MRM (m/z) 177 → 113 for DCBQ, 191 → 127 for DCMBQ, 211 → 175 for TCBQ, and 267 → 79 for DBBQ were used to quantify their concentrations in treated water samples. Detection limits (3σ, calculated using data acquired for 11 sample replicates over three days) of the LC–ESI–MS/MS (MRM) method for DCBQ, DCMBQ, TCBQ, and DBBQ were 1.0, 1.9, 1.7, and 0.5 ng/mL, respectively.

8.4.4 SPE-LC-ESI-MS/MS (MRM) method performance

The combination of the SPE with the LC-ESI-MS/MS (MRM) method enabled the determination of the four HBQs at ng/L levels. We validated the SPE-LC-ESI-MS/MS method through the analysis of a series of water samples spiked with the standards at 1, 5, 50, and 100 ng/L. Table 8.4 presents the recovery, the limit of quantification (LOQ), and precision of the SPE–LC–MS/MS method. With the presence of sample matrices, the LOQs of the SPE–LC–MS/MS method (at S/N of 10) were as low as 1–6 ng/L, and the recoveries of DCBQ, DCMBQ, TCBQ, and DBBQ from the chlorinated water samples were 84 ± 1%, 69 ± 3%, 59 ± 9%, and 78 ± 3%, respectively. The precision (relative standard deviation) of the method was 1–15%.

Table 8.4 Recovery, LOQ and precision of the SPE-LC-ESI-MS/MS (MRM) method.

	Recovery [a] (%)	LOQ [b] (ng/L)	Precision (RSD, %)
DCBQ	84 ± 1	3	1
DCMBQ	69 ± 3	5	4
TCBQ	59 ± 9	6	15
DBBQ	78 ± 3	1	4

[a]Recovery of the analytes, spiked in treated water at 1, 5, 50 and 100 ng/L concentrations, from the SPE pretreatment.
[b]LOQ of the SPE coupled with the LC–MS/MS method for analysis of the HBQs in the treated water samples. 10σ, calculated using data acquired for 11 replicates of water samples spiked with the HBQs, 5 ng/L each.

8.5 SURVEY OF WATER TREATMENT PLANTS

8.5.1 Sampling

Sixteen sets of raw and treated (finished water leaving plant) water samples were collected from nine water treatment plants (WTPs) in the USA and Canada between May 2009 and April 2010. At each WTP, one to four sets of sampling campaigns were carried out. The nine WTPs use different disinfection treatments, as summarized in Table 8.5.

8.5.2 Confirmation of HBQs as DBPs

Figure 8.9 show the MRM chromatograms of individual HBQs extracted from the treated water samples using chlorination (A, WTP #5), chloramination (B, WTP #2), chlorination with chloramination (C, WTP #3), and ozonation with chloramination (D, WTP #8) in comparison with their raw water samples. All four HBQs were detected in the chlorinated drinking water, but only DCBQ was detected in chloraminated water. In the drinking waters using chlorination/chloramination and ozonation/chloramination, DCBQ, DCMBQ, and DBBQ were detected but not TCBQ. No HBQs were detected in any of the raw water samples and the blanks. These results confirmed that the four HBQs are produced as DBPs during the disinfection processes.

Table 8.5 Summary of water treatment processes used in the nine water treatment plants (WTP 1–9).

WTP #	1	2	3	4	5	6	7	8	9
$KMnO_4$					×	×			
H_2CO_3						×			
Powdered Activated Carbon	×	×			×	×			
Aluminum Sulfate Flocculation	×	×	×	×	×	×	×	×	×
Polymer	×	×			×			×	×
Sedimentation	×	×	×						
Sand filtration	×		×	×			×		
Carbon filtration	×		×				×		
Membrane filtration	×	×			×	×		×	×
Biofiltration									×
Lime/soda ash		×	×					×	×
Fluoridation	×	×	×	×	×	×		×	×
Orthophosphate			×					×	×
UV	×	×				×			
Sodium hypochlorite					×	×			
Cl_2			×	×					
O_3							×	×	×
Chloramines ($Cl_2 + NH_3$)	×	×	×	×			×	×	×

8.5.3 HBQ occurrence

The concentrations of DCBQ ranged from 3.3 to 274.5 ng/L, with a mean concentration of 68.9 ng/L and a median concentration of 23.0 ng/L. The concentrations of DBBQ ranged from below the limit of detection 0.5 ng/L to 37.9 ng/L, with a mean concentration of 3.6 ng/L and a median concentration of below the detection limit of 0.6 ng/L. The DCMBQ concentrations ranged from <0.9 ng/L (below the detection limit of 0.9 ng/L) to 6.5 ng/L with mean 1.5 ng/L and median <0.9 ng/L (below the detection limit of 0.9 ng/L), and the TCBQ concentrations ranged from <1.5 ng/L (below the detection limit of 1.5 ng/L) to 9.1 ng/L with mean 2.4 ng/L and median <1.5 ng/L (below the detection limit of 1.5 ng/L), respectively. The total concentrations of the four HBQs in the 16 treated water samples ranged from 9.2 to 290.8 ng/L, with mean 76.5 ng/L and median 24.4 ng/L.

DCBQ and DBBQ were the most frequently detected among the four HBQs detected in the present study: 16 and 11 out of the 16 treated samples from the nine WTPs, respectively. The occurrence frequencies were 100% for DCBQ, 69% for DBBQ, 38% for DCMBQ, and 19% for TCBQ.

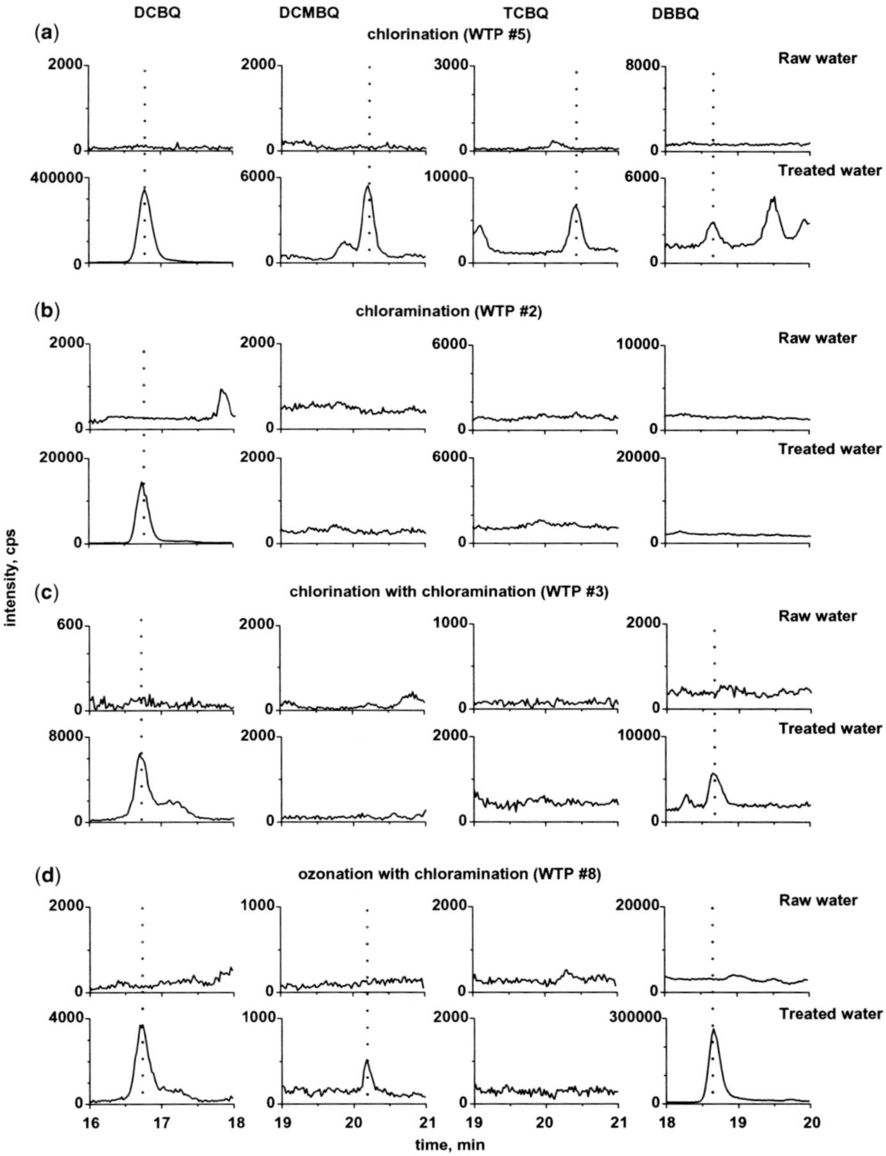

Figure 8.9 Confirmation of the formation of DCBQ, DCMBQ, TCBQ, and DBBQ resulting from chlorination (A, WTP #5), chloramination (B, WTP #2), chlorination with chloramination (C, WTP #3), and ozonation with chloramination (D, WTP #8), demonstrated by the comparison of MRM chromatograms of individual HBQs in treated and raw water extracts

8.6 CONCLUSION

An SPE-LC-ESI-MS/MS (MRM) method was developed for the characterization and determination of chloro- and bromo-benzoquinones in drinking water. The method provides low limits of quantification of 3 ng/L for DCBQ, 5 ng/L for DCMBQ, 6 ng/L for TCBQ, and 1 ng/L for DBBQ. Using this method, we studied the occurrence of the four HBQs in nine drinking water systems in Canada and the USA.

- HBQs are confirmed as DBPs in drinking water
- HBQs are found in authentic water samples from operating water treatment plants
- Monitoring for HBQs has required a substantial research investment to develop valid analytical, sampling, and storage methods
- The predicted low threshold for HBQ toxicity suggests that these DBPs warrant further investigation, both in terms of their formation/occurrence and in terms of their toxicological properties.

REFERENCES

Anichina J., Zhao Y., Hrudey S. E., Le X. C. and Li X.-F. (2010). Electrospray ionization mass spectrometry characterization of interactions of newly identified water disinfection byproducts halobenzoquinones with oligodeoxynucleotides. *Environ. Sci. Technol.*, **44**(24), 9557–9563.

Bull R. J., Reckhow D. A., Rotello V., Bull O. M. and Kim J. (2006). Use of Toxicological and Chemical Models to Prioritize DBP Research. AWWA Research Foundation and American Water Works Association, Denver, CO.

Heasley V. L., Fisher A. M., Herman E. E., Jacobsen F. E., Miller E. W., Ramirez A. M., Royer N. R., Whisenand J. M., Zoetewey D. L. and Shellhamer D. F. (2004). Investigations of the reactions of monochloramine and dichloramine with selected phenols: examination of humic acid models and water contaminants. *Environ. Sci. Technol.*, **38**(19), 5022–5029.

Hoq M. F., Indu B., Ernst W. R. and Neumann H. M. (1991). Kinetics of the reaction of chlorine with formic acid in aqueous sulphuric acid. *J. Phys. Chem.*, **95**(2), 681–683.

Li X.-F., Hrudey S. E., Bull R. J., Reckhow D. A., Humpage A., Joll C. and Heitz A. (2011). Analytical Methods for Predicted DBPs of Probable Toxicological Significance. Water Ressearch Foundation, Denver, CO, U.S.A. and Drinking Water Inspectorate, London, England, 155.

Qin F., Zhao Y. Y., Zhao Y., Boyd J. M., Zhou W. J. and Li X.-F. (2010). A toxic disinfection by-product, 2,6-dichloro-1,4-benzoquinone, identified in drinking water. *Angew. Chem. Int. Ed.*, **49**(4), 790–792.

West H. L. and Rollefson G. K. (1936). The photochemical reaction of chlorine with formic acid. *J. Am. Chem. Soc.*, **58**(11), 2140–2144.

Zhao Y., Qin F., Boyd J. M., Anichina J. and Li X.-F. (2010). Characterization and determination of chloro- and bromo-benzoquinones as new chlorination disinfection byproducts in drinking water. *Anal. Chem.*, **82**(11), 4599–4605.

Chapter 9

Mutagen X: The evolving story of an extremely potent mutagen, its toxicology and human health risk assessment

A. R. Humpage

9.1 INTRODUCTION

9.1.1 General description

The organic compound known as MX ("Mutagen X", 3-chloro-4-(dichloro-methyl)-5-hydroxy-2(5H)-furanone) was identified following investigations of an extremely mutagenic fraction of chlorinated wood pulp mill effluent. MX was subsequently identified in chlorinated humic acid solutions and in chlorinated drinking water; (Hemming *et al.* 1986; Holmbom *et al.* 1981; Holmbom *et al.* 1984). No other sources of MX are known.

Other halogenated hydroxyfuranones have been identified in drinking water, for which there is limited toxicological information. These are shown in Figure 9.1.

The brominated forms occur in significant quantities when source waters containing naturally high levels of bromide (>100–200 μg/L) are chlorinated to yield reactive bromine (Krasner *et al.* 2006). Most of these compounds occur at physiological pH as open-ringed forms and the toxicological importance of this feature is discussed below (see Section 9.2.4 Quantitative Structure-Activity Relationships).

9.1.2 Occurrence worldwide

Most early studies in the US, Finland, Australia, China, Japan, Russia, Spain, and the UK generally found very low levels of MX in drinking water (<80 ng/L; reviewed in McDonald and Komulainen, 2005), although Cherepovets in Russia was reported to have average concentrations of 160 ng/L (Egorov *et al.* 2003). More recently, higher concentrations have been detected, the highest level reported being in Massachusetts (310 ng/L; Krasner *et al.* 2006; Weinberg *et al.* 2002). The other halogenated hydroxyfuranones in Table 9.1 have been reported to occur at similar concentrations as MX (Krasner *et al.* 2006; Smeds *et al.* 1997; Weinberg *et al.* 2002), meaning that concentrations of total halogenated total hydroxyfuranones can often be over 1000 ng/L, the highest reported concentration being 2380 ng/L (Krasner *et al.* 2006). Halogenated hydroxyfuranones have generally been found to be stable, or in some cases to increase in concentration, in distribution systems (Krasner *et al.* 2006; Weinberg *et al.* 2002). The limited data for Australia indicate only low levels occur in drinking water. Sixteen drinking water samples from 8 Australian towns and cities were found to contain MX from undetectable levels (<0.5 ng/L) to 33 ng/L (Simpson and Hayes, 1998).

Name of Halogenated Furanone	Structure
MX: 3-chloro-4-(dichloromethyl)-5-hydoxy-2(5H) – furanone	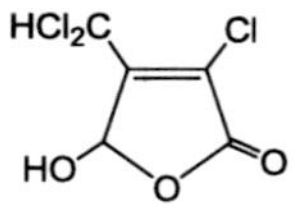
ZMX: (Z)-2-chloro-3-(dichloro-methyl)-4-oxobutenoic acid	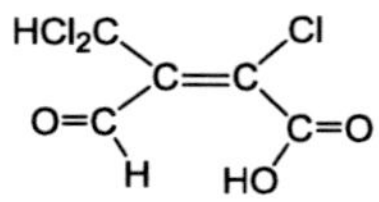
EMX: (E)-2-chloro-3-(dichloro-methyl)-4-oxobutenoic acid	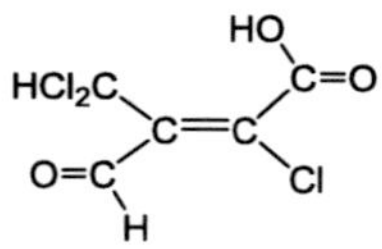
red-MX: 3-chloro-4-(dichloro methyl)-2-(5H)-furanone	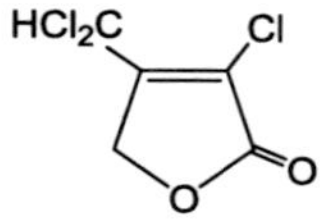
ox-MX: (E)-2-chloro-3-(dichloro methyl) butenedioic acid	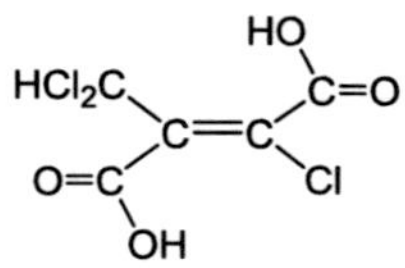

Figure 9.1 Structures of the halogenated furanones (adapted from Hrudey, 2009)

Name of Halogenated Furanone	Structure
MCA: 2,3-dichloro-4-oxobutenoic acid	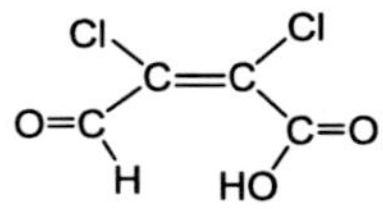
BMX1: 3-chloro-4-(bromochloromethyl)-5-hydoxy-2(5H) – furanone	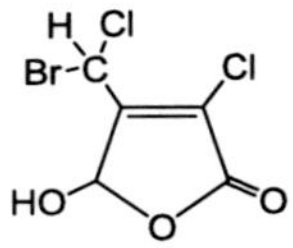
BMX2: 3-chloro-4-(dibromomethyl)-5-hydoxy-2(5H) – furanone	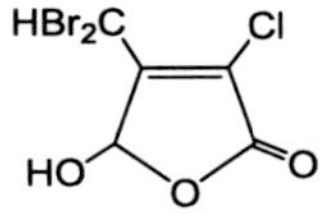
BMX3: 3-bromo-4-(dibromomethyl)-5-hydoxy-2(5H) – furanone	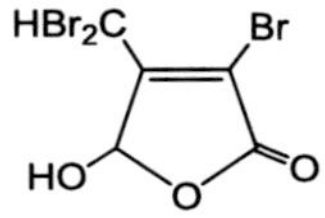
BEMX1: (E) 2-chloro-3-(bromochloromethyl)-4-oxobutenoic acid	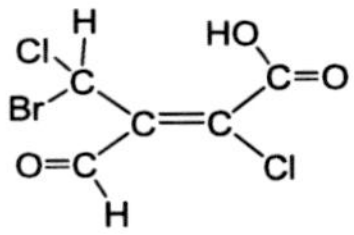
BEMX2: (E) 2-chloro-3-(dibromomethyl)-4-oxobutenoic acid	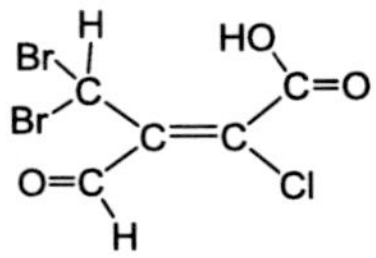
BEMX3: (E) 2-bromo-3-(dibromomethyl)-4-oxobutenoic acid	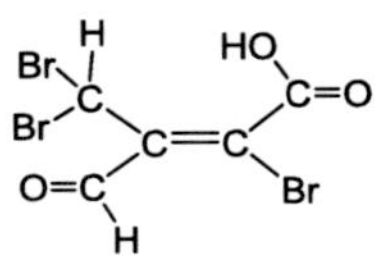

Figure 9.1 Continued

9.2 TOXICOLOGY

As explained by Hrudey (2009, online supplemental data), it is an unfortunate statistical reality that low incidence of cancer in large human populations cannot be replicated by treating a few animals in a carcinogenicity trial at the concentrations that humans are exposed to. The dose of the toxic agent given to the animals must be increased to a level that the incidence of the effect becomes statistically observable. All studies of MX have used concentrations that are 1000× or more above those likely to occur in human blood after exposure to MX in drinking water. However, it is not appropriate to simply dismiss the animal studies as not being relevant to the real world. Instead, it is important to try to understand whether the mechanisms (i.e. mode of action, Chapter 10) that are thought to underlie the effects seen in these studies at high doses would also be likely to apply at the lower doses to which people are exposed. If such mechanisms do apply, then reduction of the exposure concentrations may be required through mandating safe levels in drinking water guidelines.

For compounds that occur at ng/L concentrations such as MX, a key question is whether it is genotoxic *in vivo*. Genotoxic compounds change the structure of DNA and so can cause cells to behave abnormally. Sometimes a cell that has been damaged by a genotoxin can grow out of control and give rise to cancer. Since, at least in theory, a single change to a cell's DNA could cause it to become cancerous, genotoxins have traditionally been considered to have some small but non-zero likelihood of having a harmful effect even at the lowest exposure concentrations. For these cases a cancer slope factor (CSF) will typically be used (Chapters 1, 10 and 11). In contrast, non-genotoxic agents are considered to have a toxicity threshold below which the body's defences can metabolise or otherwise deal with the toxin to make it effectively safe.

In reality it may not be as simple as described above because science has moved well ahead of regulatory policy. For example, these classifications do not easily take account of: (1) multiple modes of action (Melnick *et al.* 1996) in which potency of a genotoxic mode of action may be much lower than non-genotoxic modes; (2) the greater tolerance of DNA damage than was originally recognised (Waters *et al.* 2009), which may provide a pseudo-threshold, or modify the response slope at low doses, even for direct-acting genotoxins; (3) such recently discovered modes of carcinogenesis as perturbation of microRNA function (Costinean *et al.* 2006; Gonzalez and Shah, 2008) or modification of DNA methylation or chromatin arrangement (Jaenisch and Bird, 2003; Patra *et al.* 2008), both of which have the potential to cause epigenetic changes that last for the life of the organism even though they may be caused by exposures that occur only during critical developmental stages; (4) and the roles of DNA repair capacity and a better understanding of comparative endogenous vs. exogenous DNA damage.

Nevertheless, this dichotomous classification of genotoxin or non-genotoxin still underpins regulatory human health risk assessment worldwide. Thus the key questions for current drinking water regulation are: (1) Is MX genotoxic *in vivo*? (2) What is its potency as a carcinogen? and (3) Does it occur in drinking water at concentrations calculated from the carcinogenicity potency data to be a potential risk to consumers? Unfortunately, even if all of this information was available, the subsequent interpretation may not be clear-cut. Add to this considerable variation in regulatory policy among regulatory regimes, and, as we shall see, the conclusions regarding public health risk can vary widely.

The toxicology of MX has been extensively studied and has been reviewed on a number of occasions (Anon, 2001; MacDonald, 2000; McDonald and Komulainen, 2005; Richardson *et al.* 2007; UKWIR/WRc, 2006; WHO, 2004). McDonald and Komulainen (2005) cite more than 80 publications on MX genotoxicity and note that it has been shown to cause DNA damage in over 45 strains of bacteria as well as a range of other organisms. Only a summary of the most pertinent results will be provided here to give a flavour of how an understanding of its potency and mode of action has changed as the science has developed.

9.2.1 Mechanism or mode of action

As explained in the previous section, determining whether a compound has a genotoxic or non-genotoxic mode of action is important for determining how the potential risk will be assessed. But understanding mode of action is also very important for assessing the likelihood that effects seen in cells in culture (*in vitro*) or in whole animals (*in vivo*) might also apply in humans. The better we can define the specific biochemical mode of action of a compound (usually studied in isolated *in vitro* models), the more targeted we can be in looking for those same specific effects in exposed animals or humans, and hence, the more confidence we can have that the effects seen *in vivo* are really due to that compound.

Mutagenicity is a subtype of genotoxicity in which single DNA bases are substituted or deleted, changing the "code" contained in the sequence of DNA bases. The well-validated Ames test is designed to specifically detect mutagens and is based on detection of base changes at specific sites in genetically constructed tester strains of *Salmonella typhimurium.* The mutagenicity of MX and other halogenated furanones has been extensively studied with this test. MX is one of the most potent agents tested in this bacterial system, and MX concentrations in drinking water samples can account for 50% or more of the total mutagenicity found in these samples. Addition of a rat liver extract containing a mix of metabolic enzymes (so called S9) considerably reduced the potency of MX, suggesting either efficient metabolism to a non-mutagenic product or reaction with nucleophiles in the S9 that reduced exposure of the cells (Meier *et al.* 1987). In contrast to bacterial systems, the genotoxic potency of MX is lower in assays that use mammalian cells, probably due to differences in cellular uptake, metabolic detoxification and DNA repair capabilities of the test organisms. For example, Plewa *et al.* (2002) compared the relative potencies of 6 haloacetic acids (HAAs), MX and potassium bromate in the Ames test and the SCGE genotoxicity test in mammalian Chinese Hamster Ovary CHO cells. MX and bromoacetic acid (BA) were the two most potent mutagens in the Ames test, with MX being 150× more potent than BA. However, in the CHO test BA was 31 × more cytotoxic and 14× more genotoxic than MX.

The mechanisms underlying MX mutagenicity have been investigated, with two mechanisms appearing to contribute: DNA adduct formation, mainly of guanine, and non-covalent abstraction of an electron from a target nucleotide. Either mechanism would allow mistakes in DNA repair that would cause various mutations to occur. Other halogenated furanones are also mutagenic, although the spectrum of mutations induced is often different to those induced by MX.

The genotoxicity of MX has also been investigated in whole animals. Rats receiving oral gavage of MX for 14 days showed no increase in mutagenic compounds in their urine except at the highest dose of 64 mg/kg/day, which increased mutagenicity 400-fold. This could suggest that a protective mechanism that converted MX to non-genotoxic metabolites at lower doses had been saturated (Meier *et al.* 1996). Studies conducted *in vitro* have shown that MX conjugates with the cellular antioxidant glutathione (GSH) and that this reduces or eliminates its activity as a mutagen (Clark and Chipman, 1995; Fekry *et al.* 2011; Ishiguro *et al.* 1987; Meier *et al.* 1990). In addition, Clark and Chipman (1995) chemically depleted GSH in isolated gut mucosa and showed that this increased the translocation of MX-derived mutagenicity from one side of the mucosa to the other. However, other studies have not shown this effect in liver or testicular cells (Brunborg *et al.* 1991; Nunn *et al.* 1997). In contrast, Fekry *et al.* (2011) proposed a thiol-dependent activation mechanism for compounds similar to MX, which relies on transitory formation of an episulfonium ion and results in preferential formation of strand breaks at guanidine groups in artificial oligonucleotides. They suggest that GSH is a less effective activating agent compared to other (microbial) thiols. Thus, the role for GSH as a protective agent is still uncertain. Furthermore, Komulainen *et al.* (1997) demonstrated MX carcinogenicity in rats at a dose ~100-fold less than the supposed saturation dose found by Meier *et al.* (1996). This could mean that mutagenicity is not the only, or even the main, mechanism that leads to tumour formation during long-term exposure to MX.

In vivo DNA damage (micronucleus formation, sister chromatid exchange, unscheduled DNA synthesis) has been detected in the stomach, lung, intestine, liver, spleen and kidneys of exposed mice or rats (reviewed in McDonald and Komulainen, 2005). However, results have been consistently negative in bone marrow, which is a site that is very commonly used to test for *in vivo* genotoxicity because of the relative ease of visualising the genotoxic effects. Sasaki *et al.* (1997) showed that DNA damage in the liver caused by acute oral exposure to 100 mg/kg MX was subsequently repaired within hours, whereas tissues of the urinary and gastrointestinal tract still exhibited DNA damage 24 h later. A number of reviews have noted that, in general, short-term studies find no evidence for *in vivo* genotoxicity whereas longer-term studies do (McDonald and Komulainen, 2005; Richardson *et al.* 2007). This again suggests that direct reaction of MX with DNA is not enough to initiate lesions in DNA that can lead to tumour formation.

MX has also given positive results in both *in vitro* and *in vivo* tumour promotion assays, and a non-genotoxic mechanism that may account for this has also been observed *in vitro*. Using a two-stage cell transformation assay, Nakajima *et al.* (2005) showed MX was active in both the initiation and promotion stages (this confirms earlier findings of Laaksonen *et al.* (2001), see McDonald and Komulainen, 2005). MX, at nM concentrations, was shown to be a potent inhibitor of gap-junction intercellular communication (GJIC) (Hakulinen *et al.* 2004), a finding that was soon confirmed using a different assay methodology (Nakajima *et al.* 2005; Nishikawa *et al.* 2006). Other chlorinated furanones are considerably less potent in this assay. However, in studies comparing the potency of MX, CMCF, MCA, and MCF for inhibition of GJIC it was found that potency (order and relative magnitude) varied between mouse 3T3 fibroblasts and rat WB-F344 hepatocytes, that the connexin 43 protein in gap junctions is a specific target in hepatocytes, and that protein levels rather than mRNA were reduced by the furanones (Hakulinen *et al.* 2006). Since exchange of low MW chemical messengers via GJIC is thought to be very important for maintaining coordination between normal cells in a tissue, and cancerous cells very often have lost this ability, these findings could provide an explanation of the non-mutagenic processes indicated by the *in vivo* studies mentioned above.

9.2.2 Carcinogenicity

In order to predict the potential risk to humans from a lifetime exposure to a compound, we need to investigate the quantitative relationship between exposure to the compound and the adverse effects it produces in the whole animal. It is important to take into account all of the modifying effects of uptake of the compound, its distribution throughout the body, metabolism by various enzymes in a range of tissues, and excretion of the parent compound and its metabolites. Historically, the only means of doing this has been to conduct studies that expose animals (generally rats and mice of both sexes) via the relevant route of exposure for as near to their full life as possible.

A long-term animal carcinogenicity study has been reported (Komulainen *et al.* 1997) in which Wistar rats were exposed to MX in drinking water at doses of 0.4, 1.3 or 5.5 mg/kg/d (males) or 0.6, 1.9, or 6.6 mg/kg/d (females) for 104 weeks. The mean concentrations of MX were 5.9 μg/mL (low dose), 18.7 μg/mL (mid dose), and 70.0 μg/mL (high dose). A detailed post-mortem examination included histopathology of 47 organs and all lesions detected. No overt toxicity was induced by the treatment. However, large statistically significant increases were observed in the incidence of cholangioma (cancer of the bile duct cells) in the liver, follicular adenoma and follicular carcinoma in the thyroid gland, and cortical adenomas of the adrenal gland in both genders. Incidence of cholangioma was up to 66% in high dose females but only 8% in high dose males, whereas follicular adenoma of the thyroid was seen in up to 70% of both males and females. Adrenal tumours were found in up to 30% of both genders. Mammary tumours (adenocarcinoma and fibroadenoma, 20% and 68%, respectively, in the highest dose) were found in females, as was a significant incidence of lymphoma or leukaemia (8%). In males, significant increases were seen in basal cell carcinomas of the skin (6%), and adenomas in the lungs (14%), as well as a non-significant increase in lymphoma/leukaemia (6%). These results indicate that, at the high concentrations tested, MX is a multi-site carcinogen in the rat.

Thyroid cancer in the rat is known to be efficiently induced by elevated levels of thyroid stimulating hormone (TSH; IARC, 1999) making this endpoint questionable for extrapolation to human cancer risk. Consequently, the authors measured TSH, and the thyroid hormones T3 and T4 in the serum of the same strain of rats used the Komulainen *et al.* (1997) study after exposure to MX. They also measured prolactin and growth hormone levels. The concentrations of these hormones were not increased by MX treatment, leading the authors to conclude that the rat-specific TSH-dependent mechanism was not operating under the test conditions and so MX may have a more general human-relevant carcinogenic effect in this organ (Komulainen *et al.* 2000). This is important because, as the most sensitive organ to MX-induced carcinogenesis, the thyroid cancer incidence would normally be used as the basis for a cancer risk assessment unless there was evidence that the underlying mechanism does not apply in humans.

In a study using renal cancer-prone Eker rats, exposure to 70 mg/L MX in drinking water for 10 months significantly increased the numbers of renal adenomas and renal carcinomas per animal (Hooth *et al.* 2002). Neither potassium bromate, chloroform, nor bromodichloromethane significantly increased tumour incidence when administered individually, and the tumour incidence in animals treated with a mixture of all four DBPs was not significantly different from that seen in animals treated with MX alone.

However, the evidence for carcinogenicity is not clear-cut. The *in vivo* carcinogenicity of MX and MCA in the gastrointestinal tract has been studied (Steffensen *et al.* 1999). MX or MCA at 0.2 mg/mL in the drinking water of C57Bl/6-Min (multiple intestinal neoplasia)/+ mice (male and female), either alone or with co-administration of 1,2-dimethyl hydrazine or azoxymethane, did not increase the incidence of aberrant crypt foci (ACF) or tumours above the high background. Intra-rectal administration of MX or MCA had only a slight promoting effect of ACF growth. Similarly, no genotoxicity (as measured by mutation of a reporter gene in transgenic mice), nor overt tumours, were seen after 78 weeks treatment with MX at concentrations from the same range to higher than used by Komulainen *et al.* (1997). Cellular proliferation rates were not increased in the stomachs of the mice (Nishikawa *et al.* 2006). These results again suggest that there are both species and tissue differences in the carcinogenic response to MX.

9.2.3 Toxicokinetics

In toxicokinetic studies of MX given orally (Komulainen *et al.* 1992; Ringhand *et al.* 1989), at least 20–40% of the MX was rapidly absorbed into the body and widely distributed. It was also rapidly eliminated mainly in the form of uncharacterised metabolites, with less than 1% appearing in the urine as the parent compound. Studies using radio-labelled MX found the highest levels in the stomach and small intestine as well as in the kidneys and urinary bladder along the main excretion pathway (Lebetkin *et al.* 2002 – see McDonald and Komulainen, 2005). This was taken to indicate high chemical reactivity at sites nearest to the route of exposure and excretion, which would be consistent with a mechanism involving direct reaction without metabolic activation. However, these are not the sites where the most tumours are found in carcinogenicity studies, although urinary bladder cancer is the most consistent cancer outcome associated with disinfection by-products in human epidemiology studies (Chapters 10 & 11).

9.2.4 Quantitative structure-activity relationships

Woo *et al.* (2002) reviewed the structure-activity relationship data for 10 MX congeners and came to the following conclusions: (1) While the open ring forms are the most prevalent at physiological pH and may be the actual mutagenic form, the closed ring forms, which are more hydrophobic, seem to be required for efficient penetration into the cell. The 5-OH group is important for cyclisation and so congeners that lack this group have lower potency; (2) brominated forms appear to be of similar potency to the chlorinated counterparts; (3) carcinogenic potency, based on the thyroid tumour data from Komulainen *et al.* 1997, is lower than might be expected from the mutagenic potency of MX in bacteria, suggesting significant detoxification activity *in vivo*. Based on their analysis of 209 known DBPs, Woo *et al.* 2002 rated MX and the 3 brominated congeners as being of "high-moderate" concern, although this is tempered by their occurrence at only ng/L concentrations. Bull and co-workers came to a similar conclusion regarding MX, but noted that structurally related compounds also predicted to form during chlorination could add significantly to the risk posed by this group of compounds (Bull *et al.* 2006; Bull *et al.* 2011).

Taken together, the various genotoxicity and carcinogenicity studies suggest that MX is mutagenic, and a potent carcinogen in some tissues in animal experimental studies, for example, liver and some endocrine organs but not tissues of the gastrointestinal tract or bone marrow. It may be more potent in rats than in mice but both species have been shown to be affected in at least some studies. The studies also suggest that mechanisms exist *in vivo* that modulate MX's direct reactivity with cellular DNA and that these, as well as epigenetic processes such as inhibition of GJIC, may be important for its overall carcinogenicity.

9.3 RISK ASSESSMENT METHODOLOGY FOR GENOTOXIC CARCINOGENS

The relative contribution to the observed *in vivo* carcinogenicity of the diverse potentially carcinogenic mechanisms that have been proposed is still a matter of some contention in the literature. For example, Richardson *et al.* (2007) state that "*MX is the most potent* in vitro *mutagen in Salmonella and the most potent rodent carcinogen of all DBPs; it also induces tumours at more organ sites than does any other DBP*". However, Nakajima *et al.* (2005) suggest that MX has only weak genotoxic activity in mammalian systems *in vivo* and give more weight to MX exhibiting a cancer promotion effect. Hence Nakajima *et al.* (2005) suggest that regulation via threshold models should be considered. But the authoritative WHO body, the International Agency for Research on Cancer (IARC, 2004), have concluded that: "*MX is a potent, direct-acting mutagen that induces primarily GC to TA transversions in both bacterial and mammalian cells. It induces DNA damage in bacterial and mammalian cells as well as in rodents* in vivo. *MX is a chromosomal mutagen in mammalian cells and in rats, and it induces mammalian cell transformation* in vitro. *The MX-associated thyroid gland tumours in rats are caused by mechanisms other than TSH-mediated hormonal promotion*" (IARC, 2004, p. 442). Overall, IARC classified MX as Group 2B: Possibly carcinogenic to humans, which is one level down from the highest rating they use based on animal evidence of carcinogenicity (Group 2A: Probably carcinogenic to humans). As we shall see, to date, all risk assessments performed by health regulators have used the basis of a direct-acting genotoxic mechanism.

9.3.1 Low dose extrapolation methodology

Once the cancer incidence has been determined in animal studies using high doses of a chemical, the risk to human consumers who are exposed to very low doses must be estimated. This obviously involves adopting a range of assumptions about how effects seen at high doses are likely to manifest at low doses, about how animal physiology relates to that of humans, and about what is an acceptable risk. This process involves a number of steps (see Hrudey, 2009 Online Supplementary Material 1 for a more detailed treatment):

(a) the most appropriate tumour type is chosen to provide the "point of departure". This is usually the tumour type showing the highest incidence unless there are toxicological reasons for thinking that the mechanism that gave rise to these tumours in the animals may not apply in humans. For MX, the data from the Komulainen study for cholangioma incidence in females, the thyroid follicular adenoma incidence, the combined follicular adenoma and carcinoma incidence, or the combined incidence of all tumours have variously been used.

(b) allowance may be made for the fact that animals and humans have different metabolic rates and physiology by applying a body weight scaling factor, usually bodyweight raised to the 2/3 or 3/4 power (McDonald and Komulainen (2005, p 200) state that the choice is "a matter of science policy as current evidence cannot distinguish between the two").

(c) the tumour incidence versus dose data for the chosen tissue is fitted to various mathematical models to test goodness of fit. The slope of the dose-response curve to be used for the extrapolation is selected based on fitting various mathematical models to the data from carcinogenicity studies and choosing the slope of the best-fitting model. The most frequently used is the so-called Linearized Multistage, or LMS, model.

(d) this slope is used to calculate a concentration of the compound that represents a given increased lifetime risk of cancer. It is not practical, and in fact would be counter-productive, to mandate that none of the compound should be present in water, so the slope is used to calculate a concentration that represents a very low "acceptable" increased risk of cancer, usually in the range 1 in 10,000 to 1 in 1,000,000. An additional element of safety is built into the calculation by using the upper bound on the 95% confidence interval (95% CI) of the slope, the CSF. Originally, upper bound lifetime cancer risks above 1 in 10,000 usually required regulatory action while upper bound lifetime cancer below 1 in 1,000,000 were regarded as negligible (*de minimus*). In practice, an upper bound lifetime cancer risk of 1 in a 1,000,000 has become a *de facto* default target for maximum contaminant levels in the U.S.

(e) the concentration so-derived is compared to occurrence concentration data for the compound of concern and if it has been shown to occur near the concentration of concern, a regulatory level may be implemented to promote management actions to reduce its occurrence levels.

Note that in Australia, an alternative methodology was proposed called the modified BMD (mBMD) method (NHMRC, 1999) in which, rather than selecting a single slope, the slopes of all of the well-fitting models were used to calculate doses that would cause a significant increased risk (usually 5%) not far from what can actually be measured in an animal experiment. These were averaged, and the mean (not the upper 95% CI bound) was used as the point of departure for application of various modifying factors in much the same way as threshold toxins are treated. This method avoids the need to stipulate a specific very low acceptable increased risk and extrapolate to a dose that corresponds to it a long way below the doses used in the carcinogenicity experiment. A guideline calculation for MX based on this method has been published (Fitzgerald and Robinson, 2009, described below). However, the Australian National Health and Medical Research Council (NHMRC) has since withdrawn, without explanation, the report including the mBMD method.

9.3.2 Available risk assessments

All of the quantitative risk assessments cited below have been based on the experimental data from the Komulainen *et al.* (1997) study. Thus the toxicological basis for all of them is the same. The often marked differences between their conclusions arise from what are essentially regulatory policy decisions made during the process just outlined.

The first such assessment was published as an editorial piece on the implications of the Komulainen *et al.* (1997) study in the same June 1997 issue of the Journal of the National Cancer Institute in which the Komulainen study was published (Melnick *et al.* 1997). An LMS extrapolation from the female cholangioma incidence data (without body weight scaling) was cited with the conclusion that "the estimated upper-bound excess cancer risk for lifetime consumption of water containing MX at 67 ppt is two in one million". Sixty-seven ppt was then the highest recorded occurrence of MX in drinking water. More recently, a value of 310 ng/L has been reported (Krasner *et al.* 2006; Weinberg *et al.* 2002), which would correspond to an upper bound lifetime cancer risk closer to 10 in 1,000,000 (10^{-5}) if such exposure continued over a lifetime. Melnick *et al.* 1997 used the cholangioma data rather than the thyroid tumour data because Komulainen *et al.* (1997) did not report the combined incidences of follicular adenomas and carcinomas in their paper.

The World Health Organisation reviewed the toxicological information on MX in 2004 (WHO, 2004). Again using the unscaled cholangioma data from Komulainen *et al.* (1997) they derived a concentration in drinking water likely to represent a 1 in 100,000 upper bound lifetime cancer risk. This concentration was 1.8 μg/L. Because this concentration is considerably higher than any concentration reported to be found in drinking water, it was not considered necessary to propose a formal guideline value for MX in drinking water. This assessment was not revised or updated in the 2011 WHO Drinking Water Guidelines (WHO, 2011).

The California EPA's Office of Environmental Health Hazard Assessment produced a draft report on the Carcinogenicity of MX (MacDonald, 2000) and a May 2001 report from the same office details the derivation of a No Significant Risk Level (NSRL) for MX (Anon, 2001) based on a Monte Carlo analysis of all of the tumour incidence data including the thyroid tumour data (supplied by Komulainen at the request of the authors). The derived slope factor was corrected for interspecies differences by multiplying by the ratio of human to animal bodyweights raised to the 1/3 power. The NSRL derived is 0.11 μg/d for an increased upper bound lifetime cancer risk of 1 in 100,000. Assuming consumption of 2L of water per day, this would imply an acceptable concentration in drinking water of 55 ng/L.

McDonald and Komulainen (2005), based on the earlier Californian EPA reports, also employed a Monte Carlo analysis of the incidence of all tumours (scaled using the 3/4 power of bodyweight, animal to human ratio) to derive a combined cancer potency of MX in each gender. Since the calculated potency was similar in each gender, these were averaged. The combined thyroid adenoma and carcinoma data from the Komulainen *et al.* 1997 was published in this paper. The drinking water concentration likely to represent a 1 in 1,000,000 increased lifetime cancer risk was calculated to be 7.8 ng/L, or 230x less than the acceptable level in the WHO (2004) document, albeit for a 10 fold more stringent cancer risk criterion. Unlike the conclusions in the WHO document, an acceptable concentration of 7.8 ng/L suggests a pressing need to set a regulatory limit.

The U.S. EPA has not adopted any risk estimation for MX, but a draft US EPA assessment (marked "Draft – Do Not Cite" and not available through the EPA website) proposed a risk assessment that is consistent with McDonald and Komulainen (2005).

Fitzgerald and Robinson (2009) took a different approach to the problem. Using the NHMRC 1999 mBMD methodology, they applied various models to the unscaled combined thyroid follicular adenoma and carcinoma data (Komulainen *et al.* 1997; McDonald and Komulainen, 2005) to calculate an average dose associated with a 5% increased risk of cancer (0.03 mg/kg/d). This was divided by the sum of modifying factors (to account for inter-species and intra-species differences, database adequacy, malignancy and genotoxicity, product = 6000) to derive a tolerable daily intake of 5.0 ng/kg/d. Assuming consumption of 2 L of drinking water per 70 kg person per day, and 50% of MX coming from drinking water, a provisional guideline of 87 ng/L was suggested.

Thus the available human health risk assessments for MX suggest that a drinking water concentration with an associated low acceptable risk (upper bound lifetime cancer risk of 10^{-5} to 10^{-6}) is probably somewhere in the range 8 ng/L to 1800 ng/L. Since all of the risk assessments were based on the same carcinogenicity study, the breadth of this range is a measure the variability inherent in the risk assessment process, and is in fact much greater than the statistical uncertainty in the experimental animal data upon which the assessments are based. This finding is not unusual because the knowledge uncertainty associated with translating experimental animal data to human health risk is inherently large.

9.4 CONCLUSIONS

MX was initially described as the most potent mutagen ever tested in the Ames Test. However, it has since been found to be much less potent in tests based on mammalian cells in culture. While MX has been shown to be mutagenic and genotoxic in treated animals, carcinogenicity studies have generally found tumours in different tissues to the main sites of genotoxicity. This has led to the conclusion that other, non-genotoxic, mechanisms may be important in MX carcinogenicity. Nevertheless, current risk assessment methodology makes a distinction between genotoxic carcinogens and threshold carcinogens. MX has been classified as genotoxic based on the weight of experimental evidence. Despite this general consensus, and the fact that all published risk assessments of MX used data from the same carcinogenicity study, these quantitative assessments of the cancer risk posed by MX in drinking water vary considerably. Variation is due to differences in selection of the tumour type used to derive the point of departure, body-weight scaling factor, the mathematical models used to estimate risk at very low dose, and differences in what is considered to be acceptable upper bound lifetime cancer risk. In the absence of good toxicological data, these effectively become decisions of science policy rather than scientific fact.

This variability in the calculated risk suggests considerable randomness in the procedure and uncertainty in the conclusions, which might lead the public (and the water industry) to question which level is *really* safe. The saving grace of all of the derivations is their inherent caution, although it is obvious that some limits must, in practice, be more cautious than others. But disproportionate caution may come at the cost of unnecessarily increased water treatment requirements, potentially diverting resources from issues that actually represent a greater threat but are better quantified and so have lower attendant uncertainty. Treating the drinking water criteria generated as hard and fast limits of safety is not accurate or wise.

This means that the risk assessment for MX must be done in the context of other drinking water contaminants and with an appreciation of the costs as well as the benefits of regulation. Data presented by Bull *et al.* 2006 suggest that MX poses about the same risk as NDMA and some THMs and HAAs. Hence, one could argue that if these are regulated, then why not MX? On the other hand, these compounds collectively do not account for more than one tenth the apparent risk from drinking chlorinated water calculated from the epidemiological studies (see Bull *et al.* 2006; Bull *et al.* 2011, and Chapters 7, 10 and 11). It could therefore be argued that, lacking knowledge of the real DBP "bad actors", regulation of THMs is adequate as a surrogate for all DBPs. Adding another DBP to the regulations would just impose unnecessary monitoring costs on the water industry without significant public health benefit. The one thing that everyone agrees about is that infectious diseases pose a much greater risk than all DBPs combined, and so any regulation of DBPs must not be at the expense of effective disinfection.

In summary:

- MX is a potent genotoxin and carcinogen that occurs at very low concentrations in chlorinated drinking water. Estimating the expected magnitude of the risk to human health has proven to be problematic.
- Derivation of human health risk assessment from animal data is not an exact science. The methods used are inherently cautious but can vary considerably in their conclusions.

- In practice, the risk calculated for a particular compound must be assessed in the context of other health risks, and of the relative costs and benefits of alleviating these risks.

REFERENCES

Anon (2001). No Significant Risk Level (NSRL) for the Proposition 65 carcinogen MX (3-chloro-4-(dichloromethyl)-5-hydroxy-2(5H)-furanone), Reproductive and Cancer Hazard Assessment Section, Office of Environmental Health Hazard Assessment (OEHHA), California Environmental Protection Agency, Oakland, California. http://www.oehha.ca.gov/prop65/law/pdf_zip/MX_NSRL.pdf (accessed 9 March 2012).

Bull R. J., Reckhow D. A., Li X., Humpage A. R., Joll C. and Hrudey S. E. (2011). Potential carcinogenic hazards of non-regulated disinfection by-products: Haloquinones, halo-cyclopentene and cyclohexene derivatives, *N*-halamines, halonitriles, and heterocyclic amines. *Toxicology*, **286**, 1–19.

Bull R. J., Reckhow D. A., Rotello V., Bull O. M. and Kim J. (2006). Use of Toxicological and Chemical Models to Prioritize DBP Research, Report 2867, American Water Works Association Research Foundation, Denver, CO, USA.

Brunborg G., Holme J. A., Soderlund E. J., Hongslo J. K., Vartiainen T., Lotjonen S. and Becher G. (1991). Genotoxic effects of the drinking-water Mutagen 3-Chloro-4-(Dichloromethyl)-5-Hydroxy-2[5h]-Furanone (Mx) in mammalian-cells *in vitro* and in rats *in vivo*. *Mutation Research*, **260**(1), 55–64.

Clark N. W. E. and Chipman J. K. (1995). Absorption of 3-chloro-4-(dichloromethyl)-5-hydroxy-2-[5H] furanone (MX) through rat small intestine *in vitro*. *Toxicology Letters*, **81**(1), 33–38.

Costinean S., Zanesi N., Pekarsky Y., Tili E., Volinia S., Heerema N. and Croce C. M. (2006). Pre-B cell proliferation and lymphoblastic leukemia/high-grade lymphoma in E(mu)-miR155 transgenic mice. *Proceedings of the National Academy of Science*, **103**(18), 7024–7029.

Egorov A. I., Tereschenko A. A., Altshul L. M., Vartiainen T., Samsonov D., LaBrecque B., Maki-Paakkanen J., Drizhd N. L. and Ford T. E. (2003). Exposures to drinking water chlorination by-products in a Russian city. *International Journal of Hygiene and Environironmental Health*, **206**(6), 539–551.

Fekry M. I., Price N. E., Zang H., Huang C. F., Harmata M., Brown P., Daniels J. S. and Gates K. S. (2011). Thiol-activated DNA damage by alpha-Bromo-2-cyclopentenone. *Chemical Research in Toxicology*, **24**(2), 217–228.

Fitzgerald D. J. and Robinson N. I. (2009). Application of the modified benchmark dose methodology to derive a tolerable daily intake for 3-chloro-4-(dichloro-methyl)-5-hydroxy-2(5H)-furanone. *Environmental Health*, **9**(1 & 2), 11–16.

Gonzalez F. J. and Shah Y. M. (2008). PPARα: Mechanism of species differences and hepatocarcinogenesis of peroxisome proliferators. *Toxicology*, **246**, 2–8.

Hakulinen P., Maki-Paakkanen J., Naarala J., Kronberg L. and Komulainen H. (2004). Potent inhibition of gap junctional intercellular communication by 3-chloro-4-(dichloromethyl)-5-hydroxy-2(5H)-furanone (MX) in BALB/c 3T3 cells. *Toxicology Letters*, **151**(3), 439–449.

Hakulinen P., Rintala E., Maki-Paakkanen J. and Komulainen H. (2006). Altered expression of connexin43 in the inhibition of gap junctional intercellular communication by chlorohydroxyfuranones in WB-F344 rat liver epithelial cells. *Toxicology and Applied Pharmacology*, **212**(2), 146–155.

Hemming J., Holmbom B., Reunanen M. and Kronberg L. (1986). Determination of the strong mutagen 3-chloro-4-(dichloromethyl)-5-hydroxy-2(5H)-furanone in chlorinated drinking and humic waters. *Chemosphere*, **15**(5), 549–556.

Holmbom B. R., Voss R. H., Mortimer R. D. and Wong A. (1981). Isolation and identification of Ames mutagenic compounds present in kraft chlorination effluents. *Tappi*, **64**(3), 172–174.

Holmbom B., Voss R. H., Mortimer R. D. and Wong A. (1984). Fractionation, isolation and characterization of Ames-mutagenic compounds in kraft chlorination effluents. *Environmental Science & Technology*, **18**, 333–337.

Hooth M. J., McDorman K. S., Hester S. D., George M. H., Brooks L. R., Swank A. E. and Wolf D. C. (2002). The carcinogenic response of Tsc2 mutant Long-Evans (Eker) rats to a mixture of drinking water disinfection by-products was less than additive. *Toxicological Sciences*, **69**(2), 322–331.

Hrudey S. E. (2009). Chlorination disinfection by-products, public health risk tradeoffs and me. *Water Research*, **43**(8), 2057–2092.

IARC (1999). Species Differences in Thyroid, Kidney and Urinary Bladder Carcinogenesis. World Health Organisation Internationa Agency for Research on Cancer, Lyon, France.

IARC (2004). Some Drinking-water Disinfectants and Contaminants, including Arsenic. World Health Organisation International Agency for Research on Cancer, Lyon, France.

Ishiguro Y., LaLonde R. T., Dence C. W. and Santodonato J. (1987). Mutagenicity of chlorine substituted furanones and their inactivation by reaction with nucleophiles. *Environmental Toxicology and Chemistry*, **6**(12), 935–946.

Jaenisch R. and Bird A. (2003). Epigenetic regulation of gene expression: how the genome integrates intrinsic and environmental signals. *Nature Genetics*, **33**(Suppl), 245–254.

Komulainen H., Kosma V. M., Vaittinen S. L., Vartiainen T., KalisteKorhonen E., Lotjonen S., Tuominen R. K. and Tuomisto J. (1997). Carcinogenicity of the drinking water mutagen 3-chloro-4-(dichloromethyl)-5-hydroxy-2(5H)-furanone in the rat. *Journal of the National Cancer Institute*, **89**(12), 848–856.

Komulainen H., Tuominen R. K., Kosma V. M. and Huuskonen H. (2000). 3-Chloro-4-(dichloromethyl)-5-hydroxy-2(5H)-furanone (MX), a rat thyroid gland carcinogen, does not affect serum levels of TSH and thyroid hormones. *Environmental Toxicology and Pharmacology*, **8**(4), 267–273.

Komulainen H., Vaittinen S. L., Vartiainen T., Lotjonen S., Paronen P. and Tuomisto J. (1992). Pharmacokinetics in rat of 3-chloro-4-(dichloromethyl)-5-hydroxy-2(5H)-furanone (MX), a drinking water mutagen, after a single dose. *Pharmacology and Toxicology*, **70**(6 Pt 1), 424–428.

Krasner S. W., Weinberg H. S., Richardson S. D., Pastor S. J., Chinn R., Sclimenti M. J., Onstad G. D. and Thruston A. D. (2006). Occurrence of a new generation of disinfection byproducts. *Environmental Science & Technology*, **40**(23), 7175–7185.

Laaksonen M., Maki-Paakkanen J. and Komulainen H. (2001). Enhancement of 3-methylcholanthrene-induced neoplastic transformation by 3-chloro-4-(dichloromethyl)-5-hydroxy-2(5H)-furanone in the two-stage transformation assay in C3H 10T1/2 cells. *Archives of Toxicology*, **75**(10), 613–617.

Lebetkin E. H., Chen L. J. and Burka L. T. (2002). Disposition of 3-chloro-4-(dichloromethyl)-5-hydroxy-2(5H)-furanone (MX) in B6C3F1 mice and F344 rats. *Journal of Toxicology and Environmental Health-Part A*, **65**(24), 2101–2118.

MacDonald T. (2000). Evidence on the Carcinogenicity of MX (3-chloro-4-(dichloromethyl)-5-hydroxy-2(5H)-furanone). Reproductive and Cancer Hazard Assessment Section, Office of Environmental Health Hazard Assessment (OEHHA), California Environmental Protection Agency, Oakland, California.

McDonald T. A. and Komulainen H. (2005). Carcinogenicity of the chlorination disinfection by-product MX. *Journal of Environmental Science and Health Part C-Environmental Carcinogenesis & Ecotoxicology Reviews*, **23**(2), 163–214.

Meier J. R., Blazak W. F. and Knohl R. B. (1987). Mutagenic and clastogenic properties of 3-chloro-4-(dichloromethyl)-5-hydroxy-2 (5H)-furanone: a potent bacterial mutagen in drinking water. *Environmental and Molecular Mutagenesis*, **10**(4), 411–424.

Meier J. R., DeAngelo A. B., Daniel F. B., Doerger J. U., Chang L. W., Kopfler F. C., Robinson M. and Ringhand H. P. (1990). Genotoxic and carcinogenic properties of chlorinated furanones: important by-products of water chlorination. In: Genetic Toxicology of Complex Mixtures, M. O. Waters, F. B. Daniel and J. Lewtas (eds.), Plenum Press, New York, pp. 185–195.

Meier J. R., Monarca S., Patterson K. S., Villarini M., Daniel F. B., Moretti M. and Pasquini R. (1996). Urine mutagenicity and biochemical effects of the drinking water mutagen, 3-chloro-4-(dichloromethyl)-5-hydroxy-2[5H]-furanone (MX), following repeated oral administration to mice and rats. *Toxicology*, **110**(1–3), 59–70.

Melnick R. L., Boorman G. A. and Dellarco V. (1997). Water chlorination, 3-chloro-4-(dichloromethyl)-5-hydroxy-2(5H)-furanone (MX), and potential cancer risk. *Journal of the National Cancer Institute*, **89**(12), 832–833.

Melnick R. L., Kohn M. C. and Portier C. J. (1996). Implications for risk assessment of suggested nongenotoxic mechanisms of chemical carcinogenesis. *Environmental Health Perspectives*, **104**(Suppl 1), 123–134.

Nakajima M., Shimada S., Nagai M., Mizuhashi F., Sugiyama C., Masuda S., Hayashi M. and Kinae N. (2005). 3-Chloro-4-(dichloromethyl)-5-hydroxy-2(5H)-furanone [MX] shows initiating and promoting activities in a two-stage BALB/c 3T3 cell transformation assay. *Mutagenesis*, **20**(5), 375–379.

NHMRC (1999). Toxicity Assessment for Carcinogenic Soil Contaminants. Office of the National Health and Medical Research Council, Health Advisory Unit, Canberra, Australia.

Nishikawa A., Sai K., Okazaki K., Son H. Y., Kanki K., Nakajima M., Kinae N., Nohmi T., Trosko J. E., Inoue T. and Hirose M. (2006). MX, a by-product of water chlorination, lacks *in vivo* genotoxicity in gpt delta mice but inhibits gap junctional intercellular communication in rat WB cells. *Environmental and Molecular Mutagenesis*, **47**(1), 48–55.

Nunn J. W., Davies J. E. and Chipman J. K. (1997). Production of unscheduled DNA synthesis in rodent hepatocytes *in vitro*, but not *in vivo*, by 3-chloro-4-(dichloromethyl)-5-hydroxy-2[5H]-furanone (MX). *Mutation Research-Fundamental and Molecular Mechanisms of Mutagenesis*, **373**(1), 67–73.

Patra S. K., Patra A., Rizzi F., Ghosh T. C. and Bettuzzi S. (2008). Demethylation of (cytosine-5-C-methyl) DNA and regulation of transcription in the epigenetic pathways of cancer development. *Cancer and Metastasis Reviews*, **27**(2), 315–334.

Plewa M. J., Kargalioglu Y., Vankerk D., Minear R. A. and Wagner E. D. (2002). Mammalian cell cytotoxicity and genotoxicity analysis of drinking water disinfection by-products. *Environmental and Molecular Mutagenesis*, **40**(2), 134–142.

Richardson S. D., Plewa M. J., Wagner E. D., Schoeny R. and DeMarini D. M. (2007). Occurrence, genotoxicity, and carcinogenicity of regulated and emerging disinfection by-products in drinking water: a review and roadmap for research. *Mutation Research-Reviews in Mutation Research*, **636**(1–3), 178–242.

Ringhand H. P., Kaylor W. H., Miller R. G. and Kopfler F. C. (1989). Synthesis of 3-14C-3-chloro-4-(dichloromethyl)-5-hydroxy-2(5H)-furanone and its use in a tissue distribution study in the rat. *Chemosphere*, **18**(11–12), 2229–2236.

Sasaki Y. F., Nishidate E., Izumiyama F., WatanabeAkanuma M., Kinae N., Matsusaka N. and Tsuda S. (1997). Detection of *in vivo* genotoxicity of 3-chloro-4-(dichloromethyl)-5-hydroxy-2[5H]-furanone (MX) by the alkaline single cell gel electrophoresis (Comet) assay in multiple mouse organs. *Mutation Research-Genetic Toxicology and Environmental Mutagenesis*, **393**(1–2), 47–53.

Simpson K. L. and Hayes K. P. (1998). Drinking water disinfection by-products: An Australian perspective. *Water Research*, **32**(5), 1522–1528.

Smeds A., Vartiainen T., MakiPaakkanen J. and Kronberg L. (1997). Concentrations of Ames mutagenic chlorohydroxyfuranones and related compounds in drinking waters. *Environmental Science & Technology*, **31**(4), 1033–1039.

Steffensen I. L., Paulsen J. E., Engeset D., Kronberg L. and Alexander J. (1999). The drinking water chlorination by-products 3,4-dichloro-5-hydroxy-2[5H]-furanone (mucochloric acid) and 3-chloro-4-(dichloromethyl)-5-hydroxy-2[5H]-furanone do not induce preneoplastic or neoplastic intestinal lesions in F344 rats, Balb/cA mice or C57BL/6J-Min mice. *Pharmacology & Toxicology*, **85**(2), 56–64.

UKWIR/WRc (2006). Datasheet on MX, United Kingdom Water Industry Research/Water Research Council, London, UK.

Waters L. S., Minesinger B. K., Wiltrout M. E., D'Souza S., Woodruff R. V. and Walker G. C. (2009). Eukaryotic translesion polymerases and their roles and regulation in DNA damage tolerance. *Microbiology and Molecular Biology Reviews*, **73**(1), 134–154.

Weinberg H. S., Krasner S. W., Richardson S. D. and Thruston A. D. Jr. (2002). The Occurrence of Disinfection By-Products (DBPs) of Health Concern in Drinking Water: Results of a Nationwide DBP Occurrence Study. U.S. EPA, Athens, GA. www.epa.gov/Athens/publications/reports/EPA_600_R02_068.pdf (accessed 3 March 2012).

WHO (2004). MX in Drinking Water. Background Document for Development of Who Guidelines for Drinking-Water Quality. WHO/WSH/03.04/108, World Health Organisation, Geneva, Switzerland.

WHO (2011). Guidelines for Drinking Water Quality. 4th edn, World Health Organisation, Geneva, Switzerland.

Woo Y. T., Lai D., McLain J. L., Manibusan M. K. and Dellarco V. (2002). Use of mechanism-based structure-activity relationships analysis in carcinogenic potential ranking for drinking water disinfection by-products. *Environmental Health Perspectives*, **110**, 75–87.

Chapter 10

Toxicological evaluation of experimental data that informs the magnitude of cancer risk from disinfection by-products

Richard J. Bull, Ph.D.

10.1 INTRODUCTION

Concern related to possible cancer risks associated with disinfection by-products arose with the publications of Rook (1974) and Bellar *et al.* (1974) that four trihalomethanes are produced in the chlorination of drinking water (Chapter 1). This temporally coincided with knowledge of the results of a cancer study of chloroform by the National Cancer Institute (NCI, 1976) in which chloroform was shown to induce liver cancer in mice and kidney cancer in rats. These findings were followed by the development of a total trihalomethane (TTHM) regulations or guidelines (National Health and Welfare Canada, 1978; U.S. EPA, 1979) that address the sum (by weight) of the four most commonly found THMs, chloroform, bromodichloromethane (BDCM), chlorodibromomethane (CDBM), and bromoform (TBM).

From its conception the TTHM regulation was assumed to be a surrogate for DBP production. While much attention was paid to the actual regulation, little attention was paid to the complex array of precursors that occurred in source waters or the diverse nature of by-products that were certain to result from reactions of disinfectants. The complexity of DBP mixtures was repeatedly discussed at early conferences organized by Jolley *et al.* (1978–1990), symposia conducted by the U.S. EPA (*Environmental Health Perspectives, 1982 vol **46** and 1986, vol. **69***), and, most recently, in the organization of Gordon Research Conferences (www.grc.org). Despite the communications of the past, the THMs remained the major focus and THMs have been widely used as a surrogate for the formation of other chlorination by-products, implicitly suggesting they are valid indicators for the cancer risk associated with chlorinated water. The validity of this assumption has never been confirmed and, in fact, existing data suggest that THMs are not good surrogates for other chlorination by-products (Bull *et al.* 2009a)

The list of chlorination by-products has expanded and now includes several hundred chemicals (Richardson *et al.* 2007). Along with the expansion of the list of DBPs, several of the more common DBPs were studied as carcinogens by the National Toxicology Program, U.S. EPA Health Effects Laboratories, and academic scholars. Parallel to these activities, epidemiological studies were conducted to determine the extent to which there were associations between the occurrence of various cancers in the human population and exposure to chlorinated drinking water. Gradually, the emphasis of the epidemiological studies migrated from associations with the practice of disinfection in water supplies to focus specifically on the THMs as the exposure metric. This evolution has resulted in papers that appeared to suggest that the observed associations were with the TTHM or single members of that class. Observed associations, most notably with bladder cancer, have come to be thought of as causal in some quarters.

This chapter is concerned with the mistaken view that a reduction of THM concentrations will necessarily result in substantial decreases in bladder cancer risk as has been suggested by the U.S. EPA. The practical outcome of acting on this simplistic notion has led to utilization of combinations of two or more disinfectants whose benefits are at least partly justified by the reduction of cancer risk proportional to the reduction of THM concentrations. The thesis of this chapter is that pursuit of alternative methods of disinfecting water without a much better understanding of the causal relationships involved in producing bladder cancer risk brings to mind the old caution against looking only under a lamppost for items lost in the dark.

The specific approach of this chapter will be to utilize both descriptive and mechanistic toxicological data to examine the issue of causality of cancer associated with the chlorination of drinking water. Clearly a lack of action to address a possible public health risk is a sure way to lose public confidence. However, that reality does not justify the perpetuation of incomplete or inaccurate understanding of a public exposure from drinking water, which is so central to the public health.

10.2 THE NATURE AND SCOPE OF A TOXICOLOGICAL EVALUATION

Toxicological evaluations are frequently viewed as simple examination of the dose-response for adverse effects for chemicals in humans or experimental animals. The transition into risk assessment is viewed as a simple mechanical process of identifying a level at which no significant adverse outcome would be anticipated. This impression arises from observing the superficial approaches that are described by risk assessors in explaining uncertainty factors to adjust observed data to account for lack of information about interspecies and intraspecies variability, or use statistical models to extend estimates of the probability of adverse effects to levels of exposure encountered by humans. This practice has been quite common over the first 30+ years in which data started to become available for chemicals for which little prior toxicological data existed. What is often missed is that where initially there was little application of mechanistic and mode of action (MOA) data to the interpretation of descriptive toxicological data, MOA is now taking a more central role in risk assessment modeling (U.S. EPA, 2005).

There are limits to how descriptive toxicological data can be used as the template for understanding the development of human disease. First, the differences in approach taken in experimental toxicology and human medicine must be recognized. Descriptive data from experimental toxicology most commonly focuses on the development of pathology[1] where variables other than the treatment are carefully controlled. In contrast, human medicine begins with the identification of symptoms that are coupled with a patient history and, in some cases, with an environmental assessment. In the end, an adverse outcome in human medicine may be coupled to a well characterized pathological outcome, but the actual relationship of an outcome to an exposure to a candidate chemical can be highly uncertain, especially where low doses are involved and the patient exists in a very complex environment of other exposures and risk factors. Second, it cannot be taken for granted that parallels in pathological outcomes are indicative of a common etiology.[2] A common pathology can have a multitude of etiologies and sometimes possible etiologies can include a variety of chemical exposures. Third, although it is often assumed to be the case, many pathologies seen with drugs are poorly detected in preclinical toxicology studies. In this case, an investigator has a good estimate of exposure in both cases, but an adverse effect that is only produced in a sensitive population of humans (e.g., autoimmune hepatitis) will rarely be detected in standard descriptive toxicological tests in animals or even in preclinical studies in humans because they tend to be rare and would go undetected in these small studies. Conversely, there are pathologies that develop in animals that do not appear even at high doses in humans (e.g., liver tumor induction by phthalates). Fortunately, these are exceptions to the rule. Fourth, experimental toxicology is the method to identify the specific mechanistic linkages that are responsible for adverse outcomes in human as well as experimental animal studies. This extends to identifying the reasons why one species is affected by a chemical and another is not. Descriptive toxicological studies have prevented the introduction of dangerous chemicals into commerce and provided the evidence to remove older products or reduce exposures to health threats that were unrecognized in the past. However, these

[1]Pathology is the anatomic and physiological deviations from the normal that constitute disease or characterize a particular disease

[2]Etiology is the cause or causes of a disease or abnormal condition

examples demonstrate why it is important to understand mechanisms of toxicity and modes of action in both experimental animals and humans.

Descriptive toxicological data developed from treatment of humans or experimental animals under controlled conditions still plays an important role in estimating risk, largely because of challenges in comparing and scaling toxicological responses observed *in vitro*[3] to predict adverse effects that result from *in vivo*[4] exposures. Only in the past 5–10 years have research programs in regulatory agencies begun to tap into large databases to determine how such predictions can be made (Dix *et al.* 2007; Knudsen *et al.* 2009; Martin *et al.* 2009). In the interim (the next decade or two), we will continue to struggle with uncritical attempts to simply translate findings *in vitro* to *in vivo* exposures. The obvious approach needed is to quantify linkages between both the doses applied and the responses measured in the two types of experiments.

Much of this chapter focuses on how mechanistic and mode of action data has and should be employed to resolve attributes of DBP risks. These analytical approaches can identify genuine paths to the mitigation of risks. Some of the information that will be discussed is basic biomedical science and may not be recognized as toxicological data for those unacquainted with the training and practice of the discipline. Much of the time, these data are thought by laypersons to be irrelevant information of importance only to those interested in doing research. This chapter seeks to dispel some of that suspicion by illustrating how those analyses can be brought to bear in risk assessment in as clear a way as possible.

The nature of the data that will be discussed will be diverse. It is hard to generalize about mode of action research in lay terms, but the essence of the analysis is to take the processes apart to confirm that they do indeed contribute to the adverse outcome. The processes involved can be as different as the classes of DBP that have been identified. To be useful, descriptions of MOAs must be imbedded in a quantitative and valid dose-response framework. A first step in this process is understanding the relationship between the external dose and the systemic concentrations that produce adverse effects in humans and/or the experimental animals (i.e. pharmacokinetics). Without this information, it is very difficult to judge whether the concentrations added to a cell culture (*in vitro*) have any meaning for toxicological effects observed *in vivo*. If the chemical produces effects in cell culture only at concentrations that are substantially greater than concentrations achieved in blood and/or tissues *in vivo* that are associated with adverse effects when given to animals, the response seen *in vitro* is unlikely to be of importance. This is illustrated with some conclusions that can be drawn with haloacetic acids for which extensive pharmacokinetic and MOA data are available both *in vitro* and *in vivo*.

---

[3] *in vitro* refers to experiments conducted outside the living body and in an artificial environment

[4] *in vivo* refers to experiments conducted in the living body of an animal

10.3 OVERVIEW OF THE WEIGHT OF EVIDENCE THAT DBPs ARE A CANCER RISK

In addition to a large volume of descriptive screening and toxicological studies that identify individual DBPs as carcinogens, there is a large body of epidemiological data addressing associations of various adverse health effects with the use of disinfectants (primarily chlorine). These studies are being reviewed in Chapter 11. Therefore, the present chapter will only address aspects of epidemiological work that can be informed by experimental data obtained in humans or experimental animals or *in vitro* assessments of mechanisms involved in the development of toxicological responses. For brevity, only epidemiological investigations of associations of disinfectant use with cancer are considered. However, the principles would apply to epidemiological studies of other health endpoints.

Bladder cancer has been the most consistent observed risk associated with chlorination of drinking water in epidemiological studies (see Chapter 11). As the THMs are frequently used as a surrogate measure of chlorination by-products, in part because THMs are the most commonly measured DBPs, most of the more recent studies have focused on THMs. However, a conclusion that an association between bladder cancer and THMs exposure is causal is not justified because there are several hundred by-products in chlorinated drinking water and a least a dozen others have been shown to be carcinogenic and many more to be mutagenic (Bull *et al.* 2011; Richardson *et al.* 2007). There has been no attempt to rule out these other compounds as being causal. There are no experimental data that support the rationale that THMs should be isolated from this very complex mixture as the causal agents. Biological plausibility was one of the key criteria cited by the U.S. EPA (2005) as determining whether the observed epidemiological associations were causal.

One important question of biological plausibility is whether the THMs are sufficiently potent as carcinogens to account for a risk of the magnitude observed in the epidemiological studies (Bull *et al.* 2011). The lack of any evidence that the THMs are capable of producing bladder cancer in any relevant organism is also of concern. Further, experimental animals studies of chloroform (Jorgenson *et al.* 1985) and BDCM (NTP, 2006) using a drinking water vehicle rather than corn oil raise the question of whether these compounds should even be considered carcinogens. These issues are more thoroughly explored in the following section.

10.4 EVALUATION OF RISKS ASSOCIATED WITH THMs

10.4.1 Toxicological data

The THMs have been extensively studied. All four have been subject to chronic cancer studies (Jorgensen *et al.* 1985; NCI, 1976; NTP, 1985, 1986, 1989, 2006).

These studies all documented carcinogenic responses in either rats, mice or both species when administered by gavage at high doses (50 to 200 mg/kg day^{-1}) when dissolved in a corn oil vehicle. Subsequent studies administering chloroform (Jorgensen *et al.* 1985) and BDCM (NTP, 2006) in drinking water resulted in a much smaller tumor induction in one species and produced no tumors in the other in the former case and a complete lack of carcinogenic effect in the latter case. Bioassays with CDBM and TBM have not been conducted using drinking water as the vehicle. Whether the animal studies of these two DBPs should be used in human risk assessment is questionable as they only produced cancer in one of the two species in which they were tested. Therefore, the determination that THMs are carcinogenic in animals hangs almost entirely on studies that utilized corn oil gavage studies (IARC, 1991, 2004; U.S. EPA, 2003a,b) an experimental approach that raises considerable concern that the observed carcinogenic responses were artifacts of the experimental method.

In addition to the descriptive studies described above, there have been extensive studies of the mechanisms by which the THMs produce toxicity, including studies of how they are metabolized and how that relates to their toxic and carcinogenic effects. A most convincing line of evidence was developed which indicates that chloroform produces cancer in the two target sites associated with it is a result of killing liver and kidney cells (Larson *et al.* 1994a,b, 1995a,b, 1996). The cell killing results in the replacement of the dead cells by a process referred to as reparative hyperplasia. As chloroform toxicity in the liver and kidneys becomes more severe, the rate of cell division increases proportionally, stimulating the outgrowth of abnormal cells that inevitably arise spontaneously in an animal which experiences repeated cell killing. Coupled with the large database that indicates that chloroform does not directly induce DNA damage or produce mutations, it was concluded that the cancer risk associated with chloroform is not genotoxic and does not warrant estimating low dose cancer risk through the use of a modeled cancer slope factor that predicts a positive risk associated with any dose above zero (U.S. EPA, 1999). Therefore, chloroform poses essentially no cancer risk at the concentrations that result from the chlorination of drinking water.

The issue with BDCM is a little more complex. It has long been known that chloroform is metabolized by the enzyme coded for by the CYP2E1 gene referred to as cytochrome P450 2e1 (P450 2e1) to produce a toxic intermediate, phosgene. BDCM metabolism results in this same intermediate, but more of it. Phosgene (or brominated analogs with CDBM and TBM) are quite reactive with various biochemicals, including proteins, but it does not appear to bind directly to DNA. In the past decade, however, it was found that the brominated THMs are also metabolized by a different enzyme, glutathione transferase theta 1–1 (GSTT1). It was shown that a product results that is mutagenic (Ross and Pegram, 2003, 2004), raising the issue that the brominated DBPs may be initiating cancer through a genotoxic mechanism. This metabolic path is not important for chloroform (Pegram *et al.* 1997), so it does not alter the conclusion that it is unlikely to cause cancer at low doses.

Among the THMs, BDCM dissolved in corn oil was the most potent carcinogen when assessed by treating mice and rats via stomach tube. In addition, it produced cancer in several target organs, including the large intestine and in the kidney of mice as well as liver tumors in mice and kidney tumors in rats that result from treatments with chloroform. CDBM only produced liver tumors in mice (NTP, 1985) and TBM only produced tumors in the large intestinal of rats (NTP, 1989). Therefore, BDCM is the only THM that clearly met the IARC and EPA criteria as possible carcinogens in humans (although TBM was granted that status because the induction of large intestinal tumors, which are relatively rare in rats and they appeared to be a relevant tumor site in humans given epidemiological results which suggested that colorectal cancer might be associated with chlorinated drinking water exposure, but this cancer site has not been consistent across studies). Therefore, the weight of the toxicological evidence for carcinogenic activity of the THMs depends heavily on the results of BDCM studies using corn oil gavage (NTP, 1986). In fact, WHO (60 µg/L) and Canada (16 µg/L) developed lower more restrictive guidelines for BDCM than the other THMs because of this observation. Canada has since withdrawn its BDCM guideline in 2009, following a review of the new evidence provided (NTP, 2006).

The NTP conducted a second study of BDCM carcinogenicity in mice and rats where it was given via drinking water. The results of that study were negative; in fact, the incidence of spontaneous tumors observed in the control animals were actually depressed by BDCM treatment (NTP, 2006). Aside from the difference in the mode of BDCM administration in the two studies, the only substantive difference was that the highest daily dose administered in the drinking water study was slightly lower than the lowest dose administered in the corn oil gavage study (lowest dose by corn oil gavage was 25 mg/kg day^{-1} in mice and 50 mg/kg day^{-1} in rats, whereas the highest dose was 36 mg/kg day^{-1} in mice and 25 mg/kg day^{-1} to rats in the drinking water studies). If simple linear extrapolation were applied, statistically significant cancer effects should have been observed in the water dosing experimtents. A pharmacokinetic model was developed, but never published, to determine the actual delivered dose (THMs are more rapidly absorbed from water than corn oil). In any case, these data do call into question the determination that the THMs are likely to be carcinogenic in humans exposed to them at concentrations found in ambient air and drinking water. At the very least, the projected risk of the THMs will be reduced if these data are bridged by pharmacokinetic modeling to combine the results of the two studies.

Additional experimental data have been developed that need to be taken into account in assessing the possibility of a causal relationship between THM exposure and bladder cancer risk suggested by epidemiological studies. This is discussed in the next section.

10.4.2 Risk estimates based upon epidemiological data

Epidemiological studies relating disinfection of drinking water to various adverse health outcomes are reviewed in Chapter 11. The present chapter dwells on the following points: 1) a more explicit comparison of the risks identified with DBPs from toxicological research in comparison to the magnitude of the risk that can be derived from epidemiology studies, 2) the relative carcinogenic potency of other chlorination by-products, and 3) a short discussion of certain non-genotoxic carcinogens which do not contribute to risk at doses obtained from drinking water.

The magnitude of risks calculated from a series of meta-analyses of epidemiological studies of bladder cancer risk used by the U.S.EPA in the development of the Stage 2 Disinfection By-Product Rule (DBP-II) regulations (U.S. EPA, 2003a) are displayed in Table 10.1. The table reports the annual and lifetime risk of bladder cancer in the U.S. population through 2001. A review of the meta-analyses of the case-control epidemiological studies provided as background for the (DBP-II) (U.S.EPA, 2003a) estimated that between 2 and 17 percent (the upper and lower 95% confidence limits of the estimate) were attributable to chlorinated drinking water. Multiplying the lifetime probability of developing bladder cancer among men and women results in estimates of the lifetime risks these population attributable risks (PAR) represent in the population. Combining the individual risks between the sexes, these meta-analyses indicate that there is an approximately one in 1000 lifetime risk for developing bladder cancer from consuming chlorinated drinking water.

Table 10.1 Risks of bladder cancer attributed to chlorination of drinking water.

	Males	**Females**
Bladder cancer incidence (annual)	0.000377[a]	0.000096[a]
Lifetime probability for developing bladder cancer	0.0380[b]	0.0116[b]
PAR[c] for chlorinated drinking water applied to lifetime probability above		
2%	0.0008	0.0002
17%	0.006	0.002

[a]Age adjusted incidence for years 2003–2007, Cancer Facts 2011, American Cancer Society (http://www.cancer.org)

[b]Years 1999–2001, Cancer Facts, 2005, American Cancer Society (http://www.cancer.org)

[c]PAR, population attributable risk, is a statistical estimate of what proportion of a total outcome, like bladder cancer, can be attributed to a specified cause based on epidemiological data, like exposure to chlorinated drinking water. The 2% and 17% represent the lower and upper bounds of the meta-analyses (USEPA, 2003a)

10.4.3 Risk based upon toxicological data

Those DBPs that have been shown to be carcinogenic in at least one animal species are listed in Table 10.2. The lower bound of the dose that produced a 10% increase in cancer (LTD_{10}) in animals or the benchmark dose estimated to produce the same result (BMDL) is listed for each DBP. The reciprocal of these values provides a measure of the relative potency[5] of these DBPs as carcinogens. This value was multiplied by 1000 simply to express the range in whole numbers. The relative potency ranges from a low of 7.6 with benzaldehyde to a high of 71,400 for NDMA. Thus, the characterized DBPs have a 10,000-fold range in potency. The last column provides a judgment of whether the DBP is a genotoxic carcinogen or not. Some of these carcinogens are classified as genotoxic carcinogens based upon *in vitro* genotoxic data in the absence of appropriate *in vivo* studies. Therefore, the characterization of some of these compounds as genotoxic carcinogens could change with additional toxicological study.

Table 10.2 List of DBPs whose carcinogenic activity has been confirmed in chronic animal studies.

DBP	LTD_{10} or BMDL (mg/kg/d)	Potency ($1000/LTD_{10}$)	Genotox (G) / non-genotox (NG)
Chloroform	11.8[a]	85	NG
BDCM	5.24[a]	190	G
CDBM	12.9[a]	78	G
TBM	54.3[a]	18	G
DCAA	12[a]	83	NG
BCAA	2.6[b,c]	305	G
DBAA	5.2[b,d]	192	G
TCAA	51.3[a]	19.5	NG
MX	0.0691[a]	14,471	G
2,4,6-Trichlorophenol	41.9[a]	24	NG
Dibromoacetonitrile (DBAN)	7.4[b,e]	135	G
Chloroacetaldehyde	2.9[a]	345	G
Formaldehyde (inhalation risk only)	0.145[a]	6,900	G

(Continued)

[5]potency for a drug or chemical (such as a carcinogen) is the reciprocal of the dose of a chemical that produces a common increase in response. In other words, reciprocal dose results in compounds with greater potency to be characterized with a larger number. As presented in the table, the potencies allow direct comparison of different carcinogens. For example, MX is 14,471/190 = 75 times more potent than BDCM.

Table 10.2 List of DBPs whose carcinogenic activity has been confirmed in chronic animal studies (*Continued*).

DBP	LTD_{10} or BMDL (mg/kg/d)	Potency ($1000/LTD_{10}$)	Genotox (G) / non-genotox (NG)
Benzaldehyde	132[a]	7.6	NG
Bromate	0.77[a]	1,300	G?
N-Nitrosodimethylamine (NDMA)	0.014[a]	71,400	G

[a]LTD_{10}, Gold *et al.* (2011)
[b]$BMDL_{10}$ calculated by BMD software, U.S. EPA. Note figures derived from the $BMDL_{10}$ calculation include surface area correction to human equivalent dose ($BW^{3/4}$).
[c]NTP, 2009
[d]NTP, 2007
[e]NTP, 2010

There are several other genotoxic DBPs that have been identified as carcinogens based on shorter term *in vivo* studies that are not included in Table 10.2. These include bromochloroacetonitrile (BCAN)(Bull *et al.* 1985a), 1,3-dichloroacetone (Bull *et al.* 1985b), 2-chloropropenal and 2-bromopropenal (Robinson *et al.* 1989). Nitrosamines other than NDMA have also been detected as by-products, but at a much lower frequency (see Chapters 5, 6 and 7). Most nitrosamines that have been studied in chronic studies are clearly carcinogenic. Some nitrosamines approach NDMA in potency (e.g. *N*-nitrosodiethylamine or *N*-nitrosomorpholine) (Gold *et al.* 2011), but their concentrations in drinking water are so much lower that their contribution to the average carcinogenic risk from drinking water is negligible. *N*-Nitrosodiphenylamine has been identified in drinking waters, but it is very weak as a carcinogen (Gold *et al.* 2011), as well as being present at low concentrations (Li *et al.* 2011).

Caution is warranted with respect to the assigned risks in Table 10.3. The primary intent of this chapter was to compare risks from different DBP classes, so it was necessary to utilize a study that was inclusive of the DBPs for which there were toxicological data. Thus, the literature source for drinking water concentrations for most of the DBPs (Krasner *et al.* 2006) was selected primarily because it provided data on a diverse set of DBPs which were not included in prior studies in the same water systems as well as the more frequently measured DBPs. The survey involved only 12 water systems. The use of disinfectants among the systems included in the study was diverse. One system used O_3 prior to NH_2Cl, a second used $O_3 + Cl_2$, one system used Cl_2 only, five systems used Cl_2 followed by NH_2Cl, two systems employed ClO_2 followed by Cl_2 and then NH_2Cl, one used NH_2Cl followed by O_3 and one system used NH_2Cl only. Therefore, these quantitative data are unlikely to be truly representative of DBPs from use of any given disinfectant and the study is clearly too small for valid statistical analysis of the occurrence of DBPs using different disinfectants. Neither was the sample size large enough or designed properly for estimating national exposures to DBPs in drinking water treated with different disinfectants. Nevertheless, the data are reasonable when compared to large studies that surveyed fewer DBP classes.

Table 10.3 Carcinogenic risk (lifetime – 70 year – 95% upper confidence limit) from chlorination by-products that have been chronically studied in rodents.

DBP	Median conc., μg/L	Risk × 10^{-6} @ median exp.	Max. conc., μg/L	Risk × 10^{-6} at Max. Conc.	Conc. (μg/L) required to produce 10^{-3} risk[h]
BDCM	8.4[a]	8.4[d]	19[a]	19	1000
TBM	1.4[a]	0.2[d]	3.8[a]	0.5	4600
THM total		8.6		19.5	1000
BCAA	5.7[a]	6.3[f]	18[a]	19.8	910
DBAA	3.3[a]	1.8[f]	11[a]	6.1	1800
DHAA total		8.1		25.9	910
DCAN	1.2	0.48[f]	5.0	2.0	6000
BCAN	0.8	0.32[f]	2.0	0.8	6000
DBAN	0.5	0.19[e]	2.0	0.8	6000
DHAN total		0.99		3.6	6000
Chloroacetaldehyde	0.42	0.06[e]	1.9	0.27	148
MX[b]	0.053	8.8[e]	0.18	30	6
BMX-1[b]	0.04	6.7[f]	0.17	28	6
BMX-2[b]	0.002	0.33[f]	0.02	3.3	6
BMX-3[b]	0.041	6.8[f]	0.20	33	6
MX analogs total		22.6		94	6
NDMA	2.5[c]	0.8[g]	0.03[c]	10	3
Grand total		41		153	

[a]Krasner *et al.* (2006). Mean concentrations calculated based on supporting data for paper rather than median concentration reported by the authors
[b]Onstad *et al.* (2008)
[c]Valentine *et al.* (2006). Detected in 59% chloraminated and 33% chlorinated supplies. No median concentration given. 42/47 chloraminated systems <2.5 ng/L. Maximum concentration measured was 30.
[d]Risks based on U.S.EPA (2003b)
[e]Risk estimated from LTD_{10} or $BMDL_{10}$ by linear extrapolation (i.e. human equivalent dose divided by 100,000).
[f]Values estimated from the congener with a chronic cancer bioassay.
[g]Unit risk from Cal EPA PHGs (OEHHA, 2011).
[h]When combining risks within a class, the potency of the most prevalent and most potent congener is used.

The drinking water concentrations and volume ingested of these DBPs are also inadequate measures of total exposure to individual DBPs. It has been shown that some DBPs of low molecular weight, low polarity, and high partial pressures are absorbed to a significant extent through the skin and by inhalation. To estimate the actual mean dose received at these concentrations requires the use appropriate pharmacokinetic models.

Finally, some of the risks presented in Table 10.3 include DBPs for which no chronic studies have been conducted. If only one chemical within a class was subjected to a chronic cancer study and others have not, the risk calculated/μg L^{-1} was assigned to other members of the class on a relative weight basis. Using this approach, risks calculated for DBAN were assigned to dichloroacetonitrile (DCAN) and bromochloroacetonitrile (BCAN) and the risk calculated for MX was extended to the three brominated analogs. This allowed inclusion of all the classes for which there were both chronic cancer studies and evidence of genotoxicity to be included. However, the assigned risks must be taken as approximations that have yet to be verified by actual data. It is probable that these approximations overestimate the likely contribution of the dihaloacetonitriles, but underestimate that of MX and its brominated analogs. The definition of DBP classes here is narrower than used by U.S. EPA. Compounds with differing numbers of halogens are not considered of the same class because their mechanisms and MOAs have been found to be quite different (e.g., Bull, 2000).

Given all the shortcomings of the exposure data, MX and its brominated analogs appear to be the dominant risks, accounting for 23 cases per million of population exposed per lifetime at average concentrations and 94 at the highest concentrations reported in the survey.[6] This compares to an attributable risk of approximately 8.6 and 19.5 per million at average and high reported concentrations of the THMs. The dihaloacetic acids, BCAA and DBAA, contribute approximately the same risk as the THMs at average concentrations and somewhat higher risks at the highest reported concentrations. If the risks for all compounds that have been shown to produce cancer in animals and are genotoxic approximates 41 per million population at average concentrations and a total of 153 at the highest concentrations reported. The key is that THM risk based upon toxicological data ranges from less than 10 per 1,000,000 population per lifetime at median concentrations and less than 20 at the extreme concentrations. None of this cancer risk estimate is attributable to chloroform. Most of that risk is attributable to BDCM. The difficulties with the cancer evidence upon which the BDCM risk assessment is based were identified in the previous section. Therefore, the THM risk is probably a substantial overestimate.

[6]These number of cancer cases would not be detectable by public health surveillance of chronic disease against the background of cancer, even in a population of 100 million (i.e. an average of 2300 cases per 70 years would be only about 30 cases a year)

The dihaloacetonitriles (DHANs) also appear capable of contributing to carcinogenic risk. Only one member of this class has been studied chronically in rodents, dibromoacetonitrile (DBAN) (NTP, 2010). It is approximately equivalent in potency to the dihaloacetic acids, but its concentrations are generally lower. Its unit risk was also used to assign a carcinogenic risk of the two members of this class that have not been subject to chronic cancer studies. All three DHANs are mutagenic (Bull *et al.* 1985a).

DBPs were excluded from this list based on substantial evidence indicating that a genotoxic mechanism is not involved in producing cancer. The nature of the evidence upon which these exclusions were made follows:

A large body of literature, summarized in the modeling paper of Miller *et al.* (2000), shows that dichloroacetic acid (DCA) does not produce cancer by a genotoxic mechanism even though the U.S. EPA IRIS database persists in reporting a cancer slope factor for DCAA. Its MOA is selective stimulation of spontaneously arising tumor cells and suppression of the division of normal hepatocytes (Stauber and Bull, 1997). This latter effect is associated with suppressed apoptosis (Snyder *et al.* 1995), a process that increases the probability that spontaneous mutations will occur in a tissue, some of which could give rise to tumors (Bull, 2000; Miller *et al.* 2000). This was elegantly illustrated by treating hepatocytes isolated from B6C3F1 mice with DCAA and producing growth of a subpopulation of cells of the same phenotype that was observed *in vivo* (Stauber *et al.* 1998). The differential effect on tumor cells and normal liver apparently arises from disturbances in insulin signaling (Linghor *et al.* 2001, 2002). Therefore, cancer risk from DCAA is not appropriately modeled by linear extrapolation to low doses (i.e. using a cancer slope factor) as was done by the U.S. EPA (2003b). Effectively, risk from DCAA is zero at both mean and extreme concentrations identified in chlorinated drinking water.

TCAA is specifically not identified with the dihaloacetic acids because of its mode of action (Bull, 2000). It is not genotoxic, produces liver tumors in mice, but not in rats, and contributes no cancer risk at the concentrations found in drinking water. Other trihaloacetic acids have not been subject to chronic cancer studies. This is an important data gap since the brominated trihaloacetic acids are much more extensively metabolized than TCAA (Austin and Bull, 1997; Hu *et al.* 1995; Merdink *et al.* 2000; Saghir *et al.* 2011) broadening the possibility that they would contribute to a cancer risk. Chloral hydrate (trichloroacetaldehyde) is also excluded because it has only produced liver cancer in mice (Daniel *et al.* 1992) and this appears dependent upon its metabolic conversion to TCAA.

Bromochloroacetic acid (BCAA) and dibromoacetic acid (DBAA) remain in Table 10.3 as DBPs that can contribute to cancer at low doses. This is based upon evidence of mutagenic activity *in vitro* and non-hepatic cancer, as well as liver cancer, results from treatment of rodents (NTP, 2007, 2009). This is clearly different than the case of DCAA, which only produces liver cancer by the mode of action described above.

In summary, the available toxicological evidence indicates that the THMs represent a risk that is two orders of magnitude less (i.e. 100 fold less) than has been associated with chlorinated drinking water in epidemiological studies. Based on animal studies (which are the only studies conducted under controlled conditions), THMs contribute significantly less risk than other classes of chlorination by-products for which there are appropriate data to estimate carcinogenic risk. As data become available to assess the risks associated with other DBPs, the importance of the THMs as cancer risks should recede completely into the background. At best, THMs are minor players with respect to the risks that are associated with chlorinated drinking water.

10.4.4 Uncharacterized chlorination by-products

It is apparent that the vast majority of DBPs have not been toxicologically characterized. Not all DBPs represent health hazards as some are simply normal physiological substrates. Many are simple acids, fatty acids, and aldehydes that are readily assimilated by animals as well as microbes as substrates for intermediary metabolism. Nevertheless, it is important to reflect upon the kinds of chemicals that might be present as chlorination by-products and whether there is some probability of their contributing to cancer risk.

Bull *et al.* (2006, 2011) undertook a study to examine most of the chemicals that were recognized as DBPs and attempted to predict whether certain toxicologically significant by-products might be formed. In selected instances, these predictions were confirmed in a subsequent study (Li *et al.* 2011). Among these chemicals some were identified that have metabolic characteristics that parallel those of known bladder carcinogens. The groups that were selected as they were potentially potent carcinogens were halobenzoquinones (HBQs), halo-cyclopentene, and -cyclohexene derivatives (see Chapters 8 and 9). Others of interest are N-halamine, longer chain halonitriles, and evidence was provided for the occurrence of a heterocyclic amine.

Figure 10.1 displays a number of HBQs, halocyclopentane and halofuranone derivatives that were predicted to occur (Bull *et al.* 2006). Some that were subsequently shown to occur in chlorinated drinking water are identified in the Figure. The discussion in this chapter will be confined to the HBQs for the sake of brevity.

Quantitative structure toxicity relationship (QSTR) predictions of the lowest adverse effect level (LOAEL) for several of the HBQs and related chemicals in Figure 10.1 approximate 0.05 mg/kg day^{-1} (Bull *et al.* 2006). These predictions can be compared to a LOAEL of 15 mg/kg day^{-1} used by the U.S. EPA in developing water quality criterion for chloroform, indicating these chemicals are roughly 300 times more potent as toxicants. By comparison, the concentrations of individual HBQs that have been found in chlorinated water are up to 0.2 μg/L and could be significant because of their much lower LOAELs.

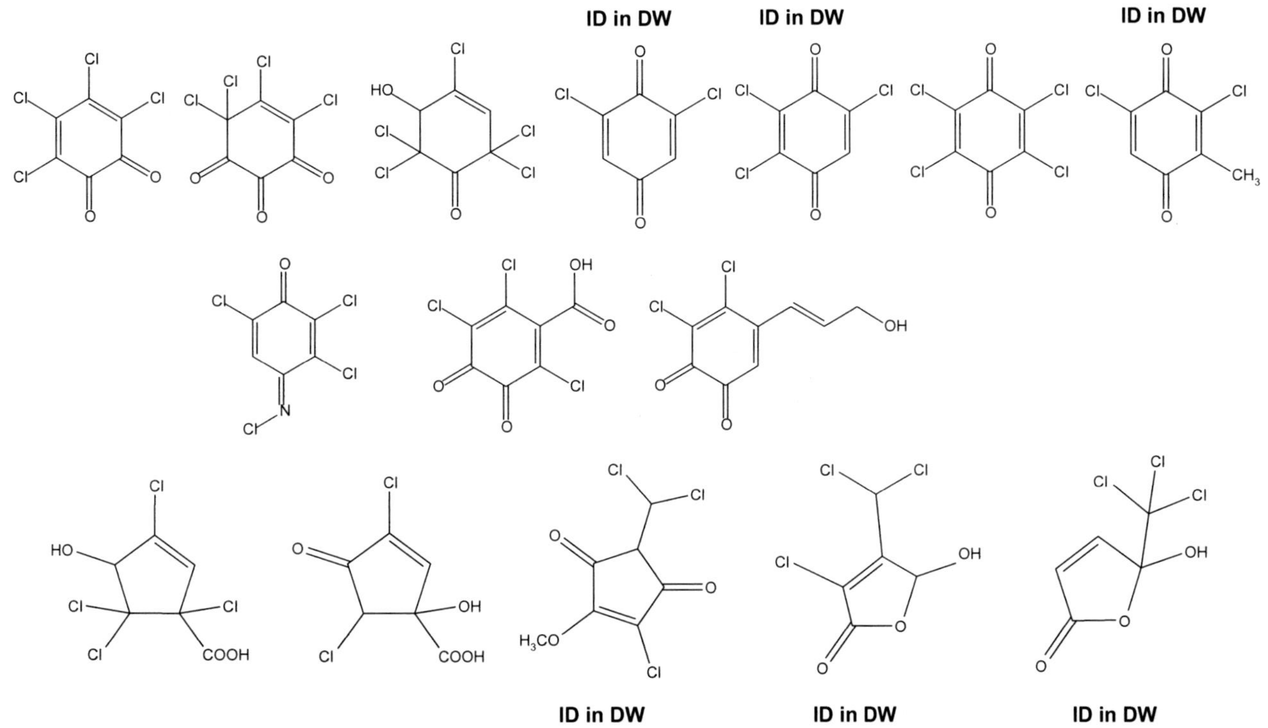

Figure 10.1 Selected haloquinone, halocyclopentane and halofuranones predicted or which are known to occur in chlorinated drinking water (Bull *et al.* 2006, 2011). ID in DW indicates chemicals that have been detected and measured in drinking water (some of brominated analogs have also been identified, but not depicted).

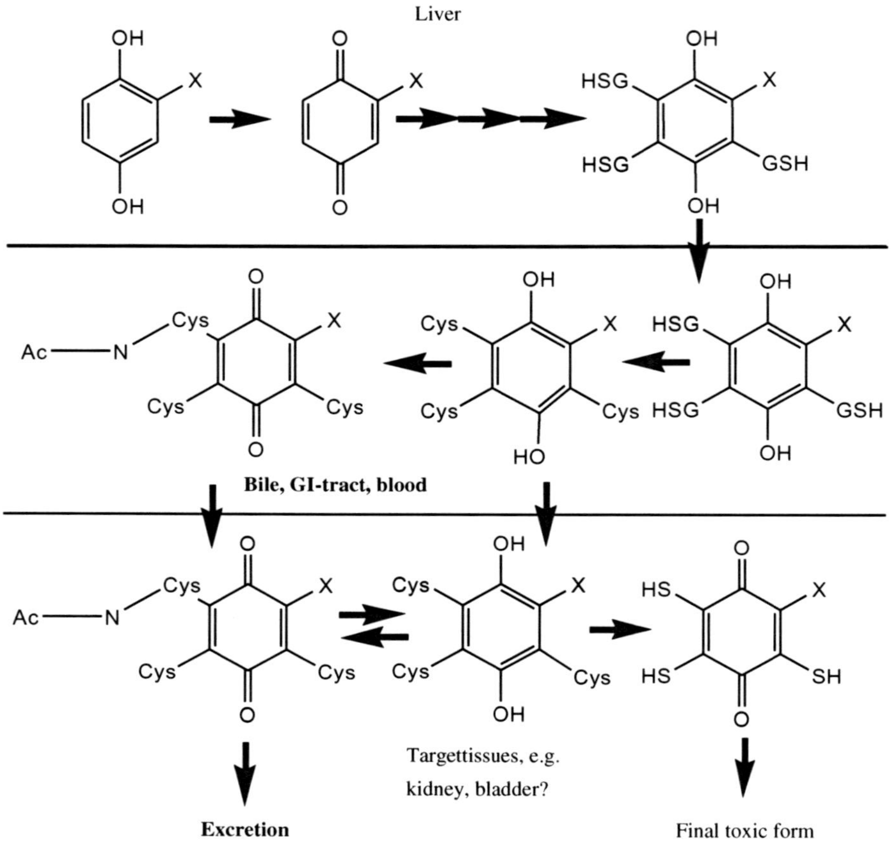

Figure 10.2 Mechanisms by which HBQs may contribute to cytotoxicity and/or genotoxicity. A key to toxicity induced by quinones is that the glutathione and cysteine conjugates, produced largely in the liver, are transported via blood and bile to tissues where they may be concentrated in cells that express γ-glutamyl transferase (γ-GT) or glutathione transporters. Once in these tissues, they are further metabolized to thiol compounds that may be directly toxic or that are metabolized to reactive intermediates. This is illustrated in the γ-GT-expressing cells. Similar metabolic processing can occur in cells that concentrate glutathione (not shown). *N*-acetylation of the cysteine conjugate detoxifies the compound, but this is an easily reversed reaction. Acetylation takes place on the amino group of one or more cysteine residues. The deacetylated form is further processed to a free thiol (shown because the actual reactive metabolite has not been established) that is converted to a reactive product. The glutathione conjugates and subsequent metabolites still give rise to reactive oxygen species and can do damage, but the activation of the thiol results in an alkylating agent. The acetylated metabolite is eliminated in the urine.

The metabolism of the HBQs is expected to parallel that of other quinones and HBQs that have been studied (Lin *et al.* 1999; Bolton *et al.* 2000). The initial metabolic activation of these compounds is through conjugation to glutathione in the liver (Figure 10.2). As the body has mechanisms for recovering glutathione and its breakdown product cysteine, there are transport systems that concentrate them in secondary tissues. Most of the compounds which have been studied have been found to localize in the proximal tubule cells of the kidney, but this can occur in other tissues as well. Further metabolic processing in these secondary tissues have been shown to greatly increase their toxicity in that tissue (Bolton *et al.* 2000). There are additional enzymatic systems that can inactivate the active compound and others that regenerate the toxic metabolite. In the figure these latter steps are acetylation to inactivate and deacetylation to reactivate to the toxic form.

10.4.5 Evaluation of recent data on enzyme variants that metabolize DBPs

Recently, Cantor *et al.* (2010) published a paper which associates the genotypes of the enzymes that are involved in the metabolism of THMs and the haloacetic acids with increased risk of bladder cancer among individuals consuming chlorinated drinking water. This paper does provide substantive evidence that chlorinated drinking water contributes to bladder cancer. It is an elegant paper, but many have taken it as strong evidence that THMs are causally involved in the production of bladder cancer. This is inaccurate for several reasons:

(1) These enzymes are involved in the metabolism of many compounds including other DBPs that are in chlorinated and/or chloraminated water.
(2) These enzymes have roles in the metabolism of lipids that have important physiological functions that could contribute to carcinogenesis.
(3) At least one transcription factor has been shown to be modified by a GST and this modifies the activity of the transcription factor. This transcription factor often plays a role in carcinogenesis.
(4) Kinetic constants for interaction of these THMs for P450 2e1 and DCA for GSTZ1 are such that the rate of THM metabolism at the low blood levels from human dermal and oral exposure will be independent of the isoforms.
(5) The GSTT1 null genotype is of particular interest from a risk assessment perspective as it should be protective (i.e. there is no enzymatic activity with the null phenotype, therefore there is no activation of the chemicals the enzyme metabolizes). However, it is probable that other DBPs may well be better substrates for the genotypes that express active isoforms of the enzyme than the THMs. These will not be toxic if that is main way they are activated toxicologically.
(6) The differences in the level of expression of these enzymes and the catalytic efficiency among the experimental species (mice and rats) and humans are

small. As a consequence this is an unlikely basis for interspecies differences in sensitivity.

(7) The human GSTT1 has comparatively low activity towards electrophilic substrates compared to the enzyme from mice and rats (Shokeer and Mannevik, 2010). This could be the basis for species differences detoxifying intermediates of DBPs other than the THMs

Table 10.4 is a short list of other environmental and endogenous chemicals that are metabolized by these enzymes. In addition, there are a number of DBPs that are known to be metabolized by glutathione transferases, but the specific isoform responsible has not been established. Some of the non-DBP compounds are included in the list as they may be models for these uncharacterized DBPs. An example is benzene, which is metabolized to phenol and quinone intermediates by P450 2e1 and conjugated to glutathione (Fisher *et al.* 2011). Apparently, in the case of glutathione conjugation to phenol metabolites is catalyzed by GSTT1 as the GSTT1-null phenotype was associated with increased susceptibility to toxicity (Chen *et al.* 1996; Wan *et al.* 2002). The GST responsible for glutathione conjugation of the benzoquinone metabolites has not been established.

Table 10.4 Disinfectant by-products and common environmental contaminants metabolized by enzymes identified in Cantor *et al.* 2010.

Compounds	P450 2e1	GSTT1	GSTZ1	Other GSTs
NDMA[a]	×			
DCAA, BCAA, DBAA[b]			×	
2-halopropionic acids[b]			×	
Endogenous fatty acids[c]	×			
Chloroform and other halocarbons[a]	×			
Brominated THMs, 1,2-dibromoethane, methylene chloride, dibromomethane[d,i,k]	×	×		
Ethanol	×			
Benzene[e]	×	×		
Haloacetonitriles[f]				×
MX and brominated analogs[g]				×
Halobenzoquinones[h]				×
1,3-Dichloroacetone, 1,3-dibromoacetone[k]		×		
Dichloropropanol[j]	×	×		
c-Jun N-terminal kinase-1[l]				(GSTP1–1)

(*Continued*)

Table 10.4 Disinfectant by-products and common environmental contaminants metabolized by enzymes identified in Cantor *et al.* 2010 (*Continued*).

Compounds	P450 2e1	GSTT1	GSTZ1	Other GSTs
Brominated trihaloacetic acids[m]	×		×	

[a]Yamazaki *et al.* (1992); Razdan *et al.* (1994)
[b]Tong *et al.* (1998a); Board and Anders, (2011)
[c]Porubsky *et al.* (2008)
[d]Pegram *et al.* 1997; Ross and Pegram (2004);
[e]Wan *et al.* (2002)
[f]Lin and Guion (1989)
[g]Indirect evidence of interaction with glutathione transferase (Heiskanen *et al.* 1995)
[h]Bolton *et al.* (2000)
[i]Shimada *et al.* (1996)
[j]Hammond and Fry, (1997)
[k]Thier *et al.* (1996)
[l]Wang *et al.* (2001)
[m]Saghir *et al.* (2011); Austin and Bull (1997)

The metabolism of endogenous fatty acids involved in cellular signaling and modification of transcription factors by enzymes involved in DBP metabolism raises the possibility that metabolism of the DBPs may not be the basis of the hypothesized interactions between DBP exposure and carcinogenic outcomes. Another example, is that c-Jun-N-terminal kinase-1 is modified by another GST (GSTP1) which modifies activity of the c-Jun transcription factor. The normal cells of the liver of mice that have been treated chronically with DCAA accumulate large amounts of c-Jun in the cytosol (or perhaps lysosomes)(Stauber and Bull, 1997). Abnormalities in the expression of this transcription factor is frequently associated with carcinogenesis. Therefore, metabolism of exogenous chemicals are not the only function of GSTs.

In addition to the chemicals identified in Table 10.4, some DBPs are metabolized by the implicated enzymes in a concerted fashion (i.e. the product of one enzyme is the substrate for another, or several others. The brominated trihaloacetic acids are metabolized by both P450 2e1 and GSTZ1. Bromodichloroacetic acid (BDCAA) is used as the example in Figure 10.3. Unlike TCAA, BDCAA and the other brominated trihaloacetic acids are extensively metabolized. At doses of 20 mg/kg to B6C3F1 mice, about 0.2% of the dose of BDCAA is eliminated in the urine unchanged (Hu *et al.* 1995), whereas at least 70–80% of TCAA is eliminated unchanged. P450 2e1 has been shown to be a major contributor to the degradation of BDCAA, chlorodibromoacetic acid (CDBAA), and tribromoacetic acid (TBAA) (Saghir *et al.* 2011). This step occurs in the endoplasmic reticulum, whereas that by GSTZ1 occurs in the cytosol. Decarboxylation to the corresponding THM is a possibility, but the amounts measured from microsomal metabolism are very low

(Saghir *et al.* 2011). However, the possibility of non-microsomal conversion does not necessarily rule out their formation *in vivo*. The main reason for assigning a role for GSTZ1 in the initial degradation of BDCAA is that pretreatment with DCAA sharply inhibits the metabolism of DCAA and has also been shown to substantially inhibit the metabolism of BDCAA (Austin and Bull, 1997). Although data reporting DCAA as a product of BDCAA metabolism measured in urine was challenged (in a second paper from the same laboratory), subsequent experiments did demonstrate that the DCAA radical was produced from BDCAA in microsomal incubations (Merdink *et al.* 2001) and DCAA is indeed produced as the product of BDCAA degradation by P450 2e1. Therefore, it can be easily visualized that differing activities of P450 2e1 and GSTZ1 could interact in the metabolism of BDCAA in a way suggested by the genotypic patterns associated with higher risks of bladder cancer and differing genotypes for these two enzymes. These data provide an additional indication of why the influence of these enzyme variants on bladder cancer risk is not simply attributable to THM exposure.

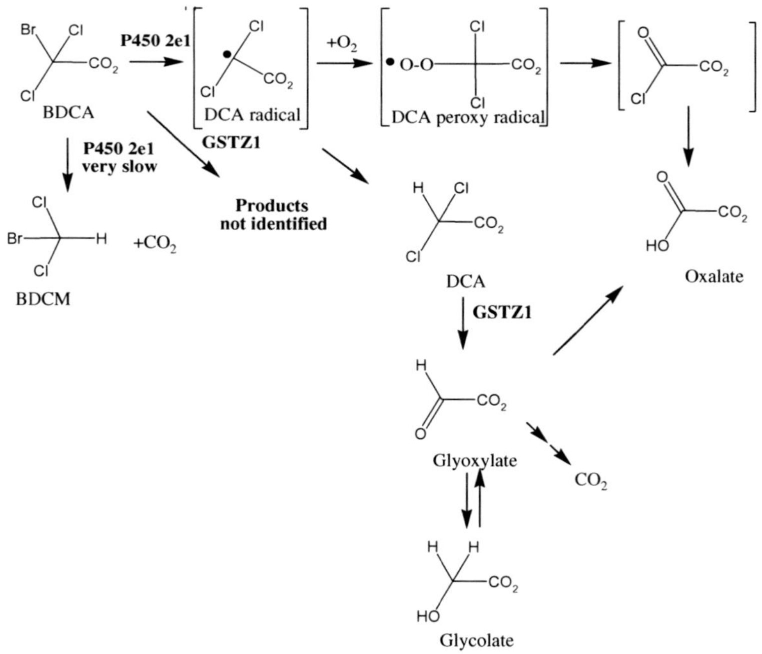

Figure 10.3 Metabolic scheme for trihaloacetic acids, using bromodichloroacetic acid as the example. DCA means the same as DCAA elsewhere in text.

Another consideration for interpreting risk is the differential metabolism of one stereoisomer of a DBP over the other. Bromochloroacetic acid (BCAA) is an example. The (−),(+)-BCAA have been shown to be metabolized at much different rates in F344 rats (Schultz and Sylvester, 2001). The (+)-BCAA is metabolized with a half-life of 0.4 h, while that of the (−)-BCAA was 0.07 h. Moreover, only the metabolism of (−)-BCAA was affected by pretreatment with DCAA, indicating its metabolism was primarily effected by GSTZ1, while that of (+)-BCAA was not. The metabolism of (+)-BCAA appears to be by another GST, that has not been identified. While the effects of DCAA appear largely attributable to the parent compound, this alternative pathway could yield some intermediates from the trihaloacetic acids that could bind to nucleic acids.

Figure 10.4 illustrates how the rate of catalysis by an enzyme is governed by the affinity of the substrate for the enzyme active site and the concentration of the substrate. The affinity of the substrate for the enzyme is reciprocally related to the Michaelis constant (K_m). The K_m is the concentration of the substrate in the media at which 50% of the active sites of the enzyme are occupied by the substrate. As illustrated in Figure 10.4, the rate of catalysis approaches zero at concentrations that are 2 orders of magnitude below the K_m. Conversely, at approximately two orders of magnitude above the K_m, the enzyme's active sites are essentially all occupied (i.e. the enzyme is saturated) and the rate of catalysis can no longer be increased by higher concentrations of the chemical.

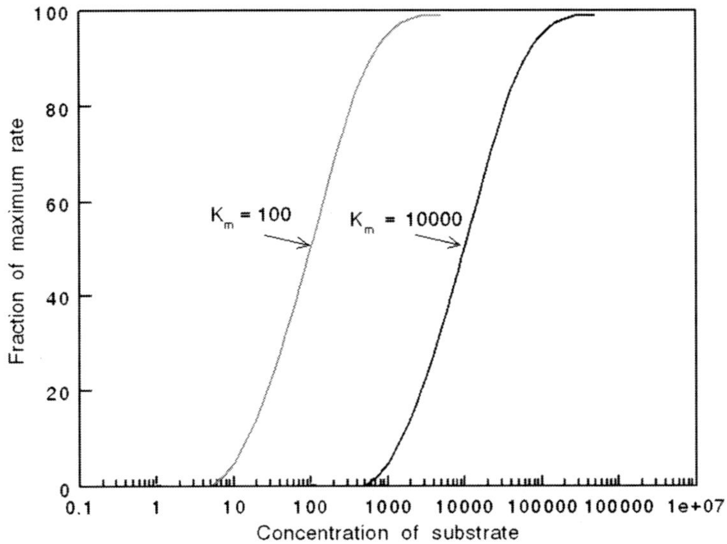

Figure 10.4 Comparison of rates of metabolism of a substrate at concentrations that are below its K_m for the enzyme. K_m is the concentration at which the enzyme active site is 50% occupied. Note that the rate of reaction falls to negligible rates when the concentration is two orders of magnitude below the K_m.

There are two curves presented in Figure 10.4 which differ in K_m by two orders of magnitude. If these enzymes have the same substrate, the enzyme with the lower K_m will consume almost all of the substrate until the substrate approaches concentrations that saturate all of the active sites of the enzyme with the lower K_m. As this happens, the concentrations approach those which are efficiently metabolized by the second competing enzyme.

Table 10.5 provides reported K_m values for the more active forms of the enzymes for which individuals were genotyped in the Cantor *et al.* (2010) study. The main point is that the K_m values for metabolizing BDCM with P450 2e1 vs. GSTT1 appear to differ by 2 orders of magnitude. The concentrations of BDCM achieved in humans under controlled exposure conditions are very low relative to the K_m of either enzyme. These data indicate that until blood concentrations begin to exceed the K_m of P450 2e1 that virtually all the BDCM is going to be metabolized by this enzyme. Metabolism via GSTT1 will be less than 1% of that metabolized via P450 2e1 simply based upon their relative K_m values. Consideration of the relative catalytic efficiency of the two enzymes may modify this relationship somewhat. Nevertheless, this comparison indicates that at the low blood concentrations that occur in human blood as a result of exposure to THMs in drinking water, the production of the mutagenic metabolite of BDCM by GSTT1 will be effectively zero. Until BDCM concentrations exceed 10 μM, the risk resulting from that mechanism will be negligible.

Table 10.5 External dose and blood level concentrations produced by defined exposures to BDCM or DCAA in humans and kinetic constants for their metabolism by cytochrome P450 2e1 or glutathione transferases.

DBP	Dose	Plasma C_{max}, μM	Km, μM	Enzyme
BDCM			2.45[c]	P450 2e1
	72 μg/kg oral	0.000016[a]	ca. 200[d]	GSTT1
	72 μg/kg dermal	0.00076[a]	ca. 200[d]	GSTT1
DCAA	15 μg/70 kg oral	0.009[b]	47.3[e]	GSTZ1

[a] BDCM concentrations measured in studies in human volunteers (Leavens *et al.* 2007)
[b] DCAA blood concentrations that are model derived from human pharmacokinetic data (Schultz and Shangraw, 2006)
[c] Lilly *et al.* 1997
[d] K_m for BDCM is estimated from concentration/mutation response curve in Figure 3A of Ross and Pegram (2004)
[e] K_m reported is for DCAA and the human GSTZ1 cytosolic protein (Tong *et al.* 1998a)

Backer *et al.* (2008) essentially confirmed the predictions related to P450 2e1. They found no significant differences in the peak values or clearance of THMs associated with the activity of P450 2e1 or with the genotype for CYP 2E1 in human volunteers after showering. The result is also consistent with modeling of the impact of P450 2e1 activity (which includes variation in isoform) at blood concentrations of trichloroethylene expected from environmental exposures (Lipscomb *et al.* 2003) and the minor differences in kinetic parameters of the different genotypes of this enzyme (Hu *et al.* 1997; LeMarchand *et al.* 1999). The main difference in these isoforms of P450 2e1 is the levels of expression (i.e., measurement of actual activity) as was done in Backer *et al.* (2008), is much better than genotyping. Slightly slower elimination of THMs was found in individuals with the null GSTT1 genotype following showering for human volunteers that possess this genotype (Backer *et al.* 2008). This latter result is difficult to rationalize because P450 2e1 will compete for virtually all of the THMs at the low blood concentrations of THMs shown to arise from exposure to chlorinated drinking water.

The K_m of GSTZ1 for DCAA is also very high relative to the plasma concentrations that result from exposures in the range expected from drinking water. Among the variant forms of the enzyme, the K_m values vary by only five-fold (Board *et al.* 2001). The maximal activity of this enzyme is very high, accounting for a very short half-time of low concentrations of DCAA in humans and experimental animals (approximately 10–20 min.). However, DCAA forms a suicide metabolite with certain variants of this enzyme at (Tzeng *et al.* 2000; Saghir and Schultz, 2002) and at high doses (e.g. above 10 mg/kg body weight) its half-life is significantly prolonged for this reason. This is in the range of the high doses of DCAA that are needed to produce liver cancer in rodents.

A more interesting aspect of GSTZ1 is that it is involved in tyrosine metabolism (referred to as maleylacetoacetate isomerase) (Lim *et al.* 2004). Deficiency in the enzyme produced in knock-out mice renders the animals sensitive to phenylalanine toxicity that parallels, in part, some of the characteristics of phenylketonuria. However, no null form of this enzyme has been identified in humans or animals. Therefore, none of the identified variants would limit the metabolism of dihaloacetic acids at blood levels that would result from consuming chlorinated drinking water. Similarly, the enzyme is present in great excess and none of the isoforms appear to limit phenylalanine metabolism.

10.5 A BROADER VIEW OF SOURCES OF CANCER RISK

The above technical discussion of the evidence of cancer risk from DBPs is mainly aimed at making it clear that attributing the cause of cancer(s) in humans that have been epidemiologically associated with drinking water disinfectants to specific DBPs is far from established. Additional technical discussion could ensue, but may well tax the interest of the reader. Therefore, it will be interrupted by a short discussion of why it is important to establish causality.

First, it should be noted that this chapter does not call into question whether chlorinated drinking water carries a cancer risk. Consideration of the combination of epidemiological and toxicological data, offers substantive evidence that chlorinated water does result in an increased risk of cancer. The degree of risk suggested by the epidemiological data is relatively large when compared to the risk levels that are commonly considered acceptable for contaminants of drinking water (usually set at a maximum of 10^{-4} lifetime risk). The main question is whether currently regulated DBPs can account for the risks identified in epidemiological studies. They may, but it will require evidence of unusually high human sensitivity and target organ specificity to arrive at that conclusion. So far data supporting these hypotheses have not been forthcoming (nor are they being pursued). Some apparently believe the association with enzymes involved in the metabolism of these compounds makes that case. This chapter has focused on why that argument is far from conclusive.

Second, one rational argument that could be brought up in this discussion is that the actual cause of cancer is unimportant because the epidemiological data has fairly consistently associated chlorine use with risk of bladder cancer. Therefore chlorine should simply not be used. Most researchers and public health authorities appear to believe, however, that continued use of chlorine is important from a public health perspective. In part, that is because alternative methods of providing this public health protection are more difficult to use (particularly on a small scale), involve more complicated processes, and are more expensive. The key question is whether there are not modifications of treatment that would control this risk, but still allow chlorine to be used as a disinfectant. To pursue this goal, understanding causality becomes critical.

Third, the carcinogenic risks from disinfection of drinking water have not been truly identified. The use of case control methodology provides information only for the site selected for study. Other ecological designs do address a greater variety of cancer, but these studies have limited specificity and as a result have not yielded consistent results. Given case control methodology is the most practical design for establishing causality for a specific cancer outcome, the lack of a specific cancer risk in situations where other disinfectants (e.g. chloramination or ozonation) are used does not provide evidence that risk of other cancers are not increased. Furthermore, the lack of sensitivity of epidemiological approaches cannot establish whether the cancer risks suggested for other target organs in animal studies can be ruled out as human risks.

Fourth, it is not clear that alternative forms of disinfection produce less health risk. All the chemical means of disinfection produce DBPs. Moreover, some of the alternative methods continue to employ chlorine or chloramine as part of the process to maintain residual disinfectant into the distribution system. No alternative to chlorine has been studied sufficiently to suggest their use would reduce overall cancer risk relative to chlorine.

Fifth, there are two aspects of the problem. One aspect is the target organ, the other carcinogenic potency. While it is a legitimate argument, the target organ specificity appears to be stronger than it perhaps is. Target organ differences between species are not unusual for the same chemical. Nevertheless, scientifically it is necessary to explain why bladder cancer is not found in the cancer studies of the regulated DBPs in animals to come to a conclusion. This is best accomplished through pursuit of comparative animal and human research on mechanisms of carcinogenesis, not the conduct of more descriptive studies with the hope of identifying an animal which would respond in the bladder. The issue of cancer potency is actually the critical issue. Given the weakness of both descriptive epidemiological and toxicological studies, neither descriptive approach can address the low dose cancer risk issue (i.e. 10^{-3} lifetime risk is about the lower limit of sensitivity of both methodologies).

Sixth, the viable alternative is hypothesis driven-research, not simply more descriptive work. Screening has its place, but that should not be the major thrust of this research. This means that research should focus on the prediction of what DBPs are likely to form, consideration of how these by-products are related to other toxicants that have been identified and perform the focused research to test those hypotheses. Most hypotheses should be addressable in short-term studies. Some aspects of the work can be explored with *in vitro* systems that are specifically designed for the purpose of establishing mechanisms and modes of action. It should involve research on human as well as animal tissues, but it will be absolutely necessary to understand whole body metabolism and pharmacokinetics of novel DBPs as well as some of the established ones which have been poorly studied.

10.6 PRACTICAL IMPLICATIONS

10.6.1 Impact of misidentifying causal linkages

What is the consequence of misidentifying causal linkages in the case of DBPs? The first and most important consequence from simply assuming that some single group of DBPs serves as a surrogate for risk from chlorinated water is an assumption that is unlikely to be confirmed, but worse, does not ask to be confirmed. It should be clear that discovering the increasing variety of DBPs that are significantly more potent as carcinogens than the THMs makes the assumption that controlling THMs will control cancer risk from disinfected water untenable.

Concern over the poor performance of THMs as surrogates for DBPs that are of health concern is exacerbated by the finding that they are not even good surrogates for chlorination by-products (Bull *et al.* 2009a). Very poor correlations were observed between major and minor halogenated by-products. This should be expected because it has been known for 30 years that the nature of by-products are affected by many factors, including source water characteristics, geographical differences in natural organic matter (NOM), the pH of water as it is being disinfected, and other treatment processes that are being employed (Bull *et al.* 2009b). Now it has become popular to use multiple disinfectants. The work that explores the impact of such changes is usually restricted to the occurrence and concentrations of regulated DBPs and how these modifications of process affect the mixture of DBPs that is produced, using broader measures of DBP occurrence and concentrations.

Preliminary data (X. F. Li, personal communication) indicate that chlorinated water contains substantially higher HBQ concentrations that chloraminated water. It appears that chloramine is much less able to oxidize phenol precursors to benzoquinones than chlorine. Other conditions that affect this reaction are being explored by the research team. These findings are of interest because in a case control study of two water systems that utilized the same source water, but used either chlorine or chloramine for disinfection found that the chloraminated system did not have an increased risk of bladder cancer (Zierler *et al.* 1988). An earlier study of water systems in Colorado supports the view that chloraminated systems carry less risk of bladder cancer risk than chlorinated systems (McGeehin *et al.* 1993). Coupled with the parallels in the metabolic processes that are involved with the metabolism of HBQs with that of aromatic amines, a circumstantial case can be developed that HBQs could be the bladder carcinogens. Therefore, we see one hypothesis that deserves to be tested.

If study of the toxicological properties of HBQs confirm that HBQs could indeed be human carcinogens, these preliminary data suggest that they would be very easily controlled by switching to chloramine and perhaps other less drastic changes in disinfection. This illustrates that there could be great benefit from identifying the agents in chlorinated drinking water that are responsible for elevated human risks from bladder cancer.

10.6.2 Regulations

The issue of drinking water standards is not the focus of the present chapter, but the topics discussed have clear implications for regulation. First, it is important to regulate compounds that pose toxic hazards. The THMs are toxic chemicals, but most concerns are well within hand when THM concentrations are kept below current drinking water standards. The issue is not whether they should be regulated, but whether they should be used as surrogates for other chlorination by-products. The use of surrogate measures is a popular one in the drinking water community because if surrogates are accepted by regulators they can save considerable analytical costs. However, in the water industry establishment of surrogates is not as rigorous as it is in most fields. Surrogates in biomedical fields are used only after it has been demonstrated that it is a valid surrogate to the point of establishing their statistical reliability with some rigor. This has not been done for the THMs.

Previous studies have shown that neither the THMs nor HAAs are reliable surrogates for chlorination by-products present at lower concentrations (Bull *et al.* 2009a). The case of HBQs (Chapter 8) suggests what the consequences could be if we continue to act on the assumption that THMs and HAAs are reliable surrogates for risks associated with chlorinated drinking water. Moreover, if the HBQs are ultimately determined to be important bladder carcinogens in chlorinated drinking water, one remedy could be to maximize THMs at the expense of the HBQs by simply increasing the pH of the water as it is being disinfected. The information summarized suggest that one might use chloramine to minimize formation of the HBQs as well as the THMs to reduce risk of bladder cancer. However, other by-products arise from chloramine.

Therefore, drinking water standards should concentrate on DBPs of known concern and, if it can be shown to be valid, to surrogates of other DBPs. Clearly, taking one class to represent the totality of chlorinated DBPs is not only invalid, it could have the opposite effect that is intended in not protecting the public health. Other approaches have been suggested, such as monitoring single members of DBP classes and utilize the distribution of chlorine and bromine in the THMs define the speciation in the class, (Bull *et al.* 2009a), but are not discussed in this chapter for the sake of brevity.

10.7 SUMMARY AND CONCLUSIONS

Despite the lengthy discussion, the conclusions to be drawn from it are pretty straight-forward as it relates to water treatment practice.

(1) It is very unlikely that THMs contribute to bladder cancer.
(2) Many DBPs that have been characterized toxicologically are greater cancer risks than the THMs.
(3) A recently identified DBP class (HBQs, Chapter 8) would appear to be more probable causes of bladder cancer than established DBPs.
(4) Enzyme variants being more strongly associated with bladder cancer is consistent with a chemical exposure-based etiology for the observed bladder cancer risk, but they do not provide substantive evidence that THMs are responsible. A wide variety of other DBPs are metabolized by these same enzymes.
(5) Research to clarify risks associated with DBPs must focus on establishing causal relationships and less time in generating descriptive data in animals and humans.
(6) Although not specifically discussed in this chapter, research identifying causal relationships between DBP exposure and adverse health effects must be based on newer methodologies that allow identification of modes of action. The application of these methods should not to be confused with screening methods. Rather tools used should be as direct a means of pursuing what the causal agents are as possible.

Issues as complex as DBPs and cancer risk have to be approached in a more scientific way than they have been in the recent past. This is a responsibility that rests largely on regulatory bodies and their research arms. This chapter serves to identify issues that have arisen from incomplete utilization of the biomedical literature to address hazards associated with DBPs. The objective should not simply be to find a way to set a standard, but needs to solve the problem. The uncertainty that still exists within this area 37 years after THMs were discovered is inexcusable. It reflects the focus on regulations rather than protecting the public health.

REFERENCES

Austin E. W. and Bull R. J. (1997). Effect of pretreatment with dichloroacetate or trichloroacetate on the metabolism of bromodichloromethane. *J. Toxicol. Environ. Health*, **52**, 367–383.

Backer L. C., Lan O., Blount B. C., Nuckols J. R., Branch R., Lyu C. W., Kieszak S. M., Brinkman M. C., Gordon S. M., Glanders W. D., Romkas M. and Cantor K. P. (2008). Exogenous and endogenous determinants of blood trihalomethane levels after showering. *Environ. Health Persp.*, **116**(1), 57–63.

Bellar T. A., Lichtenberg J. J. and Kroner E. C. (1974). The occurrence of organohalides in chlorinated drinking water. *J. AWWA*, **66**, 703–706.

Board P. G. and Anders M. W. (2011). Glutathione transferase zeta: discovery, polymorphic variant catalysis, inactivation, and properties of Gstz1-/- mice. *Drug Metab. Rev.*, **43**(2), 215–225.

Board P. G., Chelvanayagam G., Jermin L. S., Tzeng H.-F., Anders M. W. and Blackburn A. C. (2001). Identification of novel glutathione transferases and polymorphic variants by expressed sequence tag database analysis. *Drug Metab. Dispos.*, **29**(4), 544–547.

Bolton J. L., Trush M. A., Penning T. M., Dryhurst G. and Monks T. J. (2000). Role of quinones in toxicology. *Chem. Res. Toxicol.*, **13**, 135–159.

Bull R. J. (2000). Mode of action of liver tumor induction by trichloroethylene and its metabolites, trichloroacetate and dichloroacetate. *Environ. Health Perspec.*, **69**, 49–58.

Bull R. J., Meier J. R., Robinson M., Ringhand H. P., Laurie R. D. and Stober J. A. (1985a). Evaluation of mutagenic and carcinogenic properties of brominated and chlorinated acetonitriles; by-products of chlorination. *Fundam. Appl. Toxicol.*, **5**, 1065–1074.

Bull R. J., Reckhow D. A., Li X., Humpage A. R., Joll C. and Hrudey S. E. (2011). Potential carcinogenic hazards of non-regulated disinfection by-products: Haloquinones, halo-cyclopentene and cyclohexene derivatives, *N*-halamines, halonitriles, and heterocyclic amines. *Toxicology*, **286**, 1–19.

Bull R. J., Reckhow D. A., Rotello V., Bull O. M., and Kim J. (2006). Use of Toxicological and Chemical Models to Prioritize DBP Research. Water Research Foundation, Denver, CO.

Bull R. J., Rice G. and Teuschler L. K. (2009b). Determinants of whether or not mixtures of disinfection by-products are similar. *J. Toxicol. Environ. Health A*, **72**, 437–460.

Bull R. J., Rice G., Teuschler L. and Feder P. (2009a). Chemical measures of similarity among disinfection by-product mixtures. *J. Toxicol. Environ. Health. A*, **72**, 482–493.

Bull R. J. and Robinson M. (1985b). Carcinogenic activity of haloacetonitrile and haloacetone derivatives in the mouse skin and lung. In: Water Chemistry, Environmental Impact, and Health Effects, R. L. Jolley, R. J. Bull, W. P. Davis, S. Katz, M. H. Roberts, Jr. and V. A. Jacobs (eds.), Vol. 5, Lewis Publishers Inc., Chelsea, MI, pp. 221–236.

Cantor K. P., Villanueva C. M., Silverman D. T., Figueroa J. D., Real F. X., Garcia-Closas M., Malats N., Chanock S., Yeager M., Tardon A., Garcia-Closas R., Serra C., Carrato A., Castano-Vinyals G., Samanic C., Rothman N. and Kogevinas M. (2010). Polymorphisms in GSTT1, GSTZ1, and CYP2E1, disinfection byproducts, and risk of bladder cancer in Spain. *Environ. Health Persp.*, **118**, 1545–1550.

Chen H., Sandler D. P., Taylor J. A., Shore D. L., Liu E., Bloomfield D.C. and Bell D. H. (1996). Increased risk for myelodysplastic syndromes in individuals with glutathione transferase theta 1 (GSTT1) gene defect. *Lancet*, **347**, 295–297.

Daniel F. B., DeAngelo A. B., Stober J. A., Olson G. R. and Page N. P. (1992). Hepatocarcinogenicity of chloral hydrate, 2-chloroacetaldehyde, and dichloroacetic acid in the male B6C3F1 mouse. *Fundam. Appl. Toxicol.*, **19**(2), 159–168.

Dix D. J., Houck K. A., Martin M. T., Richard A. M., Setzer R. W. and Kavlock R. J. (2007). The ToxCast Program for prioritizing toxicity testing of environmental chemicals. *Toxicol. Sci.*, **96**(1), 5–12.

Fisher A. A., Labenski M. T., Malladi S., Chapman J. D., Bratton S. B., Monks T. J. and Lau S. S. (2011). The frequency of 1,4-benzoquinone-lysine adducts in cytochrome c correlate with defects in apoptosome activation. *Toxicol. Sci.*, **122**(1), 64–72.

Gold L. S., Ames B. N., Bernstein L., Blumenthal M., Chow K., Da Costa M., de Veciana M., Eisenberg S., Garfinkel G. B., Haggin T., Havender W. R., Kim Hooper N., Levinson R., Lopipero P., Magaw R., Manley N. B., MacLeod P. M., Peto R., Pike M. C., Rohrbach L., Sawyer C. B., Slone T. H., Smith M., Stern B. R. and Wong M. (2011). Carcinogenicity Potency Database (CPDB)[online] http://potency.berkeley.edu/pub/tables/chemical summary [cited July 11,2011].

Hammond A. H. and Fry J. R. (1997). Involvement of cytochrome P4502E1 in the toxicity of dichloropropanol to rat hepatocyte cultures. *Toxicology*, **118**(23), 171–179.

Heiskanen K., Linderstöm-Seppa P., Haataja L., Vaittinen S.-L., Vartiainen T. and Komulainen H. (1995). Altered enzyme activities of xenobiotic biotransformation in kidneys after subchronic administration of 3-chloro-4-(dichloromethyl)-5-hydroxy-2 (5H)-furanone (MX) to rats. *Toxicology*, **100**, 121–128.

Hu Y., Oscarson M., Johansson I., Yue Q. Y., Dahl M.-L., Tabone M., Arinco S., Albano E. and Ingelman-Sundberg M. (1997). Genetic polymorphism of human CYP2E1: characterization of two variant alleles. *Mol. Pharmacol.*, **51**, 370–376.

Hu G., Stevens D. K. and Bull R. J. (1995). Metabolism of bromodichloroacetate in B6C3F1 mice. *Drug Metab. Dispos.*, **23**(12), 1412–1416.

IARC (1991). Monographs on the Evaluation of Carcinogenic Risks to Humans. Chlorinated Drinking Water; Chlorination By–Products; some other Halogenated Compounds; Cobalt and Cobalt Compounds. International Agency for Research on Cancer, Lyon, France, pp. 45–296.

IARC (2004). Some Drinking Water Disinfectants and Contaminants, including Arsenic. Vol. 84, IARC Press, Lyon, France.

Jolley R. L., Bull R. J., Davis W. P., Katz S., Roberts M. H., Jacobs V. A., Condie L. W., Johnson J. D., Minear R. A. and Mattice J. S. (1978–1990). Water Chlorination: Environmental Impact and Health Effects. Volumes, 1–4, Ann Arbor Science, Ann Arbor, MI, Volumes 5–6, Lewis Publishers, Chelsea. MI.

Jorgenson T. A., Meierhenry E. F., Rushbrook C. J., Bull R. J. and Robinson M. (1985). Carcinogenicity of chloroform in drinking water to male Osborne-Mendel rats and female B6C3F1 mice. *Fund. Appl. Toxicol.*, **5**, 760–769.

Knudsen T. B., Martin M. T., Kavlokc R. J. and Judson R. S. (2009). Profiling the activity of environmental chemicals in prenatal developmental toxicity testing usng the U.S. EPA's ToxRefDB. *Reprod. Toxicol.*, **28**, 209–219.

Krasner S. W., Weinberg H. S., Richardson S. D., Pastor S. J., Chinn R., Sclimenti R. J., Onstad G. D. and Thruston A. D. Jr. (2006). Occurrence of a new generation of disinfection by-products. *Environ. Sci. Technol.*, **40**(23), 7175–1185.

Larson J. L., Sprankle C. S. and Butterworth B. E. (1994b). Lack of chloroform-induced DNA repair in vitro and in vivo in hepatocytes of female B6C3F1 mice. *Environ. Mol. Mutagen.*, **23**, 132–136.

Larson J. L., Templin M. V., Wolf D. C., Jamison K. C., Leininger J. R., Mery S., Morgan K. T., Wong B. A., Conolly R. B. and Butterworth B. E. (1996). A 90-day chloroform inhalation study in female and male B6C3F1 mice: implications for risk assessment. *Fundam. Appl. Toxicol.*, **30**, 118–137.

Larson J. L., Wolf D. C. and Butterworth B. E. (1994a). Induced cytotoxicity and cell proliferation in the hepatocarcinogenicity of chloroform in female B6C3F1 mice: comparision of administration by gavage in corn oil vs. ad libitum in drinking water. *Fundam. Appl. Toxicol.*, **22**, 90–102.

Larson J. L., Wolf D. C. and Butterworth B. E. (1995a). Induced regenerative cell proliferation in livers and kidneys of male F3-44 rats given chloroform in corn oil by gavage or ad libitum in drinking water. *Toxicology*, **95**, 73–86.

Larson J. L., Wolf D. C., Mery S., Morgan K. T. and Butterworth B. E. (1995b). Toxicity and cell proliferation in the liver, kidneys and nasal passages of female F-344 rats induced by chloroform administered by gavage. *Food. Chem. Tox.*, **33**, 443–456.

Le Marchand L., Wilkinson G. R. and Wilkens L. R. (1999). Genetic and dietary predictors of CYP2E1 activity: a phenotyping study in Hawaii Japanese using chlorzoxazone. *Cancer Epidemiol. Biomarkers Prevention*, **8**, 495–500.

Leavens T. L., Blount B. C., DeMarini D. M., Madden M. C., Valentine J. L., Case M. W., Silva L. K., Warren S. H., Hanley N. M. and Pegram R. A. (2007). Disposition of bromodichloromethane in humans following oral and dermal exposure. *Toxicol. Sci.*, **99**(2), 432–445.

Li X-.F. (personal communication). Occurrence and formation of chloro- and bromo-benzoquinones during drinking water disinfection. In review.

Li X-.F., Hrudey S. E., Bull R. J., Reckhow D. A., Humpage A., Joll C. and Heitz A. (2011). Analytical Methods for Predicted DBPs of Probable Toxicological Significance. Water Research Foundation, Denver, CO.

Lilly P. D., Andersen M. E., Ross T. M. and Pegram R. A. (1997). Physiologically based estimation of *in vivo* rates of bromodichloromethane metabolism. *Toxicology*, **124**, 141–152.

Lim C. E. L., Matthael K. I., Blackburn A. C., Davis R. P., Dahlstrom J. E., Koina M. E., Anders M. W. and Board P. G. (2004). Mice deficient in glutathione transferase zeta/maleylacetoacetate isomerase exhibit a range of pathological changes and elevated expression of alpha, Mu, and Pi class glutathione transferase. *Am. J. Pathol.*, **165**(2), 679–693.

Lin E. L. C. and Guion C. W. (1989). Interaction of haloacetonitriles with glutathione and glutathione-*S*-transferase. *Biochem. Pharmacol.*, **38**(4), 685–688.

Lin P.-H., Waidyanatha S., Pollack G. M., Swenberg J. A. and Rappoport S. M. (1999). Dose-specific production of chlorinated quinone and semiquinone adducts in rodent livers following administration of pentachlorophenol. *Toxicol. Sci.*, **47**, 126–133.

Linghor M. K., Bull R. J., Kato-Weinstein J. and Thrall B. D. (2002). Dichloroacetate stimulates glycogen accumulation in primary hepatocytes through an insulin-independent mechanism. *Toxicol. Sci.*, **68**, 508–515.

Linghor M. K., Thrall B. D. and Bull R. J. (2001). Effects of dichloroacetate (DCA) on serum insulin levels and insulin-controlled signaling proteins in livers of male B6C3F1 mice. *Toxicol. Sci.*, **59**, 178–184.

Lipscomb J. C., Teuschler L. K., Swartout J., Popken D., Cox T. and Kedderis G. L. (2003). The impact of cytochrome P450 2E1-dependent metabolic variance on a risk-relevant pharmacokinetic outcome in humans. *Risk Anal.*, **23**(6), 1221–1238.

Martin M. T., Mendez E., Corum D. G., Jusdson R. S., Kavlock R. J., Rotroff D. M. and Dix D. J. (2009). Profiling the reproductive toxicity of chemicals from multigeneration studies in the toxicity reference database. *Toxicol. Sci.*, **110**(1), 181–190.

McGeehin M. A., Reif J. S., Becher J. C. and Mangione E. J. (1993). Case-control study of bladder cancer and water disinfection methods in Colorado. *Am. J. Epidemiol.*, **138**, 492–501.

Merdink J. L., Bull R. J. and Schultz I. R. (2000). Trapping and identification of the dichloroacetate radical from the reductive dehalogenation of trichloroacetate by mouse and rat liver microsomes. *Free Rad. Biol. & Med.*, **29**(2), 125–130.

Merdink J. L., Bull R. J. and Schultz I. R. (2001). Toxicokinetics of bromodichloroacetate in B6C3F1 mice. *J. Appl. Toxicol.*, **21**, 53–57.

Miller J. H., Minard K., Wind R. A., Orner G. A., Sasser L. B. and Bull R. J. (2000). *In vivo* MRI measurements of tumor growth induced by dichloroacetate; implications for mode of action. *Toxicology*, **145**, 115–125.

NCI (1976). Report on the carcinogenesis bioassay of chloroform. NTIS PB-264-018, National Cancer Institute, Bethesda, MD.

NCI (1978). Bioassay of iodoform for possible carcinogenicity, Technical Report Series No. NCI-CG-TR-110, National Cancer Institute, Bethesda, MD.

NTP (1985). Toxicology and Carcinogenesis Studies of Chlorodibromomethane (CAS No. 124-48-1) in F344/N Rats and B6C3F1Mice. Technical Report Series No.TR 282, National Toxicology Program, Research Triangle Park, NC.

NTP (1986). Toxicology and carcinogenesis studies of bromodichloromethane in F344/N rats and B6C3F1 mice, Technical Report Series No. TRS No. 321, NIH No. 88-2537, CAS. NO. 75-27-4, National Toxicology Program, Research Triangle Park, NC.

NTP (1989). Toxicology and carcinogenesis studies of tribromomethane (bromoform) in F344/N rats and B6C3F1 mice, Technical Report Series No. 350, NIH No. 89-2805, CAS NO. 75-25-2, National Toxicology Program, Research Triangle Park, NC.

NTP (2006). Toxicology and Carcinogenesis Studies of Bromodichloromethane (CAS No. 75-27-4) in Male F344/N Rats and Female B6C3F1 Mice (Drinking Water Studies). TR-532 [Online]. Available: <http://ntp.niehs.nih.gov/index.cfm?objectid=00271EF 5-F1F6-975E-73E6FE7AEE1A1A31> 2005b [cited June 29, 2011].

NTP (2007). Toxicology and Carcinogenesis Studies of Dibromoacetic acid (CAS No. 631-64-1) in F344/N Rats and B6C3F1 mice (Drinking Water Studies). TR537 [Online] http://ntp.niehs.nih.gov/?objectid=8831333E-F1F6-975E-71D4F287C2229308. [cited 7/7/2011]

NTP (2009). Toxicology and Carcinogenesis Studies of Bromochloroacetic acid (CAS No. 5589-96-8) in F344/N Rats and B6C3F1 mice (Drinking Water Studies). TR-549 [Online] http://ntp.niehs.nih.gov/?objectid=82F002240F1F6-975E-7COA4D903E1262D [cited 7/7/2011]

NTP (2010). Toxicology and carcinogenesis studies of dibromoacetonitrile (CAS No. 3252-43-5) in F344/N rats and B6C3F1 mice (drinking water studies). National Toxicology Program Technical Report TR-544. NIEHS; National Toxicology Program, Research Triangle Park, NC.

OEHHA (2011). Public Health Goals. Available on line: http://oehha.ca.gov/water/phg/allphgs.html (accessed 15 August 2011).

Onstad G. D., Weinberg H. S. and Krasner S. W. (2008). Occurrence of halogenated furanones in U.S. drinking waters. *Environ. Sci. Technol.*, **42**, 3341–3348.

Pegram R. A., Andersen M. E., Warren S. H., Ross T. M. and Claxton L. D. (1997). Glutathione *S*-transferase-mediated mutagenicity of trihalomethanes n Salmonella typhimurium: contrasting results with bromodichloromethane and chloroform. *Toxicol. Appl. Pharmacol.*, **144**(1), 183–188.

Porubsky P. R., Meneely K. M. and Scott E. E. (2008). Structures of human cytochrome P-450 2E1. Insights into the binding of inhibitors and both small molecular weight and fatty acid substrates. *J. Biol. Chem.*, **263**(48), 33698–33707.

Razdan R., Frei E., Spiegelhalder B. and Siddiqi M. (1994). Caffeine-derived *N*-nitroso compounds IV: kinetics of mononitrosocaffeidine demethylation by rat liver microsomes. *Cancer Lett.*, **79**, 117–122.

Richardson S. D., Plewa M. J., Wagner E. D., Schoeny R. and DeMarini D. M. (2007). Occurrence, genotoxicity, and carcinogenicity of regulated and emerging disinfection by-products in drinking water: a review and roadmap for research. *Mutat. Res.*, **636**, 178–242.

Robinson M., Bull R. J., Olson G. R. and Stober J. (1989). Carcinogenic activity associated with halogenated acetones and acroleins in the mouse skin assay. *Cancer Lett.*, **48**, 197–203.

Rook J. J. (1974). Formation of haloforms during chlorination of natural waters. *Water Treat Exam.*, **23**, 234–245.

Ross M. K. and Pegram R. A. (2003). Glutathione transferase theta 1-1-dependent metabolism of the water disinfection by-product bromodichloromethane. *Chem. Res. Toxicol.*, **16**, 216–226.

Ross M. K. and Pegram R. A. (2004). *In vitro* biotransformation and genotoxicity of the drinking water disinfection by-product: DNA-binding mediated by glutathione transferase theta 1-1. *Toxicol. Appl. Pharmacol.*, **195**, 166–181.

Saghir S. A., Ghanayem B. I. and Schultz I. R. (2011). Kinetics of trihalogenated acetic acid metabolism and isoform specificity in liver microsomes. *Internat J. Toxicol.*, Article in press, **5**, 551–561, DOI: 10.1177/1091581811414213.

Schultz I. R. and Shangraw R. E. (2006). Effect of short-term drinking water exposure to dichloroacetate on its pharmacokinetics and oral bioavailability in human volunteers. A stable isotope study. *Toxicol. Sci.*, **02**(1), 42–50.

Schultz I. R. and Sylvester S. R. (2001). Stereospecific toxicokinetics of bromochloro- and chlorofluoroacetate: effect of GST-ζ depletion. *Toxicol. Appl. Pharmacol.*, **175**, 104–113.

Shimada T., Yamazaki H., Oda Y., Hiratsuka A., Watable T. and Guengerich F. P. (1996). Activation and inactivation of carcinogenic dihaloalkanes and other compounds by glutathione transferase 5-5 in Salmonella tester strain NM5004. *Chem. Res. Toxicol.*, **9**(1), 333–340.

Shokeer A. and Mannervik B. (2010). Residue 234 is a master switch of the alternative-substrate activity profile of human and rodent theta class glutathione transferase T1-1. *Bichemica Biophys. Acta*, **1800**(4), 466–473.

Snyder R. D., Pullman J., Carter J. H., Carter H. W. and DeAngelo A. B. (1995). *In vivo* administration of dichloroacetic acid suppresses spontaneous apoptosis in murine hepatocytes. *Cancer Res.*, **55**, 3702–3705.

Stauber A. J. and Bull R. J. (1997). Differences in phenotype and cell replication behavior of hepatic tumors induced by dichloroacetate (DCA) and trichloroacetate (TCA). *Toxicol. Appl. Pharmacol.*, **144**, 235–246.

Stauber A. J., Bull R. J. and Thrall B. D. (1998). Dichloroacetate and trichloroacetate promote clonal expansion of anchorage-independent hepatocytes, *in vivo* and in vitro. *Toxicol. Appl. Pharmacol.*, **150**, 287–294.

Thier R., Pemble S. E., Kramer H., Taylor J. B., Guengerich F. P. and Ketterer B. (1996). Human glutathione S-transferase T1-1 enhances mutagenicity of 1,2-dibromoethane, dibromomethane and 1,2,3,4-diepoxybutane in Salmonella typhimurium. *Carcinogenesis*, **17**(1), 163–166.

Tong Z., Board P. G. and Anders M. W. (1998a). Glutathione transferase zeta-catalyzed biotransformation of dichloroacetic acid and other α-haloacids. *Chem. Res. Toxicol.*, **11**, 1332–1338.

Tzeng H.-F., Blackburn A. C., Board P. G. and Anders M. W. (2000). Polymorphism- and species-dependent inactivation of glutathione transferase by dichloroacetate. *Chem. Res. Toxicol.*, **13**, 231–236.

U.S. EPA (1979). Federal Register, November 1979 (44 FR 68624).

U.S. EPA (1999). Chloroform Mode of Action Analysis. [Online]. http://www.epa.gov/sab/chloro00.htm 1999 [cited January 3, 2002 1999].

U.S. EPA (2003a). National Primary Drinking Water Regulations: Stage 2 Disinfectants and Disinfection Byproducts Rule; National Primary and Secondary Drinking Water Regulations: Approval of Analytical Methods for Chemical Contaminants; Proposed Rule. Parts 141, 142, and 143, 68, 49548–49681.

U.S. EPA (2003b). Health Risks to Fetuses, Infants, and Children: Proposed Stage 2 Disinfectant/Disinfection Byproducts: A review. EPA-822-R-03-010.

U.S. EPA (2005). Guidelines for Cancer Risk Assessment. EPA/630/P–03/001f. U.S. Environmental Protection Agency, Washington, DC, March 2005.

Valentine R. L., Choi J., Chen Z., Barrett S. E., Hwang C., Guo Y., Wehner M., Fitzsimmons S., Andrews S. A., Werker A. G. Brubacher C. and Kohut K. (2006). Factors Affecting the Formation of NDMA in Water and Occurrence. Awwa Research Foundation, Denver, CO.

Wan J., Shi J., Hui L., Wu D., Jin X., Zhao N. Huang W., Xia Z. and Hu G. (2002). Association of genetic polymorphisms in CYP2E!, MPO, NQQ!, GSTM1 and GSTT1 genes with benzene poisoning. *Environ. Health Perspect.*, **110**(12), 1213–1218.

Wang T., Arifoglu P., Ronal Z. and Tew K. D. (2001). Glutathione *S*-transferase P1-1 (GSTP1-1) inhibits c-Jun-*N*-terminal kinase (JNK-1) signaling through interaction with the C-terminus. *J. Biol. Chem.*, **216**(24), 20999–21003.

Yamazaki H., Inui Y., Yun C.-H., Guengerich F. P. and Shimada T. (1992). Cytochrome P450 2E! and 2A6 enzymes as major catalysts for metabolic activation of *N*-nitrosodialkylamines and tobacco-related nitrosamines in human liver microsomes. *Carcinogenesis*, **13**(10), 1789–1794.

Zierler S., Feingold L., Danley R. A. and Craun G. (1988). Bladder cancer in Massachusetts related to chlorinated and chloraminated drinking water: a case-control study. *Arch. Environ. Health*, **43**, 195–200.

Chapter 11

Epidemiological inference and evidence on DBPs and human health

S. E. Hrudey

11.1 TAXONOMY OF HEALTH RISK EVIDENCE

Science gains evidence about the possible adverse effects of various environmental exposures on human health by a diverse range of methods, each with varying sources and degrees of uncertainty (Thomas and Hrudey, 1997). The extent and nature of that uncertainty is not as widely appreciated as it should be to assure sound understanding of risk trade-offs that arise with issues like DBPs in drinking water. While the protection of public health has been effective, in part, by the exercise of precaution, knowing how cautious to be in any given set of circumstances is always a challenge, whether explicitly recognized or not (Hrudey and Leiss, 2003). In a case of risk trade-offs such as effective disinfection *vs*. formation and exposure to DBPs, over-simplified reliance on precaution will not work.

Before exploring the specifics of epidemiologic evidence regarding DBPs and human health, it is worth considering, first, the taxonomy of evidence about any adverse human health effect, then considering the key features of the epidemiologic method within that evidence hierarchy. Attribution of health effects to environmental causes is inherently predicated, whether explicitly stated or not, on a causal chain model that relates an environmental exposure to a disease outcome. This model, as depicted in Figure 11.1, is an enormous over-simplification of reality because we have ample evidence that any disease process involves a complex web of causal and contributory factors. The subject of our attention in this book, DBPs and human health, is a typical illustration of that over-simplification. We seek to understand whether DBPs (the agent or hazard) via drinking water exposure actually cause a human health outcome such as cancer.

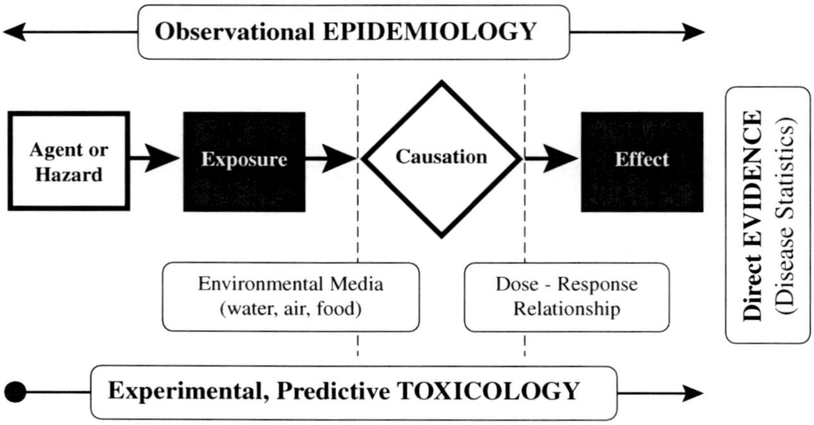

Figure 11.1 Chain of causation in environmental health (adapted from Thomas and Hrudey, 1997)

11.1.1 Evidence and uncertainty of health effects

Our tool box of methods for assessing this question often seems incredibly complex when we view the literature, such as we have outlined in this book. In spite of a complex tool box, the tools within are generally not as 'sharp' as researchers require or as the public might think. The bluntness of many of the tools is the result of many of the challenges outlined below and thus limits what can or can not be concluded from the available health risk evidence. Moreover, the tools available inevitably come back to inferences we draw from some combination of disease statistics, epidemiology studies on human populations and toxicology studies conducted on experimental animals.

How the available forms of evidence inter-relate needs to be considered in terms of the uncertainty which each source of evidence inherently entails. An attempt to consider this taxonomy of evidence and uncertainty is illustrated in Figure 11.2 (Thomas and Hrudey, 1997). This figure depicts our totality of knowledge about health impacts, as illustrated by mortality, given the certainty that death provides as an endpoint. A similar depiction for morbidity would be difficult to illustrate in a manageable illustration.

The foundation of the described barrel of knowledge in Figure 11.2 is the total number of deaths in a given time period, a data set that most countries now maintain as part of their vital statistics. The shaded core of the figure represents what we know with certainty, while the transparent space outside this core depicts our uncertainty at any vertical level in the barrel of knowledge.

There is always some uncertainty about the total number of deaths in any jurisdiction in any time period because of reporting errors, in- and out-migration and so on, but the degree of that uncertainty is trivial by comparison with the higher levels of uncertainty about knowledge where we seek understanding about causes of death.

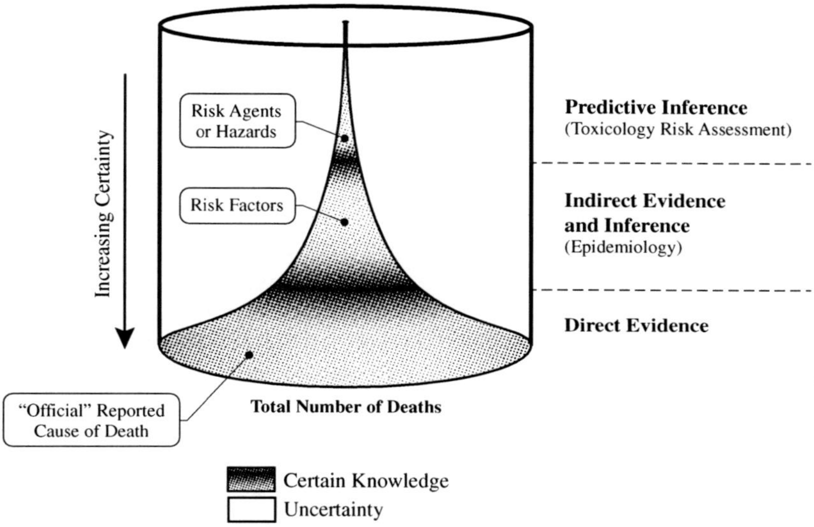

Figure 11.2 Taxonomy of health risk evidence (adapted from Thomas and Hrudey, 1997)

In the lowest portion of the barrel of knowledge, we find the reported cause of death which varies from almost certain at the bottom, up to much more uncertain at the first upper boundary/transition from direct to indirect evidence. This knowledge is intended to reflect the cause of death which is officially reported on the death certificate for any individual fatality. For a teenager who crashes a motorcycle into a brick wall and dies, a cause of death specified as "external – motor vehicle – physical trauma" is reasonably certain to be accurate. At the other extreme, a 99 year old dying in her sleep might be classified as "respiratory disease – other". While all deaths lead to respiratory failure, in the absence of any recorded disease (e.g. cancer, emphysema) under active treatment, the physician who must report the cause of death will make a judgement call with attendant uncertainty because "old age" cannot be cited as the official cause of death.

Important to note is that most developed countries maintain active cancer registries so that individuals diagnosed with cancer are most likely to have that diagnosis assigned as a cause of death for vital statistic purposes. These vital statistics provide valuable guidance about the burden of disease in a population but they do not inform us as to what factors have contributed to cause any specific disease. Unless an autopsy is done, uncertainty about cause of death can be substantial. One study comparing cause of death assignments by 68 physicians, of whom only 6 had received formal training in assigning cause of death, showed correct assignments varied according to case presented from 15% to 99%, but the overall average accuracy was under 60% (Messite and Stellman, 1996). Even when autopsies are done, errors in assignment of cause of death occur. Given the level of uncertainty illustrated for something as deceptively simple as the ultimate cause of death, imagine the level of uncertainty involved in trying to ascertain the underlying causes of a complex disease state like cancer.

11.1.2 Inference and uncertainty

To attribute causation of a specific disease to a risk factor it is necessary to move up the hierarchy into indirect evidence and inference, mostly gained through epidemiology studies. Ultimately for attribution of cause to a toxic agent, it is necessary to reach the top section of predictive inference, mostly gained by toxicology evidence. The boundaries as we move up the hierarchy are intentionally not shown as sharp in Figure 11.2. The quality of disease diagnosis, for example, may be more accurate in a well-designed epidemiology study than is routinely available for cause of death assignment in population vital statistics. However, as will be noted below, epidemiology studies are conducted on subset samples of the population, not the entire population that is covered by death certificates. As a result, any findings of an epidemiology study must be generalized to apply those findings to an entire population, thereby contributing to additional uncertainty and need for application of judgement.

11.1.3 Epidemiology

Epidemiology studies may either start with the disease and work back to find any association with exposures or may start with an exposure and work forward to find any association with disease (i.e. two-headed arrow in Figure 11.1). Either way, studies are ultimately seeking to determine an association between a disease and exposure to a risk factor. For all the reasons that will be elaborated below, this exercise invites uncertainty in the eventual findings. The nature of the findings from epidemiology are inherently more useful for managing or ultimately preventing risks that may be attributed to a specific risk factor (cause) than the direct and more certain evidence found lower down in the hierarchy. Because the objects of an epidemiology study are human, the relevance of the findings from such studies are inherently important for public health, provided the study design and quality of study performance are appropriate to the question being asked. A common criticism relative to the public health preference for precaution is that epidemiological evidence can only be gained after humans have been exposed, thereby limiting the precautionary capability of epidemiology as a means of learning about possible health risks.

11.1.4 Predictive inference (toxicological risk assessment)

The upper-most region of the barrel of knowledge makes it possible to exert the preference for precaution, as well as to gain essential evidence about mode of action (Chapter 10) for possible causative agents. The latter is essential to buttress the reality that epidemiology by itself is limited to statistical inference about association (correlation) between hypothesized cause and observed disease outcome (Angell, 1997). Assignment of causation for epidemiology studies, as we shall see in the next section, requires other evidentiary inputs and associated inference. Evidence regarding individual chemical exposures is commonly developed from controlled experiments with laboratory animals, requiring substantial interpretation for its applicability and relevance to human health. While this predictive inference in the form of toxicological risk assessment offers the advantage of being available before evidence of human exposure occurs, this approach is also subject to even more uncertainty about relevance to human health than the categories below it in the knowledge hierarchy.

Uncertainty in toxicological risk assessment arises for many reasons, but among the most important are the reliance on animals which may exhibit health outcomes different than humans because of differing anatomy and physiology and because the inferences about human health risk being sought are usually about much lower levels of exposure than are typically used in the animal experiments. As a consequence, substantial uncertainty is inevitable with toxicological risk assessment.

11.1.5 Translation of inference into policy

The foregoing challenges make the need for understanding mode of action (Chapter 10) critical, as has been illustrated for DBPs with the protracted saga associated with chloroform and predictions of cancer risk at low dose. In most jurisdictions, public policy specifies that knowledge gaps are bridged with policy assumptions and debates are generally resolved by adopting cautious assumptions. Hence, for example, there may be substantial mode of action evidence available, but until the quantity of evidence becomes compelling, default cautionary assumptions will likely be retained. This was certainly the case with chloroform (Chapter 1 & 10) where it took up to 20 years from the first publication of strong evidence that chloroform was a threshold carcinogen until regulatory policy formally accepted that evidence in some jurisdictions. Chapter 12 will provide some evidence that another decade after that, researchers publishing in the field remain unaware of these findings.

11.2 EPIDEMIOLOGIC METHODS AND EVIDENCE

Ultimately reaching valid conclusions about adverse human health effects relies upon epidemiologic evidence, that is, study of groups of humans. Evidence of individual adverse health outcomes, so-called case reports, may be useful for clinical medicine, but they are typically a result of extreme exposures (i.e. acute poisoning incidents) and lack statistical meaning for generalizing expectations of health risk for an exposed population.

11.2.1 Epidemiologic inference

Epidemiology studies on human populations seek to determine any meaningful associations between exposure to suspected causal agents and adverse health outcomes (a disease). This fundamental comparison of outcome in relation to exposure to a suspected cause requires that both outcome and exposure are known with as much clarity and to the greatest degree of accuracy possible (at least in consistent comparative terms) for every individual who will be studied. The search for evidence of causation is an exercise in determining the correlation of a specific health outcome to an exposure that is the suspected cause of that outcome. The lack of clarity on either exposure or outcome can be seriously problematic (Angell, 1997).

11.2.1.1 Correlation of outcome with exposure

Reduced to basic terms, this search for correlation consistent with causation may be represented by a 2 × 2 table where exposures and outcomes can be dichotomized as shown in Figure 11.3. In the simplest terms, evidence of a positive association between exposure to the suspected cause and the disease arises when individuals in the study population are found to be distributed more commonly in boxes *a* (exposed individuals have the disease) and *d* (unexposed individuals do not have the disease), findings that are consistent with causation, than in boxes *b* (exposed individuals do not have the disease) and *c* (unexposed individuals have the disease), findings that are inconsistent with causation.

	Disease (+)	No Disease (-)
Exposed (+)	*a*	*b*
Not Exposed (-)	*c*	*d*

Figure 11.3 Basic 2 × 2 table for epidemiologic analysis

Exposure to DBPs in drinking water is normally a continuous variable whereby the dose of DBPs is determined by the DBP concentration in water and the volume consumed (or in the case of other potential DBP exposure routes, the quantity inhaled or absorbed across the skin). The continuous dosage variable is often dichotomized for analysis or increasingly commonly is analyzed by logistic regression which accommodates continuous data (Bonita *et al.* 2006). In any case, the underlying premise of seeking an association between exposure and outcome is fundamental to the epidemiologic method.

A critical issue regarding exposure, which will be elaborated in Section 11.2.2 discussing experimental studies *vs.* observational studies, relates to the ability to control, or know with confidence, what is the exposure status of any individual. The distinction is that for an experimental clinical trial to evaluate a surgical intervention or administration of a drug, there is comparatively little uncertainty about the exposure variable (i.e. surgery or no surgery, drug or placebo) compared with observational epidemiology of something like exposure to DBPs in drinking water (i.e. what DBP dose did any individual receive?). Yet, the public is assaulted with almost daily announcements that new clinical trials have found results contrary to earlier ones where ascertainment of exposure was reasonably direct and not a major factor contributing to uncertainty. Imagine the potential for confusing results when we move to observational environmental epidemiology studies.

11.2.1.2 Confounding

Many other factors will also influence health outcomes (age, sex, nutritional status, poverty). As many of these potential confounding factors must be identified and known in as much detail as possible to allow for an assessment of the impact of confounding on observed correlation.

Confounding will arise from a failure to account in the analysis for an association between health outcomes and a suspected causal factor for the possible concurrent exposure to a true causal factor. Smoking status of individuals is a classic concern in cancer studies given the contribution of smoking to many cancer risks. Confounding factors which may modify risk of the outcome may be unevenly distributed among individuals exposed and unexposed to the suspected cause. In experimental studies, randomization is used to reduce this potential source of confounding, but observational studies cannot randomize exposures, they can only seek to understand the potential impact of any underlying differences in confounding factors between those who experience the health outcome versus those who do not.

Well conducted studies will identify as many confounding factors as possible and will mathematically adjust for the effects of identified confounders. A concern which is generally not addressed in this regard is: how accurately and precisely can the mathematical adjustment for confounders be? This concern is particularly grave where the strength of association found between exposure and outcome is small (section 11.2.2) making comparative errors in the adjustment more important to the ultimate conclusion.

11.2.1.3 Bias

Data collection will inevitably be imperfect and may be subject to bias, despite the best intentions to find the truth. There are many sources of potential bias in the collection of epidemiologic evidence which can skew the results and either hide a true association or create a spurious association where one does not truly exist. Finally, regardless of how much care is taken in the gathering of evidence, some random and sampling error is unavoidable. As a result, it is essential to always calculate a confidence interval for any estimate of an association between a suspected cause and health outcomes. To judge the statistical significance of a quantitative measure of association between exposure and outcome, typically, the 95% confidence interval for the measure of association must exclude the null value; the value that would be measured if there was truly no association. Generally, a wider confidence interval is indicative of a less stable estimate of association. All else being equal, a study with a larger sample size will have greater capability of detecting a true association and, thus, should be accorded greater weight when comparing results among various studies.

11.2.1.4 Validity and reliability

Validity and reliability are two key concepts in epidemiology that are defined differently than in the natural sciences and engineering, the primary disciplines of water professionals. Validity in epidemiology refers to whether study results accurately represent the truth, so validity and accuracy have essentially the same meaning. Reliability, as used in epidemiology, refers to the repeatability of results and is close in meaning to precision in the natural sciences. Thus, results that can be closely replicated are considered reliable in epidemiologic studies, but repeatability (precision) does not demonstrate that the results are accurate. This epidemiologic meaning for reliability does differ from the natural sciences where reliability and accuracy may be used interchangeably thereby creating a potential source of misunderstanding if water professionals read that particular epidemiology results are considered "reliable".

Validity must be considered in terms of internal versus external validity. The following discussions about study limitations and sources of error primarily relates to judging whether a study is internally valid, that is, do the results obtained accurately represent the sample group that was studied. While critically important to the judgement of epidemiologic evidence, the broader question about applying results of a study to assessing health risk for an entire population is a matter of external validity, that is, can the results obtained for a study population be generalized to the total population? External validity is often more difficult to judge for any particular epidemiologic study and this critical matter often receives less attention in regulatory policy than it deserves, particularly when the epidemiological evidence base for regulation is limited to a small number of studies with comparatively small sample-sizes in each.

11.2.1.5 Study design for causation

There are several types of epidemiology study designs with varying complexity, rigour, cost and, most important, inherent ability to test causality. A description of the different study designs is presented in Sections 11.2.2 and 11.2.3 to emphasize the importance of the study design to the ability of an epidemiologic study to test for causality in relation to widely accepted criteria for causality.

Of particular importance is that study designs better at testing causality all use exposure and outcome data for *individuals* rather than for groups or populations. The utility of a study in testing causality depends in part on whether individual exposure can be linked to individual outcome. Imagine how much uncertainty is involved when there is great uncertainty about what level of exposure to the agent under study any individual received.

The more accurately the exposure and the outcome can be characterized in each individual, the more useful a study will be in testing a hypothesis of causality. Generally, low cost, weaker study designs will be used to generate hypotheses regarding some environmental exposure causing a human illness. Once clear hypotheses have been formulated, more complex analytical study designs are needed to test a hypothesis of causation. For environmental exposures like DBPs in drinking water, there are practical limits to the type of study design which can be used.

11.2.2 Epidemiologic methods

Epidemiology study designs can be described as experimental or observational (Bonita *et al.* 2006).

Experimental studies have the investigator assign the exposure levels and follow subjects for subsequent changes in health status. Randomized control trials are generally rated the most useful study design for testing causation and these are typically used for studying drugs and medical interventions to test their efficacy. However, this type of study is rarely used in environmental epidemiology because of the ethical problems associated with experimentally exposing subjects to potentially harmful agents and the demanding logistics that would be involved for any study design that could satisfy human ethics requirements.

Observational studies include two categories of study design: analytical studies that investigate a relationship between health status and other variables, and descriptive studies that simply describe the health status of a community based on information already available, usually from public data bases. There are several types of analytical study designs including cohort studies, case-control studies, cross-sectional studies, and aggregated studies. Aggregated studies have been referred to as ecological studies but this label is misleading because these studies do not involve the science of ecology. These have also been described as correlational studies, but because all epidemiologic studies rely on testing correlation between exposure and outcome in some manner, that terminology is not informative.

11.2.2.1 Observational desciptive studies

Aggregated studies use data from whole populations to compare disease patterns between different groups within a population during the same period of time or to compare disease patterns within the same group over several time periods. The units of observation are whole population groups (i.e. aggregated) rather than individuals. Aggregated studies tend to be relatively quick and inexpensive to conduct as the information required is often already available from public records. Aggregated studies are often a first step in investigating a possible relationship between an exposure and a disease. Because aggregated studies use data for a population rather than for individuals, exposure cannot be linked to disease for any particular individual. This limitation can lead to a so-called "ecological fallacy" when inaccurate conclusions are made regarding relationships between exposures and outcomes based on aggregated data from populations. This error arises because population rate data do not allow any determination of whether the individuals who experienced an outcome were also exposed to the suspected cause, that is, the basic logic of Figure 11.3. Aggregated studies are only suitable to propose epidemiologic hypotheses, but are not capable of testing a causal hypothesis.

Cross sectional studies, also called prevalence studies, simultaneously measure disease state and exposure of individuals in a population at a point in time. This type of study provides a "snapshot" of the state of a population with respect to specific exposures and diseases at the time of the study. The major limiting factor of cross-sectional studies is that it is usually unclear whether exposure to a suspected cause has preceded, coincided with or followed the health outcome. This ambiguity in the timing of exposure necessarily limits the capacity of cross-sectional studies to test causal hypotheses.

11.2.2.2 Observational analytical studies

Case control studies involve selecting subjects based on whether they have (cases) or do not have (controls) the health outcome under study. The cases and controls are then compared for the proportion of each group who have the exposure to the suspected cause. Case control studies are comparatively less expensive and take less time to complete relative to cohort studies (described below). They offer a solution to the difficulties of studying health outcomes with long periods for development of disease such as cancer. Case control studies allow an investigation of many potential causal exposures or characteristics for a specific health outcome. However, this option must be exercised with caution to avoid data dredging (i.e. for 95% confidence intervals, 1 in 20 findings will be judged to be statistically significant, but will actually be by chance alone). In fact, Ottenbacher (1998) evaluated 173 articles in the American Journal of Public Health and the American Journal of Epidemiology in 1996 and concluded that the error rate for accepting results to be found statistically significant (i.e., other than by chance alone) was closer to 20%, rather than the 5% commonly assumed for evaluating a 5% significance level.

Case control studies are only practical for relatively common diseases because of the need to assemble an adequate number of cases for analysis. Both the exposure and the disease must have occurred at the start of the study. This reality makes case control studies subject to possible bias on the exposure status for the selection of cases *vs.* selection of controls (selection bias), or differential reporting of past exposure data based on disease status (recall bias). In the latter circumstances, those with a disease (cases) may be more likely to recall exposures to suspected causes than those who are free of the disease (controls). Exposures may be determined from public information (i.e. exposure inferred by location, e.g. drinking water system), employment records or by interview with cases and controls, or by designated responders (e.g. a relative) in the circumstances of participants being deceased. Case control studies may be the only feasible approach for assessing DBP exposures for diseases such as cancer. Although they are subject to numerous serious challenges to avoid bias, they are moderately useful for testing causal hypotheses, provided the findings are thoroughly and effectively challenged for the plausible sources of bias and confounding.

Cohort studies involve selecting a group who do not have the disease at the time of selection. This cohort must be followed over time to determine their exposure status to the suspected cause (exposed or not exposed) as well as their health outcomes. Subjects are then compared based on the proportions of exposed *vs.* non-exposed individuals who develop the outcome of interest subsequent to the exposure. Cohort studies can be retrospective or prospective. In retrospective studies, all exposures and outcomes have occurred at the initiation of the study. Exposure status is determined from a time before the outcomes occurred. Such retrospective cohort studies differ from case control studies in that an overall cohort is chosen with an expectation of covering both exposed and unexposed individuals before information is obtained on health status.

In prospective cohort studies, the exposure may or may not have occurred at the initiation of the study, but the outcomes have certainly not occurred. The exposed and non-exposed subjects are then followed for occurrence of the health outcomes of interest. Because study subjects are free from the disease at the start of the study, the temporal relationship between the exposure and the outcome can be clear. Cohort studies are very time-consuming and expensive. As subjects must be followed for many years after exposure for a disease like cancer, there is the potential for bias caused by differential loss of subjects to follow-up (i.e. any differences in ongoing full participation between those with the adverse health outcome and those without may bias the analysis). Cohort studies are not practical for very rare diseases because the size of the cohort necessary to generate sufficient cases for analysis may be unreasonably large.

The approach of case control versus cohort studies is shown in Figure 11.4 which illustrates the entry points of these two approaches to the process of epidemiologic inference.

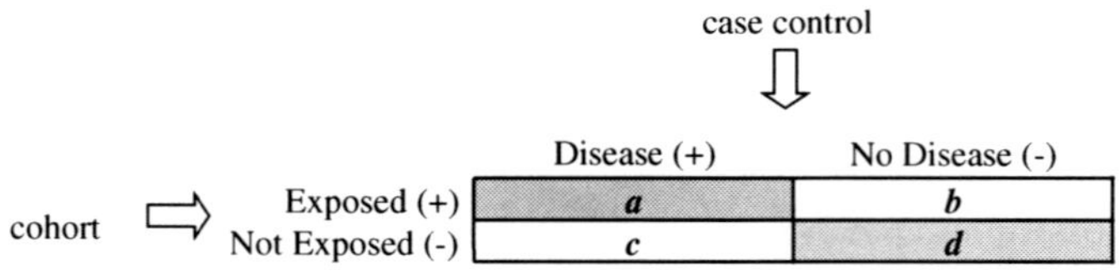

Figure 11.4 Approach of case control and cohort designs to epidemiologic inference

11.2.2.3 Experimental studies

Randomized double-blind clinical trials are the basic "gold standard" in epidemiology with regard to generating evidence on causality. This is the experimental design typically used in clinical trials of drugs or medical interventions. In these designs, individuals are recruited and randomly assigned to control or treatment (exposed) groups. Randomization of assignment to treatment or control group can substantially reduce but cannot completely eliminate bias. Likewise, blinding of both participants and researchers (double-blinding) to the exposure status of the individual for the measurement of outcomes is also commonly used to reduce sources of bias. The experimental design may allow for a cross-over, whereby those who were unexposed are switched to the exposed category and vice-versa, to further reduce the potential impacts of undetected bias among groups. These studies are very complex, time-consuming and expensive.

> An experimental design to address possible human health risks from DBPs would generally need to avoid intentionally increasing DBP exposures for any subjects above what they might reasonably otherwise experience. Consequently, an experimental design would need to find a population with comparatively high DBP exposures and then implement measures to reduce DBP exposures that were both randomized and blinded (neither participants nor study staff interacting with participants or doing data analysis would know which participants received reduced exposures). For studying a health outcome like urinary bladder cancer, which observational studies on DBP exposures have suggested may not be evident until more than 40 years of exposure, the impracticality of such an experimental design should be apparent.

11.2.3 Weighing epidemiologic evidence for causation

By its nature, epidemiologic inference provides only evidence of correlation (association) between outcome and exposure to a suspected cause (Angell, 1997), but we are interested in determining whether any correlations observed may infer that an association between a suspected cause and a health outcome is truly a cause of that health outcome. To draw conclusions about causality requires additional inference and judgement beyond the finding of a statistically significant association between exposures and outcomes. Rigorous and unbiased assessment of a body of epidemiology, toxicology and exposure assessment literature is necessary to draw any meaningful conclusions about causality.

Criteria of causality have been derived from a set of concepts set out by Sir Austin Bradford Hill (1965) and by the U.S. Surgeon General (U.S.PHS, 1964). They have been adjusted over the years by many commentators, but the original concepts remain sound. More recently the excellent and accessible introduction to epidemiology by Bonita *et al.* (2006) refers to eight elements of causal inference: temporal relationship, plausibility, consistency, dose-response relationship, strength of association, reversibility, study design and judging the evidence. All these elements are relevant to epidemiologic studies of DBPs except, perhaps, reversibility because it is limited to outcomes that are reversible, something that does not apply to many of the most important outcomes being assessed for causation by DBPs.

11.2.3.1 Temporal relationship

A suspected cause must precede the health effect if it is a true cause. If a suspected cause does not precede the health effect, then logically it cannot be a cause. This criterion demands an accurate account of the time relationship between the exposure to the suspected cause and the resulting outcome which requires accuracy in the determination of both the exposure and the outcome. Because of the importance of this temporal relationship, the most powerful epidemiology studies will be based on incident (new) disease, rather than relying on existing or prevalent disease, because the timing of disease onset can be known with confidence for incident cases.

11.2.3.2 Plausibility

Is it biologically plausible that exposure to the suspected agent will cause the hypothesized outcome? This question is best answered by mode of action toxicology studies (Chapter 10) or in the absence of specific toxicology results, an assessment must be made of how reasonable it is to presume that agent X can cause outcome Y. An exposure-outcome relationship with a biologically plausible mode of action from toxicology studies provides a stronger argument in favor of causality. If a biologically plausible mode of action is not obvious, the argument for causality is not disproven because biological plausibility may become apparent from future research. However, until biological plausibility is determined, confidence in causation will be low. An important consideration in evaluating plausibility has to be the level and mode of exposure to the suspected cause in animal experiments versus the human exposure in drinking water (e.g. massive bolus corn oil-mediated dosage of chloroform; Chapter 1).

11.2.3.3 Consistency

If several different studies with a variety of designs, carried out in different locations and under truly different conditions, which consistently report the same result, then the argument for causality is strengthened. However, a lack of consistency does not necessarily preclude an observed association from being causal, particularly if there have been a limited number of studies. The differing study designs and circumstances (such as exposure levels) could reduce the impact of the causal agent in some of the studies. Therefore, the studies with the best designs must be assigned the greatest weight when evaluating the consistency of results. A technique used for improving the statistical power of individual randomized clinical trials is meta-analysis which essentially pools the data from comparable studies to effectively increase the overall sample size. However, application of meta-analysis to observational studies, particularly where methods have varied, is controversial and findings from such meta-analyses should be considered very cautiously. For example, a widely cited meta-analysis was performed on 10 methodologically different epidemiologic studies of possible relationships between chlorinated drinking water and cancer (Morris *et al.* 1992). This study was cited as an illustration of why such meta-analyses should not be done when studies are not methodologically comparable, when their results can be seen to be heterogeneous or when there is evidence of publication bias (Poole and Greenland, 1999).

11.2.3.4 Dose-response relationship

A dose-response relationship means that the frequency and/or severity of the outcome increases with increasing frequency or magnitude of exposure to the suspected causal agent. A clear dose-response relationship can be a good indication of consistency with a causal relationship, provided care is taken to assure that no unrecognized confounding or bias could be the underlying reason. For those exposures and effects for which a dose-response relationship is valid, defining that dose-response relationship depends on defining both the dose and the response accurately. Documenting exposure accurately is a particular problem for DBPs in drinking water (Section 11.2.4).

11.2.3.5 Strength of the association

The strength of association is usually measured by the rate ratio (also commonly called relative risk, RR) or the odds ratio (OR is an approximation of RR that is generated by case control studies) for a particular exposure – outcome comparison.

The Odds Ratio compares the occurrence of exposure in the cases and the controls, with

$$\text{OR} = \frac{\text{Odds of exposure in cases}}{\text{Odds of exposure in controls}}$$

OR = 1.0 is the null value, no association between exposure and outcome

OR > 1.0 suggests that exposure is positively associated with disease (i.e. exposure is more common in cases than in controls)

OR < 1.0 suggests that exposure is negatively associated with disease (i.e. exposure is less common in cases than in controls)

The Risk Ratio (Rate Ratio or Relative Risk) for a cohort study compares the rates of incidence of disease in exposed and unexposed groups as a ratio

$$\text{RR} = \frac{\text{Incidence rate in exposed group}}{\text{Incidence rate in unexposed group}}$$

RR = 1.0 is the null value, no association between exposure and outcome

RR > 1.0 suggests that exposure is positively associated with disease (i.e. incidence of the outcome is greater in the exposed than in the unexposed group)

RR < 1.0 suggests that exposure is negatively associated with disease (i.e. incidence of the outcome in the exposed group is less than in the unexposed group)

The RR is preferred as a measure of association over OR, but it is not possible to directly determine the RR in a case-control study because the incidence rates are not known. The less common a disease is (more rare), the closer the estimate of OR will converge on the RR. For the purposes of this book where many of the disease outcomes being studied are not common, the OR provides an adequate estimate of RR.

Finding large RR argues more strongly for causality than does a small RR because it suggests a strong effect and makes it less likely that such results could be caused by an undetected bias. However, a small RR does not preclude a causal association because the size of the RR can depend on the existence of other possible causes in relation to the suspected agent. The range of the confidence interval for an estimate of RR provides a useful indicator of the stability of the estimate. High RR values with a very large confidence interval are often based on a small sample size or highly variable findings.

Among epidemiologists, it is generally accepted that if an association is causal, weak exposure assessment resulting in non-differential misclassification of exposure will bias the resulting RR towards the null value, thereby showing a weaker association than truly exists. The logic follows, therefore, that if an association is causal, performing more accurate exposure assessment should increase the apparent strength of the association. However, a weak association may be an indication of a non-causal, spurious association.

As noted above, epidemiologists often rely on the statistical reality that non-differential misclassification of exposures will have the effect of attenuating the estimated OR or RR. This means that poor exposure assessment might be acknowledged while relying on an expectation that improved exposure assessment will inevitably result in an increased OR or RR. Of course, such improvement will only happen if the suspected cause is truly a cause. This opens the prospect of pursuing improved exposure assessment for attempts at replicating or refuting earlier findings. This has been done in the case of investigating a possible association between spontaneous abortion and exposure to THMs (Section 11.4.1.2, Savitz *et al.* 2005; Waller *et al.* 1998) and for bladder cancer and chlorinated drinking water (Section 11.3.3, Amy *et al.* 2006).

Exposure assessment is almost universally acknowledged as a weakness in environmental epidemiology studies and is particularly so for DBP studies (Arbuckle *et al.* 2002). Weak exposure assessment will inevitably result in misclassification of exposures which will cause the OR or RR that is calculated to be inaccurate. In principle, if the misclassification is completely random across the study population, it will lead to a lower estimate of OR or RR compared with perfect exposure classification. Considerable advances have been made in characterizing human exposure to DBPs so that total exposure is now understood to include activities such as showering, bathing and swimming (particularly for volatile DBPs) rather than simply ingestion. This subject is discussed further in section 11.2.4.

11.2.3.6 Reversibility

This element means that if a suspected causal agent is removed, the probability of the outcome should decrease. If the removal of a suspected causal agent results in a decrease in the occurrence of the outcome, then the argument for a causal relationship is strengthened. In evaluating this element an accurate assessment of exposure or non-exposure to the potential causal agent must be determined. This criterion is limited to causal processes that involve reversible mechanisms and thus this element has limited application to some of the health effects suggested for DBPs (e.g. spontaneous abortion, birth defects).

11.2.3.7 Study design

As noted above, there are many different epidemiologic study designs, each with a different level of ability to test for causation. Generally, the best study design for testing causation is the randomized controlled trial or clinical trial. Even so, clinical trials of drugs or medical interventions will often yield conflicting results. In these cases, exposure is determined by whether an individual received the drug or medical intervention under study. Clearly, there is substantially less uncertainty about individual exposure for such studies than there could ever be for the study of exposure to DBPs in drinking water. Clinical trials are rarely done in environmental epidemiology because of the ethical limits on experimentation with humans. Notable exceptions relevant to drinking water have been the randomized intervention trials performed in Melbourne, Australia and in Iowa, U.S.A. to investigate the risk of gastrointestinal disease from drinking water exposure (Colford *et al.* 2006, 2005; Hellard *et al.* 2001). Prospective cohort and case-control studies are the next-best study designs. The latter are commonly used in environmental epidemiology. Cross-sectional studies are less able to test causation because current cross-sectional studies will not provide evidence of a temporal relationship (the cause must precede the effect).

11.2.3.8 Judging the evidence

Even with all of these elements, there is no absolute rule for judging a potential causal relationship. All available evidence must be considered when determining whether an exposure really is the cause of an outcome. In the case of conflicting evidence, the types of evidence must be weighed with respect to their relative positions within the criteria for causality. Most importantly, a temporal relationship between the cause and the effect must be established. If the effect precedes the alleged cause, then there is no argument for a causal relationship. Plausibility, consistency, and dose-response are all very important criteria. The likelihood of a causal association is increased if evidence representing several of the criteria all points to the same conclusion, that is, consistency across criteria as well as consistency of results across different studies. The strength of association must be judged by how far-removed from the null value is the estimate of RR, but also the size of confidence interval and whether it excludes a RR or OR = 1.0 is critical. Clearly, evidence on causation from epidemiologic studies of environmental exposures cannot be free of uncertainty (Figure 11.2) and as may be apparent in different chapters in this book, different experts may reach different conclusions while evaluating the same criteria to the same studies.

Taubes (1995) generated a strong reaction among epidemiologists when he published an article based on interviews with several prominent epidemiologists. Taubes summarized many of the limitations that we have discussed here. This article and the letters it generated in response (Willett *et al.* 1995) should be required reading for anyone interested in judging causal evidence in epidemiology.

Taubes (1995) addressed the importance of strength of association quoting several epidemiologists to the effect that they did not regard RR/OR values less than 3 to 4 as being credible evidence for a causal relationship unless there were several studies showing consistent results. The issue of consistency is not well defined. Taubes (1995) quoted David Sackett, a recognized pioneer in establishing evidence-based medicine[1], regarding a catch with consistency: "*It is persuasive only if the studies use different architectures, methodologies, and subject groups and still come up with the same results. If the studies have the same design and 'if there's an inherent bias' he explains, 'it wouldn't make any difference how many times it's replicated. Bias times 12 is still bias.' What's more, the epidemiologists interviewed by* Science *point out that an apparently consistent body of published reports showing a positive association between a risk factor and a disease may leave out other, negative findings that never saw the light of day.*"

A key feature which is typically missing from observational studies compared with rigorous, randomized clinical trials is a requirement for a clear, established advance protocol for data analysis and interpretation before a study is done. Taubes quotes an epidemiologist from the U.S. Food and Drug Administration on this topic: "*Although risk factor epidemiology will never be as sharp a tool as randomized clinical trials, epidemiologists could still benefit by adopting some of the scientific practices of those studies. 'The great thing about a clinical control trial,' he says, 'is that, within limits, you don't have to believe anybody or trust anybody. The planning for a clinical control trial is prospective; they've written the protocol before they've done the study, and any deviation that you introduce later is completely visible.' While agencies like NCI do insist on seeing study protocols in risk factor epidemiology prospectively, this is still not standard procedure throughout the field. Without it, says Temple, 'you always wonder how many ways they cut the data. It's very hard to be reassured, because there are no rules for doing it.'* "

The Taubes (1995) letter triggered response letters (Willett *et al.* 1995) from eight of the epidemiologists who were interviewed. These letters were critical of Taubes for not highlighting the number of beneficial contributions to health risk knowledge which have been obtained through risk factor epidemiology, but they did not dispute the main points of caution which Taubes raised. Trichopoulus (Willet *et al.* 1995) objected to two points. First he wrote: "*Taubes writes that I have expressed the view that only a fourfold risk should be taken seriously. This is correct, but only when the finding stands in a biological vacuum or has little or no biomedical credibility.*" Second, Trichopoulus states: "*Epidemiology should be evaluated in comparison to other disciplines that serve the same objective, that is, to identify the causes of human disease and facilitate their prevention.*" In support of this premise, he notes that epidemiology avoids problems with extrapolation from high dose animal experiments to low human doses. He also noted a comparison with political systems, noting all the problems with democracy, yet it is the best choice among what is available. A comparison of epidemiology with toxicology for their comparative strengths and weaknesses is provided in Table 11.1.

---

[1]Readers may be forgiven for wondering what medicine was based on before the movement to evidence-based medicine.

Table 11.1 Comparison of observational epidemiology with experimental toxicology for yielding human risk assessment insights.

Observational epidemiology	Value[a]	Experimental toxicology	Value[a]
observe human subjects	+	use animals (typically rodents)	−
adjustment for differences in absorption, distribution, metabolism and excretion generally not required	+	adjustment for differences in absorption, distribution, metabolism and excretion are required for human risk assessment	−
large sample size possible in some cases	+	sample size limited by practicality	−
wide genetic variability possible	+/−	narrow genetic variability	−/+
diverse and wide sensitivity of subjects	+/−	narrow range of sensitivity of subjects	+/−
low exposure range – realistic but insensitive	+/−	high exposure range – sensitive but commonly yields artifacts	−/+
can assess combined realistic exposure	+	combined mixtures are difficult	−
no control over exposures	−	high control over exposures	+
individual measurement of exposure generally not feasible or limited	−	individual measurement of exposure is feasible	+
no control over confounding factors – only imperfect mathematical adjustment	−	high control over confounding through experimental design	+
randomization not possible	−	randomization is normal	+
time frame for chronic diseases is long (decades)	−	time frame for chronic diseases is much shorter (typically 2 years)	+
prospective studies are limited in feasibility	−	all experiments are prospective in nature	+
capability to ascertain mechanism by post mortem investigation is rare	−	post mortem examination is normal to provide insights into mechanism	+
recall bias a problem in case-control studies	−	recall bias plays no role	+

[a]some characteristics may be valuable in some circumstances and disadvantageous in others (+/− or −/+).

A major portion of the complexity of the chlorination DBPs issue arises from the inherent limitations of our primary scientific approaches to studying the problem: toxicology and epidemiology (Figures 11.1 and 11.2). Each has its own limitations and neither is perfect for answering the questions necessary to address a complex public health issue.

Because many users of the findings of epidemiology results are not aware of all the caution that should be considered in assessing epidemiology evidence, the reliance of the DBP regulatory process on such evidence should be based on a transparent and rigorous evaluation of the strengths and limitations of that epidemiologic evidence. Review of epidemiologic evidence provided in support of the current DBP regulations and guidelines (Chapter 2) generally does not reflect such transparent and rigorous evaluation.

11.2.4 Assessing exposure

Compared to the challenges of assessing toxicity, the assessment of exposure to DBPs may seem easy. Because humans use tapwater for showering, bathing and a variety of household uses, depending on the properties of individual DBPs, it is not accurate to assume that direct ingestion is the only important route of exposure (Figure 11.5). In fact, studies with volatile DBPs like THMs have shown that exposure via showering and bathing can be substantial in relation to direct ingestion (Arbuckle *et al.* 2002; Backer *et al.* 2000). These factors, combined with the higher levels of DBPs commonly found in swimming pools, have raised the matter of swimming as an additional risk factor for DBP exposure.

Challenges of exposure assessment include: (1) limited information on the identity, occurrence, and toxicity of the many DBPs that can be formed by disinfection meaning that it is often not clear which DBPs are the suspected causal agent to be studied; (2) the complex, spatial and time-variable chemical relationships affecting DBP concentrations within a community water distribution system; and (3) practical difficulties in acquiring accurate and reliable information on personal activity determining exposure from tapwater consumption rates at various sources and non-ingestion exposure pathways.

One approach that has been suggested as a possible gold standard for DBP exposure assessment would be a biomarker (Swan and Waller, 1998). Some volatile DBPs like chloroform are metabolized far too rapidly and they can be detected in human blood for only minutes after a shower exposure, but not at all after ingestion exposure because they are metabolized so rapidly. Ingestion and uptake from the gastrointestinal tract leads directly to the liver where metabolism eliminates chloroform from the bloodstream, compared with inhalation which allows it to reach the bloodstream directly, where it can be measured until the biomarker has been circulated through the liver. Some success has been achieved in demonstrating trichloroacetic acid (TCAA) as a possible biomarker of drinking water exposure because it is metabolized relatively slowly (Froese *et al.* 2002; Kim *et al.* 1999). The half-life of TCAA following drinking water ingestion has been found experimentally to range from 2.1 to 6.3 days (Bader *et al.* 2004) compared with only minutes for THMs.

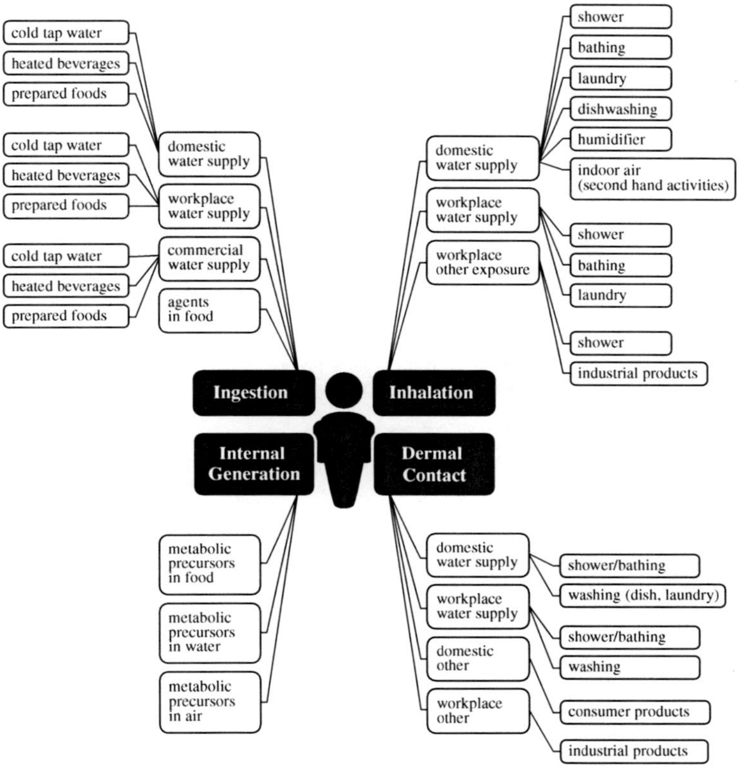

Figure 11.5 Multiple exposure sources for DBPs to be considered

A major experimental study assessed the validity and reliability of using TCAA in urine and blood as a biomarker of drinking water ingestion (Zhang *et al.* 2009a,b). This study involved a total of 46 healthy women who were provided analyzed tap water with known TCAA concentration for 15 days and provided individual urine and blood samples for TCAA measurements along with detailed fluid consumption diaries. The findings revealed that the reliability (i.e. repeatability) of measurements was excellent by using measures of TCAA ingestion, blood concentration and urinary excretion (Zhang *et al.* 2009a). However, validity is the key consideration for judging the value of a biomarker for DBP exposure (Zhang *et al.* 2009b). While the results were generally promising (a correlation coefficient of $R = 0.69$ and the data shown in Figure 11.6 establish that urinary and blood TCAA reflect TCAA ingestion), the scatter of individual results for a single exposure day shown in the log-log plot of Figure 11.6c illustrates how much variability is evident for individual data even under exceptionally carefully controlled experimental conditions.

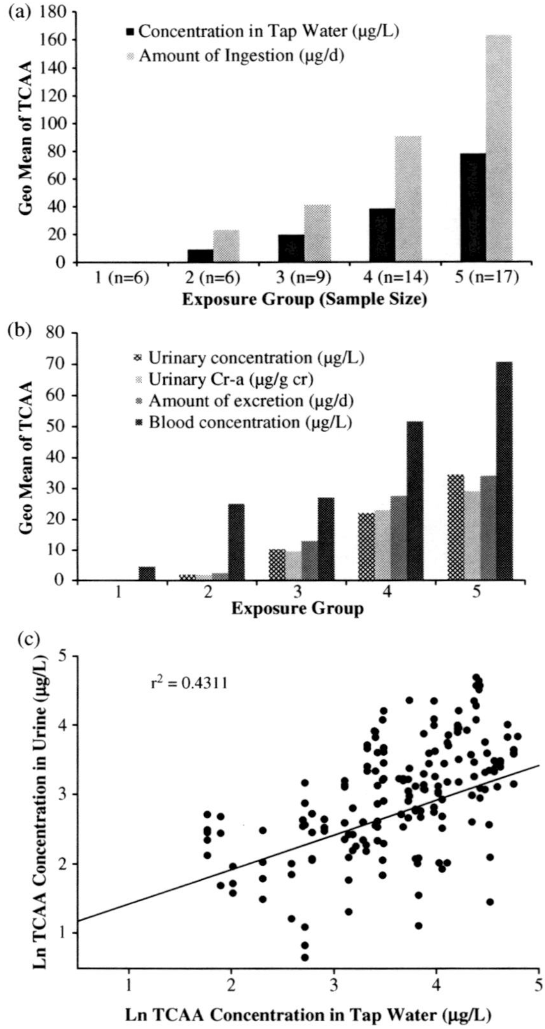

Figure 11.6 (a) TCAA concentrations in tap water and ingested amount of TCAA during 15-day exposure period in five groups; (b) levels of TCAA loading in blood and urinary excretion between exposure day 12 and 15 in five groups; (c) correlation between log-transformed TCAA concentrations in supplied tap water and log-transformed TCAA concentrations in urine in a single exposure day (after Zhang *et al.* 2009b)

Considering that these experimental data represent individual exposure assessment quality that greatly exceeds that used for any published epidemiology study addressing DBPs provides a sobering basis to be cautious about causal inference for the body of epidemiologic evidence addressing DBPs and human health effects.

> The problem of exposure assessment becomes more critical given that individual exposure assessment is required for achieving the strongest epidemiologic study designs (Arbuckle *et al.* 2002). In fact, the difficulty of accurately assessing exposure provides a major shortcoming of epidemiology studies addressing DBPs in drinking water because exposure assessment typically is neither truly individual nor measured directly. These limitations on exposure assessment need to be viewed in the context of Figure 11.3 which depicts the fundamental basis of epidemiologic inference. If we are lacking evidence or holding only weak evidence of individual exposure to DBPs, then our ability to assign individuals accurately to the appropriate exposure categories is substantially impaired. If data are seriously weak on essentially half of the basic data input required for epidemiologic inference, confidence in the overall conclusions of that inference cannot be high.

11.3 CHLORINATION DBPS AND CANCER

11.3.1 Epidemiology studies on cancer

The following summary and analyses are based primarily on a comprehensive descriptive review (Hrudey, 2008) with updates based on any relevant studies published since 2007. That reference includes summaries of 56 epidemiologic studies conducted on various cancer sites since 1977. Although there are occasional suggestive findings for various cancer sites in individual studies, overall there is no consistency or strength of association pointing to chlorination DBPs[2] as a credible source of risk for cancer sites other than urinary bladder, colon or rectum. The latter three sites are discussed in this chapter.

The International Agency for Research on Cancer (IARC, 1991; IARC, 2004) reviewed the available epidemiologic evidence on DBPs and cancer and summarized studies according to bladder, colorectal and other sites (kidney, brain, pancreatic and childhood cancer). The IARC summaries did not express conclusions about these studies, but these were used as a background for deliberations about ranking the carcinogenicity of various individual DBPs (specifically chloral hydrate, dichloroacetic acid, trichloroacetic acid and MX). In all cases, these compounds were rated as (IARC, 2004) "*I: inadequate evidence*

---

[2]Chlorination DBPs refers to any DBPs arising from chlorination or related disinfection processes like chloramination. Such DBPs need not contain chlorine nor other halogens, as noted in Chapter 5, 6 and 7, so "*chlorination* DBPs" are not necessarily "*chlorinated* DBPs".

of carcinogenicity" with respect to the characterization of human evidence, although, based on animal evidence dichloroacetic acids and MX were rated as: "*2B, possibly carcinogenic to humans*." This rating is not surprising given that the human epidemiology studies provided no exposure information on any of these specific compounds.

For bladder, colon and rectal cancers, discussion in this chapter has been limited to those analytical epidemiology studies which offered some possibility of generating evidence about causation. Consequently, analytical studies were only included that:

- were either case-control or cohort studies to assure that there was some level of individual exposure classification along with individual outcome classification.
- were based on cancer incidence rather than mortality
- provided evidence of quality control in terms of adequate sample size, high response rate, and adjustment for confounding

11.3.1.1 Colon cancer

Of all published studies found on colon and rectal cancer and summarized in Hrudey (2008), five studies (King *et al.* 2000a; Hildesheim *et al.* 1998; Doyle *et al.* 1997; Cragle *et al.* 1985; Young *et al.* 1987) were evaluated in more detail. All of these were case-control studies except for Doyle *et al.* (1997) which was a prospective cohort.

These study designs satisfied the requirements for temporality by assuring that exposures were known to have preceded outcomes. The plausibility of chlorination DBPs causing colon cancer has some limited support from the observation of large intestine tumors in rats (NTP, 1987) caused by BDCM in a corn oil gavage study, but these findings were not replicated in two later drinking water ingestion studies (George *et al.* 2002; NTP, 2006). Toxicology studies have provided some plausibility with the finding of aberrant crypt foci (pre-cancerous lesions) being formed in the large intestine of rats (analogous to the human colon), but not in mice, by high level exposure to BDCM and bromoform (DeAngelo *et al.* 2002; Geter *et al.* 2004a; Geter *et al.* 2004b; Geter *et al.* 2005).

The consistency and findings of dose response of epidemiological results for colon cancer in humans has been limited. King *et al.* (2000a) found that the OR for males in the highest category of chlorination DBP exposures was significant, but for females it was not. Hildesheim (1998) did not find any high chlorination DBP exposure to be significant. While King *et al.* (2000a) had a large number of cases (767), the total number of colon and rectal cancer cases that could be analyzed was only 43% of the total number of cases of these cancer sites identified in the cancer registry, raising concerns about selection bias occurring for the cases that were included.

Doyle *et al.* (1997) found the highest exposure category (14-287 μg/L THM4[3]) to be significantly associated with colon cancer for the female cohort studied, but not substantially more so than the next to lowest exposure category (3-13 μg/L THM4), a relatively low exposure level for THMs. Cragle *et al.* (1985) with a small study (only 200 cases) found the higher age groups (60 and above) to show significant ORs for colon cancer risk with exposure to chlorination DBPs, with higher ORs for those exposed more than 15 years versus less than 15 years. However, Young *et al.* (1987) found no support for THM4 exposure being a risk factor for colon cancer.

Nieuwenhuijsen *et al.* (2009a; 2009b) and Nieuwenhuijsen (2010) have concluded, based on a review of published studies, that the epidemiologic evidence for a relationship between DBP exposure and colorectal cancer is inconsistent and inconclusive. Rahman *et al.* (2010) undertook a meta-analysis of 13 studies (3 cohort and 10 case-control) and they concluded for colon cancer that the pooled RR estimates were 1.11 (CI: 0.73–1.70) for cohort studies and 1.33 (CI: 1.08–1.50) for case control studies providing limited evidence of an association.

Overall, the epidemiologic findings for colon cancer in relation to chlorination DBP exposure are inconsistent and the strength of association observed was modest in those studies where any association was suggested.

11.3.1.2 Rectal cancer

Of all published studies found on colon and rectal cancer and summarized in Hrudey (2008) four studies (Bove *et al.* 2007a; Hildesheim *et al.* 1998; King *et al.* 2000a; and Doyle *et al.* 1997) were evaluated in more detail for rectal cancer. All of these were case control studies except for Doyle *et al.* (1997) which was a prospective cohort.

These study designs satisfy the requirements for temporality by assuring that exposures are known to have preceded the outcomes. Again, toxicology studies have provided some plausibility for this outcome with the finding that aberrant crypt foci (pre-cancerous lesions) were formed in the rectal segment of the large intestine (analogous to the human colon) of rats (not in mice) by exposure to BDCM and bromoform (DeAngelo *et al.* 2002; Geter *et al.* 2004a).

The consistency in findings for an association between chlorination DBPs and rectal cancer has been limited. Bove *et al.* (2007a) found a significantly elevated OR for rectal cancer with exposure to bromoform in a small study (128 cases). The concentration range of this exposure group is wide for a brominated THM that is usually only present in a small proportion, if at all, in most surface water

[3]For the purposes of this Chapter, the term THM4 will be used to describe the combination of chloroform, bromodichloromethane, dibromochloromethane and bromoform. The literature will sometimes refer to THM4 as total THM or TTHM, but given the discovery of iodinated THMs (Chapter 4), TTHM is not an accurate term when it is typically limited to the chlorinated and brominated species.

supplies. Bromoform can also be an artefact in THM monitoring, so validation of high values would be warranted. In any case, this finding has limited support among the other more substantial studies. King *et al.* (2000a) found no evidence among males or females for an association of rectal cancer with chlorination DBPs. Hildesheim *et al.* (1998) found modest support for an association, but Doyle *et al.* (1997), like King *et al.* (2000a), found no evidence among a female cohort to support a causal association between exposure to chlorination DBPs and rectal cancer.

Hildesheim *et al.* (1998) found evidence of increasing risk of rectal cancer with increasing duration of exposure to chlorinated water and an estimated long term estimate of THM exposure, with the former providing a stronger measure of association. Assessment of urbanicity (population size of resident community) was correlated positively with THM exposure but negatively with rectal cancer risk.

Nieuwenhuijsen *et al.* (2009a,b) considered colorectal cancer in their review of epidemiologic evidence to conclude that evidence for a relationship between DBP exposure and colorectal cancer is inconsistent and inconclusive. Rahman *et al.* (2010) undertook a meta-analysis of 13 studies (3 cohort and 10 case-control) and they concluded for rectal cancer that the pooled RR estimates were 0.88 (CI: 0.57–1.35) for cohort studies and 1.40 (CI: 1.15–1.70) for case control studies providing only limited evidence of an association.

Overall, the epidemiologic findings for rectal cancer in relation to chlorination DBP exposure are inconsistent and the strength of association is modest in those studies where a significant association is suggested.

11.3.1.3 Urinary bladder cancer

Hrudey (2008) includes summaries of 33 epidemiologic studies that provide some measure of bladder cancer risk and 1 study (Ranmuthugala *et al.* 2003) assessing a potential early marker of bladder cancer. Of the these, 12 studies (10 case-control and 2 cohort) satisfied the quality criteria specified for having some possibility of providing meaningful evidence towards causal inference and these, plus one more recently published study (Cantor *et al.* 2010), are summarized in Figure 11.7 and Table 11.2.

Among the weaker studies capable of providing suggestive and/or hypothesis-generating evidence was one incident case-control study (Vena *et al.* 1993) that considered fluid consumption as being tapwater vs. non-tapwater, but no assessment was made for the presence of disinfection by-products in this comparison. There were another 8 case-control studies including bladder cancer (Alavanja *et al.* 1978; Brenniman *et al.* 1980; Chang *et al.* 2007; Gottlieb and Carr, 1982; Kanarek and Young, 1982; Young *et al.* 1981; Young and Kanarek, 1983; Zierler *et al.* 1988) based on cancer deaths. There were another 10 studies that were either cross-sectional or aggregated (ecologic) designs including bladder cancer (Bean *et al.* 1982; Cantor *et al.* 1978; Carlo and Mettlin, 1980; Flaten, 1992; Isacson *et al.* 1983; Koivusalo *et al.* 1994; Koivusalo *et al.* 1997; Suarez-Varela *et al.* 1994; Vinceti *et al.* 2004; Yang *et al.* 1998; Zierler *et al.* 1986; Kuzma *et al.* 1977).

Given the large number of studies performed with several being higher quality studies with some capability of generating causal evidence for bladder cancer, the discussion will be focused on the 13 studies summarized in Figure 11.7 and Table 11.2. All of these satisfy the temporality criterion.

Regarding consistency, with exception of females in Cantor *et al.* (1998); Doyle *et al.* (1997); Freedman *et al.* (1997), the associations are positive with exposure indicators of some measure of chlorinated DBPs. Eight studies provide positive indicators that are statistically significant (Bove *et al.* 2007b; Cantor *et al.* 1998; Cantor *et al.* 2010; Chevrier *et al.* 2004; King and Marrett, 1996; McGeehin *et al.* 1993; Michaud *et al.* 2007; Villaneuva *et al.* 2007). The three most recently published studies (Cantor *et al.* 2010; Michaud *et al.* 2007; Villanueva *et al.* 2007) are based on the same Spanish bladder cancer case-control study. Although these 3 papers performed different analyses of the data, the basic evidence regarding cases and THM exposures that they are analyzing is the same.

Six studies (recognizing the foregoing common data source for two of them) provide evidence of a consistent dose – response relationship (Cantor *et al.* 1998; Cantor *et al.* 2010, Chevrier *et al.* 2004; King and Marrett, 1996; McGeehin *et al.* 1993; Villaneuva *et al.* 2007). Four studies also provide evidence of a consistent duration – response relationship (Cantor *et al.* 1998; Koivusalo *et al.* 1998; McGeehin *et al.* 1993; King and Marrett, 1996).

Cantor *et al.* (2010) provides intriguing and important new evidence based upon the use of genotyping. In particular, the reported OR of 5.9 (1.8-19) for what were judged to be a group of high risk genotypes is eye-catching and offers some promise that use of this specific measure of potential susceptibility has dramatically improved the observed OR for an association between bladder cancer and chlorination DBPs. However, the discussion in Chapter 10 makes a case that the findings of Cantor *et al.* (2010) do not strengthen the case of a causal association between bladder cancer and THMs, based on the properties of the enzymes associated with these high risk genotypes. However, the results may suggest strengthened evidence of other non-THM DBPs or contaminant exposures associated with chlorinated drinking water exposure being causal for bladder cancer.

The Cantor *et al.* (2010) findings are based on very small numbers (note the wide confidence interval of 1.8 to 19), specifically only 59 cases and 47 controls for the highest THM exposure group. Likewise, the overall study was based on 1,219 cases out of 1,457 identified eligible cases and 1,271 controls out of 1,465 identified eligible controls, but the genotyping analysis was limited to 680 cases and 714 controls, only 54% and 49% respectively of the total study population raising a concern for possible selection bias. As the authors suggested, early replication of their findings in another population is warranted (Cantor *et al.* 2010).

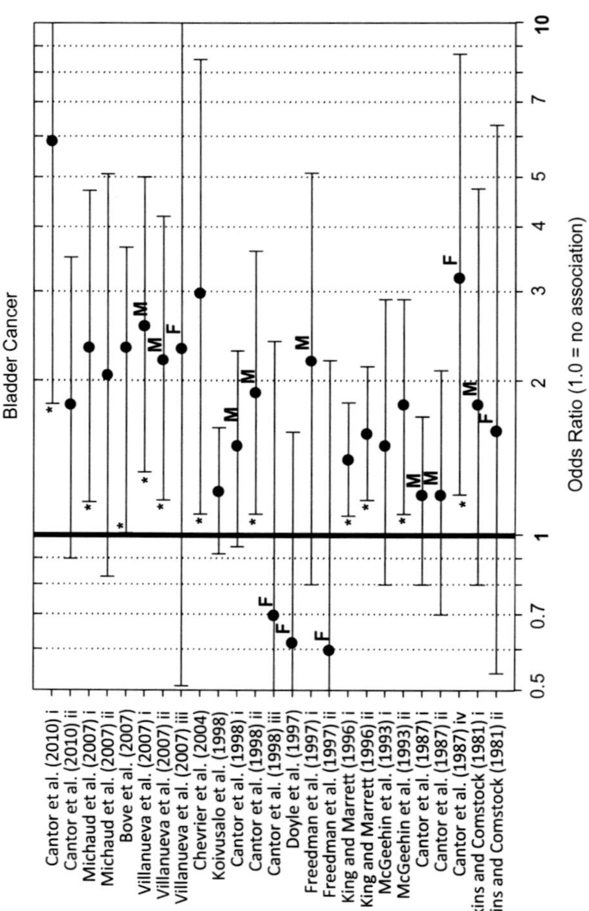

Figure 11.7 Analytical epidemiology evidence on urinary bladder cancer and exposure to chlorination DBPs

Table 11.2 DBP exposure conditions and measures of association (OR) – bladder cancer.

Reference	Exposure comparison	Adj OR (95% CI)
Cantor *et al.* (2010)i	Both: >49 µg/L THM4 vs. ≤8 µg/L; GSTT1present, GSTZ1 CT/TT	5.9 (1.8–19.0)*
Cantor *et al.* (2010)ii	Both: >49 µg/L THM4 vs. ≤8 µg/L	1.8 (0.9–3.5)
Michaud *et al.* (2007) i	Both: 26–49 µg/L THM4 vs. ≤8 µg/L	2.34 (1.16–4.71)*
Michaud *et al.* (2007) ii	Both: >49 µg/L THM4 vs. ≤8 µg/L	2.06 (0.83–5.08)
Bove *et al.* (2007)	Both: 74–352 µg/L THM4 vs. 0–38 µg/L	2.34 (1.01–3.66)*
Villanueva *et al.* (2007) i	Male: >25–30 years with chlorinated surface water[a]	2.58 (1.33–5.01)*
Villanueva *et al.* (2007) ii	Male: >30 years with chlorinated surface water[a]	2.21 (1.17–4.20)*
Villanueva *et al.* (2007) iii	Female: >30 years with chlorinated surface water[a]	2.33 (0.51–10.55)
Chevrier *et al.* (2004)	Both: >50 µg/L THM4 vs. <1 µg/L (THM4 estimated only)	2.99 (1.1–8.5)*
Koivusalo *et al.* (1998)	Both: >30 years with mutagenic chlorinated water[b]	1.22 (0.92–1.62)
Cantor *et al.* (1998) i	Male: 40–59 years with chlorinated surface water[c]	1.5 (0.95–2.3)
Cantor *et al.* (1998) ii	Male: ≥60 years with chlorinated surface water[c]	1.9 (1.1–3.6)*
Cantor *et al.* (1998) iii	Female: ≥60 years with chlorinated surface water[c]	0.7 (0.2–2.4)
Doyle *et al.* (1997)	Cohort Female: 14–287 µg/L THM4 vs. <limit of detection	0.62 (0.25–1.59)
Freedman *et al.* (1997) i	Male: >40 years with municipal (disinfected) water[d]	2.2 (0.8–5.1)
Freedman *et al.* (1997) ii	Female: >40 years with municipal (disinfected) water[d]	0.6 (0.2–2.2)
King and Marrett (1996) i	Both: >35 years with chlorinated surface water[a]	1.41 (1.09–1.81)*
King and Marrett (1996) ii	Both: >35 years with ≥25 µg/L THM4 surface water[a]	1.58 (1.17–2.14)*
McGeehin *et al.* (1993) i	Both: 21–30 years with any chlorinated water[a]	1.5 (0.8–2.9)
McGeehin *et al.* (1993) ii	Both: >30 years with chlorinated water[a]	1.8 (1.1–2.9)*
Cantor *et al.* (1987) i	Male: 40–59 years with chlorinated surface water[e]	1.2 (0.8–1.7)
Cantor *et al.* (1987) ii	Male: ≥60 years with chlorinated surface water[e]	1.2 (0.7–2.1)
Cantor *et al.* (1987) iv	Female: ≥60 years with chlorinated surface water[e]	3.2 (1.2–8.7)*
Wilkins and Comstock (1981) i	Male: Municipal (disinfected) water[f]	1.80 (0.80–4.75)
Wilkins and Comstock (1981) ii	Female: Municipal (disinfected) water[f]	1.60 (0.54–6.32)

*statistically significant; [a]compared with unchlorinated water; [b]compared with non-mutagenic water; [c]compared with unchlorinated water or chlorinated ground water; [d]compared with non-municipal water assumed to be non-chlorinated; [e]compared with unchlorinated water or chlorinated groundwater, for above-median water consumption; [f]compared with deep well water users

11.3.2 Reviews of epidemiology on bladder cancer

There have been a number of attempts to increase the power of published studies by performing meta-analyses to combine studies. Morris *et al.* (1992) performed a meta-analysis to assess the results for 10 studies including 7 that addressed bladder cancer (Cantor *et al.* 1987; Zierler *et al.* 1988; Wilkins and Comstock, 1981; Gottlieb *et al.* 1982; Young *et al.* 1981; Brenniman *et al.* 1980, and Alvanja *et al.* 1978) to determine if this technique could yield a more robust risk estimate by combining results. This analysis yielded a combined relative risk estimate of 1.21 (1.09 - 1.34). Meta-analysis has been widely practiced for enhancing interpretation of risk estimates by combining high quality consistent studies such as double-blind, controlled clinical trials.

The merits of applying meta-analysis to combine results from observational studies of widely variable character and quality have been criticized (Egger *et al.* 1997). As noted in Section 11.2.3.3, Poole and Greenland (1999) selected the Morris *et al.* (1992) paper as an example to demonstrate that performing meta-analysis is problematic when there are substantial inconsistencies among the studies combined. Poole and Greenland (1999) concluded that the overall relative risk calculations in Morris *et al.* (1992) were not valid.

Villanueva *et al.* (2003) performed a meta-analysis combining eight studies (Cantor *et al.* 1998; Koivusalo *et al.* 1998; Doyle *et al.* 1997, King and Merritt, 1996, McGeehin *et al.* 1993; Vena *et al.* 1993; Cantor *et al.* 1987; Wilkins and Comstock, 1981). This analysis reported a combined bladder cancer risk in men of 1.4 (1.1–1.9) and women of 1.2 (0.7–1.8). Villanueva *et al.* (2004) also performed a meta-analysis of six case-control studies (King and Merritt, 1996, Koivusalo *et al.* 1998; Cantor *et al.* 1998; Cordier *et al.* 1993; Lynch *et al.* 1989 and an unpublished study by Porru). The latter meta analysis found a bladder cancer risk of 1.24 (1.09–1.41) for men ever-exposed to more than 1 μg/L, but not for women who showed a bladder cancer risk of 0.95 (0.76–1.20). The credibility of a THM exposure range as low as 1 μg/L for an increased bladder cancer risk is suspect for attribution to THM. For men there was a trend to increasing risk for higher THM exposure with an OR of 1.44 (1.20-1.73) for exposure higher than 50 μg/L or longer duration of THM exposure with an OR of 1.62 (1.21-2.16) with 30-40 years exposure to chlorinated drinking water. A cumulative estimate of THM exposure (i.e. a mass of THMs derived from concentration and volume ingested) showed an increasing dose-response for all THM cumulative exposure above 15 mg.

More recently, Costet *et al.* (2011) performed another meta-analysis of two of the European case control studies included in the meta-analyses described above (Cordier *et al.* 1993; Koivusalo *et al.* 1998) plus the Spanish study (Villanueva *et al.* 2007) to analyze a total of 2831 cases and 3086 controls. They found that comparing THM exposure greater than 50 μg/L with <5 μg/L produced an adjusted OR estimate of 1.47 (1.05–2.05). However, the middle quartiles of THM exposure (>25 μg/L to 50 μg/L) and (>5 μg/L to 25 μg/L) did not yield significant OR estimates, so a consistent dose-response relationship was lacking. Their analysis of the same THM exposure quartiles applied to three North American studies (Cantor *et al.* 1998; King and Marret, 1996; Lynch *et al.* 1989) yielded significant OR for all quartiles. While these findings lend some weight to the argument for consistency of findings for bladder cancer, some key elements of their analysis were difficult to explain if THM exposure data were truly associated with a cause of bladder cancer. Most troubling was the finding that a measure of cumulative ingested total THMs (calculated from estimated concentration and consumption volume) yielded no significant OR estimates at any of the five quintiles estimated in contrast to the earlier very consistent dose-response to cumulative exposure that was reported by Villanueva *et al.* (2004). Also noteworthy was that only greater than 30 years of exposure to chlorinated surface water yielded a barely significant OR estimate of 1.57 (1.00–2.48).

These meta-analysis papers on bladder cancer (Costet *et al.* 2011; Villanueva *et al.* 2003, 2004) did not cite the Egger *et al.* (1997) concerns or the Poole and Greenland (1999) critique of Morris *et al.* (1992) having applied meta-analysis to a heterogeneous set of studies. Costet *et al.* (2011) did cite two earlier papers on meta-analysis by Greenland (1987; 1994) and also provided the best summary of the heterogeneity of exposure estimation among the studies they included. The Finnish study was based on an evaluation of the level of mutagenicity (Ames test) which was then used with an empirical equation based on mutagenicity, quality of raw water and treatment practices to estimate past total THM levels. Cases and controls were collected in 1991–92. The French study retrospectively collected information from municipalities and water providers about water sources (surface vs. groundwater) and treatment practices that were used in a predictive model to reconstruct total THM exposure. Cases and controls were collected in 1985–87. The Spanish study performed THM analyses on 105 drinking water samples and based on an assumption that measured levels remained unchanged for a constant water source, THM levels were estimated back to 1920. Cases and controls were collected in 1998–2001.

Overall, no mention was made of any expectation of changes in THM formation from water providers changing source water or treatment that might have occurred because of the discovery of THMs in 1974 or subsequent development of chloroform (WHO, 1984, 1993) or THM (Health and Welfare Canada 1978; USEPA, 1979) drinking water standards.

11.3.3 Critical analysis of epidemiology on bladder cancer

Overall, the studies investigating a relationship between bladder cancer and some measure of exposure to chlorinated DBPs have provided reasonably consistent evidence of a significant positive association across a number of different population exposure scenarios, generally in the OR range of 1.2 to 2.0. In the field of epidemiologic study, these are not strong indicators of an association, but their apparent consistency should not be dismissed. The critical article by Taubes (1995) raised concern about the ability of observational epidemiology studies to provide meaningful results for weaker associations where OR values are in the range of 3 to 4 for a single study. Where several studies are reporting consistent results, lower OR values may be viewed as meaningful.

von Elm *et al.* (2007) have reported on guidelines for reporting observational studies. Most of the recent papers address most of the checklist of items recommended, although there is certainly scope in this literature for fuller discussion of the limitations of the study "*taking into account sources of potential bias or imprecision*" and greater caution in the interpretation "*considering objectives, limitations, multiplicity of analyses, results from similar studies, and other relevant evidence.*" Finally, the published studies generally do not or do not adequately address "*the generalisability (external validity) of the study results.*"

Despite numerous valid concerns about the quality of the epidemiology results reported for DBPs and bladder cancer, a precautionary approach has been justified. Because of the extensive population exposure to chlorine-disinfected drinking water, a relatively small increase in risk that is estimated can translate into a substantive public health issue. Based on some of these studies (Cantor *et al.* 1987; Cantor *et al.* 1998; Freedman *et al.* 1997; King and Marrett, 1996; McGeehin *et al.* 1993), the U.S. Environmental Protection Agency estimated that the population attributable risk (PAR) for bladder cancer might range between 2 and 17% (Odom *et al.* 1999). For the Disinfection By-Product 2 Rule, US EPA (2006) considered three approaches for PAR, including the above range, but did the benefit calculations for cases of bladder cancer averted with a selected PAR of 16%.

According to IARC (2011) data which allows international comparisons, the U.S.A. was estimated to have 68,810 new cases of urinary bladder cancer (in 2008), while the American Cancer Society (2011) estimated 69,250 new cases for 2011. If the PAR estimates were accurate, approximately between 1400 and 12,000 cases of bladder cancer per year might be attributed to DBPs in chlorinated drinking water in the U.S. There is clearly a discrepancy between these estimates and those which arise from cancer risk assessment numbers using individual DBPs for which CSF have been estimated. The quantification of cancer risk using the CSF that were considered valid in 2005 (BDCM, bromoform, DBCM and DCAA, but no CSF for chloroform) for the benefit analysis of the Stage 2 DBP rule (USEPA, 2005) determined a maximum estimate of 92 cases of any type of cancer per year for a surface and groundwater served population of 263 million. In other words, the cancer risk assessment which is designed to be an upper bound estimate of cancer cases based on predicted carcinogenicity of individual DBPs yields an estimate that is less than 7% of the lower bound PAR number for bladder cancer and only 0.8% of the upper bound PAR estimate. There is clearly a major discrepancy between the cancer case predictions for the regulated THMs and HAAs and the epidemiologic estimates for bladder cancer alone, based on the epidemiologic PAR estimates.

The US EPA (2006, 1998) cautioned that their level of confidence in these data (PAR estimates) does not preclude the authentic number of bladder cancer cases truly caused by chlorination DBPs from being zero because causation of bladder cancer by this exposure has not been proven.

Despite the apparent consistency found in the bladder cancer epidemiology studies, there are a number of anomalies and inconsistencies that should be considered in judging the case for causation. These include:

(1) Limited evidence for early markers of bladder cancer
(2) Inconsistencies regarding smoking interaction
(3) Inconsistencies regarding female vs. male risk
(4) Inconsistencies regarding total water consumption and bladder cancer risk
(5) Limited studies including substantial urban areas free of chlorination DBPs
(6) Absence of a plausible toxicological agent among known DBPs

11.3.3.1 Evidence of early markers of bladder cancer

Ranmuthugala *et al.* (2003) studied a prospective cohort of 348 male volunteers (aged 30 to 65) for the presence of micronuclei as evidence of DNA damage in exfoliated bladder epithelial cells. About two thirds of 228 participants providing usable samples were exposed to THM4 in drinking water at levels from 38 to 157 μg/L vs. one third on an unchlorinated water supply. The results showed no difference in presence of micronuclei in the exposed vs. the unexposed groups. Villanueva *et al.* (2007) reported that they did find an association of micronuclei with THM exposure above the median (26 μg/L) vs. those below among 44 study controls who provided THM exposure information, but the difference was not statistically significant in this small sample subset.

11.3.3.2 Inconsistencies regarding smoking interaction

Smoking is recognized as an established risk factor for bladder cancer, so meaningful epidemiologic studies have to control their findings regarding exposure to chlorination DBPs for smoking status. Cantor *et al.* (1987); Freedman *et al.* (1997); McGeehin *et al.* (1993); Cantor *et al.* (1998) and Villenueva *et al.* (2007) all found that exposure to chlorination DBPs was a greater risk for smokers than for non-smokers. King and Marrett (1996) and Koivusalo *et al.* (1998) found the opposite result, smokers were at lesser risk for bladder cancer than smokers when exposed to chlorination DBPs. The expected finding of smokers being at a greater risk was more common among the studies, but the King and Marrett (1996) study is one of the strongest of all in terms of methodology and consistency of finding. The latter finding, in relation to smoking status, somewhat detracts from the overall consistency and plausibility of results given the accepted role of smoking in causation of bladder cancer. However, the King and Marrett anomalous finding on smoking status was not a strong one and it was not statistically significant.

Because smoking is recognized as a major cause of bladder cancer, adjustment for smoking status is a critical matter. Any inaccuracy or even imprecision in this correction could be a concern for studies yielding weak OR results.

11.3.3.3 Inconsistencies regarding female vs. male bladder cancer risk

Bladder cancer occurs much more frequently in men vs. women (generally at least 3 fold higher incidence). Cantor *et al.* (1998); Freedman *et al.* (1997) and Villanueva *et al.* (2004), all found that male bladder cancer risk associated with chlorination DBP exposure was much higher than that for females. On the other hand, Cantor *et al.* (1987) found female risk was much higher and Wilkins and Comstock (1981) and Villenueva *et al.* (2007) found male and female bladder cancer risk with chlorination DBP exposure to be similar. Doyle *et al.* (1997) studied a cohort of women only and found no evidence of a bladder cancer risk with chlorination DBP exposure (OR: 0.62, CI: 0.25–1.69) in contrast to the general consistency in risk findings among males or both sexes. Part of this inconsistency may be explained by the lower number of female cases of bladder cancer in all studies where both sexes were evaluated.

11.3.3.4 Fluid consumption and bladder cancer risk

Michaud *et al.* (1999) found that increased fluid consumption reduced bladder cancer risk in follow-up of the Prospective Health Professionals cohort (47,909 participants). During a 10 year period they found 252 newly diagnosed cases of bladder cancer. Consumption of water contributed to a lower risk (RR = 0.49, 0.28–0.86). Michaud *et al.* (2007) obtained a similar finding in Spain where higher water consumption yielded lower bladder cancer risk (OR = 0.47, 0.33–0.66). This trend held with bladder cancer inversely associated with water intake, regardless of THM exposure level. These findings are in contrast to Villanueva *et al.* (2006) who pooled results from six case control studies (Cantor *et al.* 1998; Cordier *et al.* 1993; Hung *et al.* 2005; King and Marrett, 1996; Koivusalo *et al.* 1998; Lynch *et al.* 1989) five of which focused on THM exposure, to find an increased bladder cancer risk with greater tap water consumption (OR = 1.46, 1.20–1.78) comparing >2L per day with <0.5 L per day.

11.3.3.5 Limited studies including urban areas free of chlorination DBPs

A critical aspect to judging the consistency of findings as a factor in supporting causation is that consistency should exist across a suitably wide range of exposures and populations to reduce the likelihood that observed associations being caused by a common, but unaccounted for, factor among the studies (see Sackett quote, Section 11.2.3.8). The troubling feature of the bladder cancer studies is that the underlying comparison involves exposures expected to be low in chlorination DBPs with those that are higher (e.g. high THMs vs. low THMs, chlorinated vs. unchlorinated, municipal disinfected vs. undisinfected).

The reality is that in most developed countries the foregoing differences will be most evident for rural communities or private farms where un-chlorinated water is commonly used. With the exception of Chevrier *et al.* (2004), the published urinary bladder cancer studies have not generally included larger urban regions using alternatives to chlorination as a substantial part of their study base. There can also be differences in THM levels because of other water quality parameters (e.g. total organic carbon in source water), but it appears likely that a substantial proportion of the low THM exposure group has likely come from small communities or private dwellings that do not practice chlorination. Chevrier *et al.* (2004) provides an intriguing exception to this concern, but this study has other limitations. The study was a rather small case-control study (281 bladder cancer cases) and while it included the feature of French communities using ozonation such that lower chlorination DBPs could be expected for a range of communities, exposure assignments in terms of THM4 was done based on an expert-opinion matrix assigning a THM4 level according to source water (ground or surface), chlorination in relation to filtration (pre- or post-) and water treatment period (1948–1966, 1967–1976, 1977–1986). The THM4 levels assigned by this matrix had some apparent anomalies, such as surface water with neither pre- nor post-filtration chlorination being assigned a THM4 level of 27.4 μg/L with no explanation of what was expected to produce the THM4.

These realities raise the concern about confounding differences that may contribute to bladder cancer that cannot be adequately adjusted for in the mathematical adjustment of the data that is done for expected confounding factors (i.e. age, smoking, etc.). Bladder cancer epidemiology studies have not taken advantage of exposure scenarios where THMs have been consistently very high for extended periods or where demographically similar communities would provide a clearer range of differences in THM levels. Likewise, there have not been major urinary bladder cancer studies done on large urban communities that do not use chlorine at all (e.g., German groundwater systems like Berlin and Dutch groundwater or riverbank filtrations systems serving urban areas) or do so only minimally in conjunction with a different primary disinfectant like ozone (e.g., urban communities in France could be studied in a larger, better resourced version of Chevrier *et al.* (2004) using actual THM4 data). Australia may provide a valuable location for finding widely differing THM exposures that still occur in major urban areas (e.g. Melbourne vs. Perth and Adelaide)

Cantor *et al.* (1998), one of the key studies showing a significant association of bladder cancer risk with chlorinated drinking water exposure, reported no association of bladder cancer risk with average lifetime population size of the city or town of residence but the data were not shown nor explained. King and Marrett (1996), another high quality study showing a significant association of bladder cancer risk with chlorinated DBPs, assessed their data for differences in response patterns among controls between urban and rural areas and were satisfied that this did not contribute bias. However, King and Marrett (1996) noted; "*While the analysis controlled for the effects of several factors, the possibility remains that results may be confounded by other factors. Of particular concern are bladder cancer risk factors which may be more common in urban areas associated with chlorinated surface-water supplies.*"

11.3.3.6 Absence of a plausible toxicological agent among known DBPs

Although all of the foregoing inconsistencies are noteworthy, the greatest concern, at least with respect to allowing the epidemiologic evidence to guide effective management of bladder cancer risk, is the absence of an identified plausible agent to explain the observed excess of bladder cancer associated with exposure to chlorination DBPs. None of the THMs, nor any other currently identified DBP, have both the capability of acting to cause bladder tumours, and sufficient potency and exposure concentration to yield bladder cancer predictions that would accord with the epidemiologic predictions.

The reality is that epidemiologic studies have been using THMs as an exposure metric for DBPs because THM monitoring data have been available and halogenated DBPs were an early focus of analytical methods development (Chapter 2), not because there is any toxicologically supported evidence to suggest that any of the THMs could be the causal agent for bladder or any other human cancer that has been predicted by those studies. This is problematic because there is no reason to expect that modifying disinfection practices to reduce THM exposures will necessarily reduce the concentration of any other agent that may be causal for bladder cancer, for which THMs have served as a surrogate.

A specific example of the problem created by the established surrogate status of THMs is that much focus has been drawn to tap water routes of exposure other than ingestion, such as showering, bathing and swimming. These alternate routes of exposure contribute substantially to the total THM exposure that humans will likely receive, but there is no reason that this additional, non-ingestion exposure is relevant to another substance which may not possess the same physical chemical properties (volatility and limited water solubility). For example, if the true causal agent (presuming that there is one) is substantially non-volatile and/or relatively unable to cross the skin barrier, predictions of importance for inhalation and dermal uptake for exposure to THMs will have no meaning for such a DBP.

Finally, we should take note that the outcome in this case, urinary bladder cancer, itself has substantial variation in rates among jurisdictions. Table 11.3 from a common IARC (2011) data base (using data provided by individual countries to IARC) suggests that the age-standardized incidence rate (ASIR) for 2008 of bladder cancer in males is almost three times higher in Spain than in Australia and Canada. Somewhat surprisingly, the ASIR of bladder cancer in males for the Netherlands is 40% higher, France is 50% higher and Germany is almost double the rates in Australia. These European countries have gone to some lengths to minimize the use of chlorine in drinking water for decades and given some of the challenging water sources used in Australia, chlorinated DBP exposures in Australia have likely been higher than typically found in North America.

Table 11.3 Bladder cancer incidence and mortality data for seven relevant countries Data from the International Agency for Research on Cancer http://globocan.iarc.fr/

Country	Male		Female		Both	
	ASIR[a]	ASMR[b]	ASIR[a]	ASMR[b]	ASIR[a]	ASMR[b]
Australia	10.2	3.0	2.7	1.1	6.2	2.0
Canada	11.2	4.1	3.1	1.3	6.8	2.5
Finland	12.2	3.0	2.7	0.9	6.9	1.8
France (metropolitan)	14.8	5.6	2.0	1.0	7.7	3.0
Germany	19.6	3.5	5.2	1.1	11.6	2.1
Spain	27.7	8.3	3.2	1.1	14.4	4.3
The Netherlands	14.1	4.8	3.8	1.7	8.4	3.0
USA	21.1	3.7	5.8	1.1	12.7	2.2

[a]ASIR is age-standardized incidence rate, incident cases per 100,000 age-standardized population
[b]ASMR is age-standardized mortality rate, deaths per 100,000 age-standardized population

The beginning of this Chapter cautioned, with good reason, against drawing substantive conclusions about causation based on population rates (e.g. limitations of aggregated studies and the "ecological fallacy"), but it is somewhat interesting that these data suggest that despite efforts to eliminate chlorination DBP exposures in European countries like the Netherlands, their population-wide bladder cancer incidence rates appear to be much higher than in Australia. Another caution is that there are differences in diagnosis and reporting (Ploeg *et al.* 2009; Wu *et al.* 2008) as well as in treatment of bladder cancer between jurisdictions that likely contributes to some of the apparent variations between countries. A particularly evident anomaly in these data is the very high ASIR for males in the U.S.A, almost double that in Canada, but the age-standardized mortality ratio (ASMR) is slightly lower in the U.S.A. than Canada.

A descriptive review of evidence on the epidemiology of bladder cancer (Murta-Nascimento *et al.* 2007) addressed to urologists mentions exposure to

DBPs in one paragraph under environmental exposures, but concludes that: "*cigarette smoking and occupational exposure to aromatic amines are the main known causes of UBC* [urinary bladder cancer]. *Phenacetin, chlornaphazine and cyclophosphamide also increase UBC risk. … Further research is needed to disentangle whether established risks / protective and susceptibility factors have a different effect on different UBC subgroups according to pathological or molecular subgroups. Similarly, an in-depth study of endogenous factors such as inflammation, oxidative stress, and hormonal status may help in identifying further causes of UBC as well as to elucidate the reasons for sex differences in incidence*." This extract indicates that there is considerable uncertainty about the causes of bladder cancer, but it also suggests that it would be a gross over-simplification to draw a direct causal connection between exposure to DBPs in drinking water and a major contribution to causation of bladder cancer.

11.4 ADVERSE REPRODUCTIVE OUTCOMES

11.4.1 Epidemiology evidence

As with the study of cancer outcomes and exposure to chlorination DBPs, the issue of exposure assessment for DBPs is a major problem for studies addressing reproductive outcomes (Arbuckle *et al.* 2002). The time exposure window for adverse reproductive effects is short compared with cancer, which typically has a lag time of 5 to more than 20 years. Theoretically this raises the prospect of obtaining better evidence of individual exposure for epidemiology studies on adverse reproductive outcomes. While there is some merit to this possibility (Zhang *et al.* 2009a,b), the downside is that the time window for causing an adverse reproductive effect is likely to be very short. That reality raises the difficulty that fluctuating water quality is likely to pose a challenge in characterizing the individual short term exposure that may be relevant to causing an adverse reproductive outcome.

The possibility of DBPs causing adverse reproductive outcomes raises a substantially different concern for DBP exposures via drinking water because, unlike cancer, a long term average exposure would not likely be the primary concern. In the case of adverse reproductive outcomes, short term, even maximum exposure values might be relevant, which raises prospects of a very different approach being needed for regulation of DBPs.

The number of possible adverse reproductive outcomes is large and many possible outcomes have been studied. There have been many published reviews considering the evidence for chlorination DBPs causing adverse reproductive outcomes. These include: Tardiff *et al.* (2006); Hwang and Jaakkola (2003); Bove *et al.* (2002); Graves *et al.* (2001); Nieuwenhuijsen *et al.* (2000a); Reif *et al.* (2000); Reif *et al.* (1996). More recently, there have been comprehensive reviews by Nieuwenhuijsen *et al.* (2009a,b,c; 2010) and a meta-analysis by Grellier *et al.* (2010).

11.4.1.1 Fetal growth and prematurity

Preterm delivery

Preterm delivery (also known as premature birth) is birth before the standard duration of pregnancy (i.e. sooner than 37 weeks after the start of the last menstrual period). A descriptive review (Hrudey, 2008) summarized findings from 14 separate studies (Lewis *et al.* 2007; Yang *et al.* 2007, 2000; Savitz *et al.* 2005, 1995; Wright *et al.* 2004, 2003; Aggazzotti *et al.* 2004; Jaakkola *et al.* 2001; Kallen and Robert, 2000; Dodds *et al.* 1999; Gallagher *et al.* 1998; Kanitz *et al.* 1996; Kramer *et al.* 1992). These, with one minor exception (Yang *et al.* 2000) provide remarkably consistent findings of no association between preterm delivery and any of the measures of DBP exposure. Hoffman *et al.* (2008a) as part of the major Savitz *et al.* (2005) study (Section 11.4.1.2) with its comparatively high quality exposure assessment, found no increase in preterm birth with high DBP exposure. A recent study in Spain (Villanueva *et al.* 2011) analyzing 2,070 newborns with a large range of THM exposures found no effect of chloroform or brominated THM on pre-term delivery. Another recent study in Crete (Paterlou *et al.* 2011) with 1359 pregnant women found no increased risk, particularly for brominated THMs for preterm delivery.

Low birth weight, very low birth weight and term low birth weight

Various measures of low birth weight have been evaluated, with the usual definition of low birth weight being less than 2500 g, very low birth weight is typically less than 1500 g and term low birth weight is less than 2500 g at 37 weeks gestation.

A descriptive review (Hrudey, 2008) summarized findings from 13 studies (Yang *et al.* 2007, 2000; Lewis *et al.* 2006; Hinckley *et al.* 2005; Toledano *et al.* 2005; Wright *et al.* 2003; Jaakkola *et al.* 2001; Kallen and Robert, 2000; Dodds *et al.* 1999; Gallagher *et al.* 1998; Kanitz *et al.* 1996; Bove *et al.* 1995; Kramer *et al.* 1992) which collectively show some consistency across these different studies for there being comparatively few significant associations between these outcomes and various measures of DBP exposure, except for Toledano *et al.* (2005), low and very low birth weight; Gallagher *et al.* (1998) and Lewis *et al.* (2006) for term low birth weight. The limitations on drawing a causal inference for Toledano *et al.* (2005) are discussed in Section 11.4.1.3, but a very large sample size used in Toledano *et al.* (2005) makes it, all else being equal, more persuasive than smaller studies.

Both Gallagher *et al.* (1998) and Lewis *et al.* (2006) are retrospective cohorts which relied upon maternal residence at birth for assigning THM exposure. Gallagher *et al.* (1998) only considered exposure during the third trimester to find significant associations with THM4 exposure whereas Lewis *et al.* (2006) considered all three trimesters for exposure and found only a significant association with THM4 exposure in the second trimester (but not the first or third). No mechanistic theory was suggested for how a second trimester effect alone would logically explain term low birth weight outcome.

A recent study in Spain (Villanueva *et al.* 2011) analyzing 2158 newborns with a large range of THM exposures found no effect of chloroform or brominated THM on birth weight. Another recent study in Crete (Paterlou *et al.* 2011) with 1359 pregnant women found no increased risk, particularly for brominated THMs for low birth weight.

In contrast, a study in Lithuania (Grazuleviciene *et al.* 2011) with 4161 pregnant women found a dose-response relationship and a significant relationship (OR = 2.17, CI: 1.19–3.98) for low birth weight associated with estimated internal chloroform dose comparing the highest tertile internal chloroform dose with the lowest tertile of dose.

Intrauterine Growth Retardation (IUGR) and Small for Gestational Age (SGA)

IUGR is typically defined by being SGA, so these concepts are intertwined. A descriptive review (Hrudey, 2008) summarized findings from 15 studies addressing these outcomes using the terminology adopted by the investigators for their study (Yang *et al.* 2007, 2000; Hinckley *et al.* 2005; Porter *et al.* 2005; Savitz *et al.* 2005; Aggazzotti *et al.* 2004; Infante-Rivard, 2004; Wright *et al.* 2004, 2003; Jaakkola *et al.* 2001; Kallen and Robert, 2000; Dodds *et al.* 1999; Gallagher *et al.* 1998; Bove *et al.* 1995; Kramer *et al.* 1992). With the exception of Kramer *et al.* (1992); Infante-Rivard (2004) there is no evidence of a significant association of IUGR and exposure to chlorination DBPs in results from four studies. Kramer *et al.* (1992) showed a significant association (OR = 1.8, CI: 1.1–2.9) for "high" chloroform ($\geq$10 μg/L vs. low $<$1 μg/L) in an early case control study with limited exposure assessment although from a water quality perspective, 10 μg/L is not a high chloroform concentration. Infante-Rivard (2004) found a very significant association with THM exposure for a subgroup of a case-control study that was found to have 1 or 2 variant alleles of the CYP2E1 gene that is implicated in metabolism of chloroform.

With the sole exception of Savitz *et al.* (2005) there is no evidence of a significant association of SGA and exposure to chlorination DBPs in nine studies. Savitz *et al.* (2005) found an isolated association of SGA with THM4 exposure only during the period of week 27 to birth (OR = 2.07, CI: 1.12–3.82) when they dichotomized exposure above and below the U.S. regulatory limit of 80 μg/L for THM4. While it is not possible to dismiss this finding, it was not found when they analyzed THM4 exposure by quintiles for THM4 concentration, THM4 ingested amount, THM4 integrated exposure or THM4 shower and bath exposure only. The failure to find an association in the latter cases does little to support the simple dichotomized analysis as being meaningful. Hoffman *et al.* (2008b) elaborated on the marginal and negative findings for SGA for this study.

A recent study in Spain (Villanueva *et al.* 2011) analyzing 2073 newborns with a large range of THM exposures found no effect of chloroform or brominated THM on small for gestational age. A study in Crete (Paterlou *et al.* 2011) with 1359 pregnant women found no increased risk, particularly for brominated THMs for

SGA. A study in Lithuania (Grazuleviciene *et al.* 2011) with 4161 pregnant women found SGA was not significantly associated with chloroform internal dose.

Summary of fetal growth and prematurity

Nieuwenhuijsen *et al.* (2009a,b) in a descriptive review summarized studies as showing some evidence of an association between THMs and SGA and to a lesser extent pre-term delivery, but evidence for low birth weight was described as inconclusive and inconsistent. Grellier *et al.* (2010) undertook a systematic review and meta-analysis concluding that there was little or no evidence for exposure to THMs and adverse birth outcomes relating to fetal growth and prematurity, with the possible exception of SGA. Nieuwehuijsen (2010) found little evidence of any effects of THMs on pre-term birth weight and low birth weight in any studies for the HiWATE DBP research program in Europe.

11.4.1.2 Spontaneous abortion

More commonly referred to as miscarriage, spontaneous abortion is the loss of the developing fetus before it is viable (pregnancy loss at $\leq$20 weeks of gestation). This is a difficult pregnancy outcome to study because it is not typically well-tracked in the healthcare system and it poses problems of verification of outcome for retrospective studies. A descriptive review (Hrudey, 2008) summarized findings from three studies which addressed spontaneous abortion (Savitz *et al.* 1995; 2005; Waller *et al.* 1998).

Savitz *et al.* (1995) performed a case-control study, the first to report on spontaneous abortion. Their use of individual questionnaires allowed for much better control of confounding than with birth certificate studies. Little detail was provided about how THM4 exposure was determined beyond obtaining data from public water suppliers whose monitoring data was apparently assigned to each case or control. Risk of miscarriage was slightly (but not significantly) increased among women who used bottled water compared with private wells while no difference was found between community (disinfected) water source vs. private well water.

Savitz *et al.* (1995) found no significant association using exposure based on an average high THM4 (81–169 μg/L) vs. lower THM4 (41–60 μg/L). This was despite noting a prediction of an association for an OR of 1.7 per 50 μg/L increase in THM4 exposure, but they discounted this finding on the basis that they had found the highest sextile of exposure had an adjusted OR = 2.8 (CI: 1.2–6.1), but the second to highest sextile had an adjusted OR = 0.2 (CI: 0.0–0.5). This is certainly indicative of there being no dose response based on THM4 for their data. An exposure metric based on THM4 concentration multiplied by the number of glasses of water consumed per day was more stable and it showed no evidence of a positive association of spontaneous abortion with THM4 exposure.

Savitz *et al.* (1995) found a consistent pattern of decreasing risk with increasing water consumption. Overall, they concluded that drinking water source was not related to risk of miscarriage. They considered only medically-treated miscarriages and they did find some indication of differential under-treatment related to social class which might bias their results.

Waller *et al.* (1998) performed the first prospective cohort study on spontaneous abortion which marked a substantial improvement in confirming the occurrence and timing of the adverse outcome. They also used personal interview questionnaires to characterize the level and nature of water consumption combined with water utility distribution system monitoring to estimate THM4 exposure levels. They found a significant OR of 1.8 (CI: 1.1–3.0) for a metric of high THM4 exposure vs. low ($\geq$75 μg/L THM4 with >5 glasses of cold tap water consumption per day vs. either <5 glasses of cold tap water with THM4 <75 μg/L or receiving water from a utility providing $\geq$95% groundwater). THM4 exposure for each individual was determined for most (77%) by averaging all the readings for the applicable water utility for the first trimester of pregnancy. For women where these data were not available for the applicable time period, readings taken within 30 days of the first trimester (4% of cohort) were used or an average for the applicable water utility (9% of cohort) was used. It was not clear from the paper what was done with the remaining 10% of the cohort. The questionnaire analysis allowed consideration of showering and/or swimming behaviour, but neither of these activities that are known to increase THM4 exposure yielded a higher risk for spontaneous abortion. A subsequent re-analysis of the study data set to refine exposure assessment failed to show any substantive advantages with other approaches to analyzing the available data (Waller *et al.* 2001a,b). When exposure was classified according to high BDCM exposure ($\geq$18 μg/L) results showed a higher OR = 2.0 (CI: 1.2–3.5) but this paper was not clear about what was the reference group.

The findings from Waller *et al.* (1998) clearly invited some follow-up study to seek validation for the findings, with improved exposure assessment being a major need. This was addressed by locating three communities with differing THM exposures, one with moderate THMs with a low proportion brominated, one with moderate THMs with a high proportion brominated and one with low THMs. The moderate THM communities used chloramination which had the effect of stabilizing THM concentrations in the distribution system. For all communities, distribution system sampling was performed to confirm consistency of THM levels so that a single sampling site could be used to characterize THM levels during the study. Weekly (biweekly at the low THM site) tap water samples were collected and analyzed for THM4, nine HAAs and total organic halogen (TOX). Several exposure indices, including a number of behavioural influences, and key variables were generated and applied during several critical time windows during pregnancy. Oddly, swimming exposure was not assessed.

The follow-up study (Savitz *et al.* 2005; 2006) was designed to be able to replicate the Waller *et al.* (1998) finding and it found essentially a null result (OR = 1.1, CI: 0.7–1.7) for high personal THM exposure ($\geq$75 μg/L THM4 with >5 glasses of cold tap water consumption per day vs. <5 glasses of cold tap water or THM4 <75 μg/L). The Savitz *et al.* (2006) study found that BDCM dichotomized in a manner similar to Waller *et al.* (1998) showed an OR = 1.6 (CI: 1.0–3.5), suggestive of a possible BDCM effect. However, when BDCM was analyzed by quintiles of either BDCM concentration or ingested amount (μg/d), it failed to show any signs of an elevated OR (ranging between 0.7 and 1.1) for any quintile referred to the lowest quintile. There were some non-significant and inconsistent signs that TOX exposure was associated with spontaneous abortion.

In an invited commentary on this paper at the time it was published Howards and Hertz-Picciotto (2006) concluded: "*Although the investigation by Savitz et al. does not preclude effects of DBPs on pregnancy, the study provides some confidence that exposure to THMs through most routes is not a threat to fetal viability during the first 20 weeks of gestation. Considering the public health value of controlling waterborne pathogens economically through chlorination, future studies of spontaneous abortion and THMs are probably not warranted, although studies of swimming may be useful.*"

11.4.1.3 Stillbirth

Stillbirth is a progression from spontaneous abortion whereby a fetus that has developed beyond 20 weeks gestation dies during birth or late stages of pregnancy. A descriptive review (Hrudey, 2008) summarized findings from five studies (Aschengrau *et al.* 1989; Dodds *et al.* 2004; King *et al.* 2000b; King *et al.* 2005; Toledano *et al.* 2005).

Aschengrau *et al.* (1989) found a significant OR = 2.2 (CI: 1.3-3.6) for stillbirth in a wide-ranging exploratory case-control study based on the limited exposure metric of determining whether cases and controls were exposed to chlorinated or chloraminated water. A similar comparison of cases and controls based on exposure to surface vs. groundwater yielded a crude OR = 0.9 (0.6–1.4). King *et al.* (2000b) studied a retrospective cohort using a Nova Scotia perinatal database with almost 50,000 singleton deliveries between 1988 and 1995. THM exposure based on public water supply monitoring was assigned to maternal address from the database. A number of significant associations between exposures and stillbirths were found, most notably for THM4 ≥ 100 μg/L vs. <50 μg/L (OR = 1.66, CI: 1.09–2.52), for the same exposure metric but with the outcome being stillbirths diagnosed as asphyxia (OR = 4.57, CI: 1.93–10.77) and for BDCM ≥ 20 μg/L vs. <5 μg/L (OR = 1.98, CI: 1.23–3.49).

Dodds *et al.* (2004) performed a case-control study with 112 stillbirth and 398 live birth controls in Nova Scotia and eastern Ontario. Subjects were interviewed and those women on a public water supply provided a residential water sample that was timed to coincide with approximately 15 weeks gestation, but one year after the stillbirth in an attempt to estimate the seasonal variation in THM levels.

Exposures were estimated based on ingestion, showering and bathing, as well as on ingestion alone. Significant results were reported for THM4 $\geq$ 80 μg/L vs. 0 μg/L (OR = 2.2, CI: 1.1–4.4) and for BDCM $\geq$ 10 μg/L vs. 0 μg/L (OR = 2.3, CI: 1.0–4.9). The referent category for these analyses contains subjects served by a private well raising a concern about possible confounding of a public vs. private individual water supply (i.e. an urban vs. rural) effect. There was a lack of a dose-response relationship for THM and BDCM for either exposure concentration or total exposure somewhat undermining the causal hypothesis. An analysis of these cases and controls for an association of stillbirth with estimates of HAA exposure found no significant associations (King *et al.* 2005).

Toledano *et al.* (2005) performed a very large retrospective cohort (a total of 869,314 live and 4852 stillbirths) in three water supply regions of England. The size of this study makes its findings more persuasive, all else being equal, compared with other smaller studies on stillbirth. Postal code of maternal residence was used to assign an exposure zone and THM4 exposure was estimated for the third trimester. In one of the three regions, the high THM4 zone showed an OR = 1.21 (CI: 1.03–1.42) compared with the low THM4 zone. The high THM4 zone in this region tended to be more socio-economically deprived than the low THM4 exposure zone. Because stillbirth is associated with socio-economic deprivation and the adjustment of the OR to account for the observed scale difference reduced the crude OR by half, there is a possibility of residual confounding in the remaining observation. The lack of any significant association of stillbirth with high THM exposure in either of the other two regions taken together with the qualified finding in one region is not supportive of THMs being causal for stillbirth.

11.4.1.4 Congenital anomalies (birth defects)

A number of studies have addressed the issue of various birth defects in relation to chlorination DBP exposure. A descriptive review (Hrudey, 2008) summarized findings from 13 studies (Hwang *et al.* 2008, 2002; Chisholm *et al.* 2008; Nieuwenhuijsen *et al.* 2008; Shaw *et al.* 2003; Cedergren *et al.* 2002; Dodds and King, 2001; Kallen and Robert, 2000; Dodds *et al.* 1999; Magnus *et al.* 1999; Klotz and Pyrch, 1999; Bove *et al.* 1995; Aschengrau *et al.* 1993). Because specific birth defects are relatively rare, a prospective cohort study design is not feasible and even case – control studies are challenging to acquire enough cases that can be traced and interviewed for detailed exposure assessment. As a result, the published studies (except for Klotz and Pyrch, 1999) are all retrospective cohort studies using birth certificates or birth defect registries. This approach makes it difficult to obtain any individual data on the birth mother beyond what is recorded in the registry.

Considering all birth defects combined, for six studies (Hwang *et al.* 2008, 2002; Chisholm *et al.* 2008; Magnus *et al.* 1999; Bove *et al.* 1995; Aschengrau *et al.* 1993) which addressed this combined measure, only Bove *et al.* (1995) and Chisholm *et al.*

(2008), both birth registry-based studies, showed a significant relationship with chlorination DBP exposure. All birth defects combined is a somewhat biologically implausible outcome. In the case of Bove *et al.* (1995), the OR reported only a 90% confidence interval so the marginal significance for the broad category of all defects is likely not meaningful. Chisholm *et al.* (2008) is only marginally significant (OR = 1.22, CI: 1.01–1.48). Hwang *et al.* (2008) found a significant relationship (OR = 1.21, CI: 1.07–1.36) for a comparison of the low THM4 (5–9 μg/L) exposure group compared with the lowest THM4 (0–4 μg/L) exposure group, but little meaning can be attached to this finding given the ORs of 0.97 and 1.00, respectively, found for the medium THM4 (10–19 μg/L) and high THM4 ($\geq$20 μg/L) exposure groups compared with the lowest THM4 exposure level.

Considering cardiovascular anomalies, for nine studies (Chisholm *et al.* 2008; Nieuwenhuijsen *et al.* 2008; Shaw *et al.* 2003; Cedergren *et al.* 2002; Hwang *et al.* 2002; Dodds and King, 2001; Kallen and Robert, 2000; Dodds *et al.* 1999; Bove *et al.* 1995) which addressed this measure, only Cedergren *et al.* (2002) and Chisholm *et al.* (2008), both birth registry-based studies, showed a significant relationship with chlorination DBP exposure. In the case of Cedergren *et al.* (2002), the study was very large (almost 72,000 live births providing 753 cases), but the exposure breakpoint of above or below THM4 of 10 μg/L is difficult to imagine as being a causal exposure. Chisholm *et al.* (2008) had a more meaningful exposure range, but the finding is only marginally significant (OR = 1.62, CI: 1.04–2.51) despite being based on 260 cases. The strongest study for this defect was Nieuwenhuijsen *et al.* (2008) which analyzed 7,823 cases and found no significant associations with various measures of THM exposure.

Considering cleft defects, for nine studies (Hwang *et al.* 2008, 2002; Nieuwenhuijsen *et al.* 2008; Shaw *et al.* 2003; Dodds and King, 2001; Kallen and Robert, 2000; Dodds *et al.* 1999; Magnus *et al.* 1999; Bove *et al.* 1995) which addressed this measure, only Bove *et al.* (1995) showed a significant relationship with chlorination DBP exposure, but the result (OR = 3.17, CI: 1.18–7.26) is neither very stable, nor meaningful, based on a 90% confidence interval. The other studies are all relatively consistent in demonstrating no relationship between cleft defects and chlorination DBP exposure.

Considering central nervous system anomalies, including neural tube defects (NTD) and spina bifida, for 10 studies (Chisholm *et al.* 2008; Nieuwenhuijsen *et al.* 2008; Shaw *et al.* 2003; Hwang *et al.* 2002; Dodds and King, 2001; Kallen and Robert, 2000; Dodds *et al.* 1999; Magnus *et al.* 1999; Klotz and Pyrch, 1999; Bove *et al.* 1995) which addressed these outcomes, Bove *et al.* (1995) showed an OR = 2.96, 90% CI: 1.26–6.62, for THM4 > 80 μg/L vs. < 52 μg/L, providing a somewhat unstable and only marginally significant finding. Dodds and King (2001) found with 77 cases of NTD, a RR = 2.5, CI: 1.2–5.1 for BDCM $\geq$ 20 μg/L vs. < 5 μg/L for NTD, while finding no relationship with chloroform exposure. In the same study, they found that this BDCM exposure comparison was significantly protective for cardiovascular defects (RR = 0.3, CI: 0.3–0.7). In contrast, Nieuwenhuijsen *et al.* (2008) found no significant association of NTD with any of three specific chlorination DBP exposure metrics, including two focused only on brominated THMs, and they analyzed 3324 NTD cases. Hwang *et al.* (2002) found an OR = 2.60, CI: 1.30–5.26 for a comparison of no chlorination with high color vs. no chlorination with low color. A comparison of chlorination with high color vs. no chlorination with low color yielded an OR = 0.68, CI: 0.24–1.95. The results from Hwang *et al.* (2002) do not support an association of NTD with exposure to chlorination DBPs while also showing the instability of predictions and the chance occurrence of spurious predictions for relatively rare defects like NTD.

Considering urinary tract defects, for six studies (Hwang *et al.* 2008, 2002; Chisholm *et al.* 2008; Nieuwenhuijsen *et al.* 2008; Magnus *et al.* 1999; Aschengrau *et al.* 1993) evaluating this outcome, two studies reported a significant association. Aschengrau *et al.* (1993) reported an OR = 4.1 (CI: 1.2–14.1) when comparing exposure to chlorinated water with chloraminated water and Magnus *et al.* (1999) found an OR = 1.99 (CI: 1.10–3.57) comparing exposure to chlorinated water with high colour to no chlorination for water with low colour. Given the ambiguous exposure metric and the wide confidence intervals observed, the Aschengrau *et al.* (1993) finding offers little evidence for a causal association of urinary tract defects with chlorination DBPs. The Magnus *et al.*(1999) study which analyzed 122 urinary tract defect cases using an imprecise exposure metric is much less convincing than Nieuwenhuijsen *et al.* (2008) which found no significant association of urinary tract defects with any of three specific chlorination DBP exposure metrics, having analyzed 5063 urinary tract defect cases.

Considering respiratory defects, for five studies (Chisholm *et al.* 2008; Nieuwenhuijsen *et al.* 2008; Hwang *et al.* 2002; Magnus *et al.* 1999; Aschengrau *et al.* 1993) which addressed this outcome, only Aschengrau *et al.* (1993) reported a significant association (OR = 3.20, CI: 1.1–9.5) when comparing exposure to chlorinated water with chloraminated water. Given the ambiguous exposure metric and the wide confidence intervals observed, the Aschengrau *et al.* (1993) finding offers little evidence for a causal association of respiratory defects with chlorination DBPs.

Summary for all congenital anomalies

Overall, the results of epidemiology studies for birth defects either in total or as major specific types are not supportive of a causal linkage between exposure to chlorination DBPs and any birth defects. The state of the evidence in this regard has been summarized in an excellent manner by Nieuwenhuijsen *et al.* (2008) the largest study to date to address the possibility of chlorination DBPs being associated with specific birth defects. Nieuwenhuijsen *et al.* (2008) concluded: "*Currently there is no plausible biological mechanism by which chlorination by-products could cause congenital anomalies, particularly at low concentrations. Nonetheless, the policy of minimizing the concentrations of chlorination by-products in the public water supply by removing natural organic precursors, while simultaneously maintaining the level of protection from disinfection, seems appropriate in view of concerns about possible adverse reproductive health effects (Nieuwenhuijsen* et al. *2000a,b,c). The WHO has continued to emphasize that high levels of protection from disinfection should never be compromised in trying to reduce disinfection by-product concentrations; our data do not detract from that view.*"

Nieuwenhuijsen *et al.* (2009a,b) performed descriptive reviews that summarized studies for congenital anomalies described the evidence as inconclusive and inconsistent. Nieuwenhuijsen (2010) found little evidence of any effects of THMs on congenital anomalies in any studies for the HiWATE program in Europe. Nieuwenhuijsen *et al.* (2009c) performed a review and meta-analysis of studies on chlorination DBPs and congenital anomalies and concluded that the meta-analysis found little evidence of an association of these outcomes with chlorination DBPs.

11.5 INTERPRETATION OF PUBLIC HEALTH RISK

11.5.1 Recognizing uncertainty in epidemiology evidence

A key issue for judging the evidence and deciding on an appropriate risk management response is how strong is the epidemiological evidence for supporting a causal association rather than merely a chance association between exposure and effect. As indicated in Section 11.2.3.8, with observational epidemiology, chance spurious associations are to be expected. For example, in a retrospective cohort, Hwang *et al.* (2002) found an apparently very high OR = 2.60 (1.30–5.26) for neural tube defects, but this was found for a comparison of mothers exposed to water having high color and no chlorination with mothers exposed to water having low color and no chlorination. In contrast, another comparison of chlorination/high color exposure yielded an apparently protective OR = 0.68 (0.24–1.95). These examples show how for observational studies, estimates of risk which clearly have nothing to do with a chlorination DBP causal hypothesis can be found to exceed an OR = 1.0 when multiple analyses are done searching for a possible association of water quality with a health outcome.

11.5.2 Exposure assessment issues

An over-riding challenge for providing better quality epidemiology studies has been the limitations of exposure assessment (Arbuckle *et al.* 2002). The default assumption among epidemiologists has been that any limitations in exposure assessment will have the effect of reducing the size of any observed association (presuming that exposure mis-classification errors will be random). So far, there are not any clear examples where studies using better exposure assessment compared with previous studies using weaker exposure assessment have yielded substantially stronger measures of association. Amy *et al.* (2006) found some slight increases in risk estimates for having improved DBP exposure assessment for the Ontario (King and Marrett, 1996) and Iowa (Cantor *et al.* 1998) case control studies. In another case discussed below, regarding spontaneous abortion, the opposite has occurred (i.e. improved measures of individual exposure have yielded weaker measures of association, a finding that is not consistent with the observed association being causal).

The possibility of chlorination DBPs causing adverse reproductive outcomes was largely one of academic and research interest before the publication of the Waller *et al.* (1998) study. Numerous previous studies had found suggestive, but inconsistent and usually not significant associations of a variety of adverse birth outcomes with chlorination DBPs. The large size and comparative strength of the prospective cohort study reported by Waller *et al.* (1998) drew justifiable attention to the reported significant association of spontaneous abortion with THM4 and even stronger association with BDCM exposure. The toxicological evidence has mainly suffered from either the lack of supporting findings or enormously high doses required with rodents to yield relevant responses (Graves *et al.* 2001; Tardiff *et al.* 2006). Typically, NOAEL or LOAEL values for evaluated DBPs range from a fraction of 1% to higher percentages of the median lethal dose (LD50) for the chemicals studied. For such comparatively steep dose-response curves (i.e. small differences in dose between a LOAEL and a median, 50%, response), there are limited grounds to expect health effects at exposures corresponding to very low fractions of the LOAEL. Considered otherwise, the exposure levels are often multiples from hundreds to tens of thousand fold higher than MAC values for the target chemicals.

There was a compelling need to confirm whether chlorination DBPs could possibly cause adverse health effects based on short-term (i.e. daily peak) exposures rather than the long-term chronic exposures of concern for bladder cancer (generally greater than 40 years of exposure needed for elevated risk). This need combined with a number of inconsistencies in the Waller *et al.* (1998) findings justified a major investment in a study to attempt replication of those findings. Waller *et al.* (2001a) undertook a re-analysis of the 1998 study to improve the individual exposure assessment using "closest site" THM monitoring data, but they did not find an increase in the strength of association compared with the original analysis using utility-wide THM data. Some increases in OR were found by applying variance-based weighting factors and data subset analyses, but these improvements resulted in smaller sample size so that the usefulness of these improvements was questioned by the authors. Overall, the lack of a "gold standard" (i.e. a valid biomarker) for establishing THM exposure of individuals remains a major limitation to these studies.

The Savitz *et al.* (2005, 2006) studies were pursued to replicate the Waller *et al.* (1998) findings. Perfect replication was not achieved because Savitz *et al.* (2005, 2006) used chloraminated systems in their study which may have introduced other unknown DBP factors and it significantly reduced variability in THM exposure levels. However, Savitz *et al.* (2005, 2006) did use substantially improved assessment of individual exposure of participants to chlorination DBPs, they found no significant association between spontaneous abortion and THM4 exposure. In the absence of strong new evidence to show that some chlorination DBPs are likely to cause spontaneous abortion, this possibility should remain a research hypothesis, but the evidence does not warrant seeking more cautionary levels for currently regulated DBPs (e.g, THMs, HAAs).

There is some limited documentation of possible exceptions to an expectation that any exposure misclassification will be non-differential. For example, Isacson *et al.* (1983) reported in a study of bladder cancer in Iowa that up to 10% of bladder cancer cases reported as resident in towns of less than a 1000 population were actually rural incident cases who were now resident in these smaller communities, possibly because of the need for healthcare services. Unless such circumstances were studied by a cohort or case-control design that ensured accurate and complete documentation of individual exposure history over most of the 40 year exposure window that has been typically found to be associated with an elevated bladder cancer risk, such transplanted rural cases would be incorrectly assigned to towns which provided disinfection exposure. The stronger bladder cancer studies have sought to assure such individualized, long term exposure histories are obtained.

11.5.3 Protecting the public's health and dealing with media

The motivation for controlling chlorination DBPs in drinking water is obviously to reduce health risk that may be associated with one or more such chemicals. This is fully consistent with the public health practice foundation that emphasizes disease prevention. Where substantial uncertainty exists, as in the case of chlorination DBPs, being suitably precautionary is justified, given the broad public exposure provided by community drinking water supplies. However, there is also a responsibility inherent among public health professionals to exercise precaution in a responsible manner that neither undermines the credibility of public health practice nor causes unwarranted fear among the public who cannot be expected to understand the nature of uncertainties involved and what levels of precaution may be warranted.

There have been a number of cases where the fear of health effects from chlorination DBPs has played a role in contributing to allowing a waterborne disease outbreak to occur (e.g. Creston/Erickson, B.C.; Walkerton, Ontario; Bramham, England; Asikkala, Finland; Transtrand, Sweden; all described in: Hrudey and Hrudey, 2004). Such cases are examples where the balance in dealing with the trade-off between the certain danger of pathogens and possible dangers of DBPs has not been achieved effectively.

11.5.3.1 DBPs and cancer

How the media covers scientific discussion of epidemiologic evidence can be illustrated from reporting of an expert panel (Mills *et al.* 1998) organized by Health Canada to review the health effects evidence for THMs. This panel, which included a number of epidemiologists involved in THM studies, concluded for the evidence available on THMs that "*it was possible (60% of the group) to probable (40%) that chlorination by-products pose a significant risk to the development of cancer, particularly, bladder cancer. The risk of bladder cancer, and possibly other cancers, poses a risk to public health; this is a moderately important public health problem.*" The expert group consisted of approximately equal numbers of epidemiologists and toxicologists. The conclusion about the severity of the public health problem being only moderate was influenced by the relatively small number of cancers expected combined with a comparatively small, 30% mortality rate for bladder cancer in Canada (CCS 2011).

In an unusual move, the Editor of the Journal which published the Expert Panel report, but who was not himself a member of that Panel, published a personal commentary in the journal along with the Expert Panel report in which he stated: "*Chloroform and several other CBPs* [chlorination DBPs] *are known to cause cancer in experimental animals … there is an urgent need to resolve this and to consider actions based on the body of evidence, which at a minimum, suggests that lowering of CBP levels would prevent a significant fraction of bladder cancers. In fact, given the widespread and prolonged exposure to CBPs and the epidemiologic evidence of associations with several cancer sites, future research may establish CBPs as the most important environmental carcinogens in terms of the attributable cancers per year.*" (Wigle, 1998).

In subsequent media commentary on this issue the Wigle commentary, dominated over the expert panel report (Driedger and Eyles, 2003). Because the official documentation associated with the drinking water criteria is largely inscrutable to all but a few specialists who follow these topics, drinking water professionals are more likely to learn about these issues from the popular media coverage which conveys a message that suggests much greater certainty about health effects than truly exists.

11.5.3.2 DBPs and adverse reproductive outcomes

Public reaction to the possibility of adverse health outcomes from chlorination DBPs has often been fearful. After the Waller *et al.* (1998) study reported an elevated risk of spontaneous abortion associated with THMs, media stories about THM risks to unborn babies were widely reported in the U.S. The Public Health Department of the municipality of Chesapeake, Virginia, which was in the midst of changing its water system to meet the federal THM standard, issued a warning to pregnant mothers to drink bottled water after the water treatment plant personnel brought the Waller *et al.* (1998) study to their attention (Huslin, 2002). One result of that attempt at informing the public of the possible health risk was that the municipality became a defendant in lawsuits from 214 plaintiffs claiming breach of contract and warranty, battery negligence, nuisance, trespass, violation of the state Consumer Protection Act and fraud (Anon, 2005). Ultimately the municipality was found to be immune to these lawsuits by the Virginia Supreme Court, citing the municipalities ongoing efforts to comply with the THM4 standard. It is not difficult to imagine the emotional toll these circumstances took on affected parties. One case was described as follows (Huslin, 2002):

> "Annette Spaven already had three children when she found out she was pregnant again four years ago.
> She and her husband were surprised but pleased by the prospect of welcoming another child into their Chesapeake, Va., home. So when she suffered a miscarriage in the first trimester, they tried again.
> Six months later, she lost another baby.
> 'I wondered if something was wrong with me,' said Spaven, 38.
> About the same time, two women on her block had miscarriages. Across town, a woman gave birth to a boy who died shortly after birth. For more than a decade, they and others wondered why they'd suddenly lost their pregnancies.
> Today, many are also wondering something else: Might they have lost their babies simply because they drank tap water while they were pregnant? It's a question that has roiled this booming port community ever since residents became aware of controversy surrounding chemicals in the public drinking supply."

An example of media coverage of an epidemiology study exploring adverse reproductive outcomes occurred in Perth, Western Australia. A scientific publication that was still 'in press' (Chisholm *et al.* 2008) at the time, was made public and the local newspaper ran a story under the headline "***Tapwater 'increases risk of birth defects'*** (Guest, 2008). The story began: "*Chemical by-products in tap water in some Perth suburbs are increasing the risk of birth defects and pregnant women may need to avoid the danger, health researchers led by the University of WA have warned.*" [emphasis added]

The newspaper article quoted the Principal Investigator stating: "*If you introduce poisons to the foetus when it's forming things go wrong, development is very complex and the slightest toxin can interrupt the normality of that development.*" While that quotation by itself might not offend a reproductive toxicologist, given our understanding of the sensitivity of the developing fetus, within the context of this story, an obvious implication that can be drawn is that DBPs are the *'poisons'* and that exposure to the slightest amount of them can make *'things go wrong'*."

Surely, any woman who has given birth to a child with a birth defect would find this news coverage distressing because of the implied message that she could be responsible for her child's defect if she had consumed Perth tapwater. Of course it is difficult to convey an accurate message within the constraints of the popular media. However, in this particular case, the certainty of the evidentiary basis for delivering a potentially guilt-inducing risk message to the public deserves extremely thoughtful consideration.

The study relied upon four years of data from a birth registry whereby birth outcomes were compared using 47 water samples analyzed for THMs. Sample locations were linked to maternal residential postal code at time of birth. Maternal age and socioeconomic status code were also obtained for each birth. Based on average THM4 analyses on samples over 6 collection dates, the study area was divided into 3 regions (low: $54 \pm 16.6\ \mu g/L$, medium: $109 \pm 28.9\ \mu g/L$, and high: $137 \pm 24\ \mu g/L$). Birth defect data were analyzed for all birth defects and 7 other categories. Significant results for this registry-based retrospective cohort study which lacked individualized exposure assessment (no questionnaire data) were only obtained for two categories, all birth defects combined (OR = 1.22, CI: 1.01–1.48) and cardiovascular system defects (OR = 1.62, CI: 1.04–2.51) as reviewed in Section 11.4.1.4.

Hrudey (2008) compared results for six studies reporting all birth defects combined. The Chisholm *et al.* (2008) study is the first to report any significant association for the somewhat biologically implausible combination of "all congenital anomalies" with chlorination DBPs. Hwang *et al.* (2002) reported a significant association between exposure groups of no chlorination – high color compared with no chlorination – low color (OR = 1.18, CI: 1.02–1.36), but this comparison clearly has nothing to do with chlorination DBPs, since neither exposure group experienced chlorination. Hwang *et al.* (2008) showed a significant OR for low TTHM (5–9 μg/L) vs. lowest TTHM (0–4 μg/L), but significance was not found when higher TTHM groups were compared with the lowest TTHM group. The results of the Chisholm *et al.* (2008) study are only marginally significant, and taken together with the negative findings from the other 5 studies and the weak study design, Chisholm *et al.* (2008) provides a very weak basis to make any credible claims about DBPs in drinking water causing birth defects.

Hrudey (2008) compared results for eight studies reporting cardiovascular anomalies. The Chisholm *et al.* (2008) study and Cedergren *et al.* (2002) are the first to report any significant association of cardiovascular anomalies with chlorination DBPs. Cedergren *et al.* (2002) was a similar, but very large, birth registry, retrospective cohort study without any individual questionnaire data to improve exposure assessment. They found an OR = 1.30, CI: 1.08–1.56 when comparing exposure groups of THM4 $> 10\ \mu g/L$ with THM4 $< 10\ \mu g/L$. These exposure categories are well below the low exposure group category in the Chisholm *et al.* (2008) study which found an OR = 1.62, CI: 1.04–2.51 in comparing the high THM4 with the low THM4. Chisholm *et al.* (2008) observed no elevated risk, OR = 1.00, CI: 0.55-1.81, when comparing the medium THM4 with the low THM4. Given that the medium THM4 in Chisholm *et al.* (2008) was much higher than the high THM4 in Cedergren *et al.* (2002), some elevated risk for cardiovascular anomalies would have been expected if chlorination DBPs were actually causal. However, of greater concern for the meaning of the Chisholm *et al.* (2008) finding are the results of Dodds and King (2001) who reported an OR = 0.7, CI: 0.5–1.0 for THM4 100 μg/L vs. <50 μg/L and Dodds *et al.* (1999) who reported an OR = 0.77, CI: 0.57–1.04 for the same exposure comparison. These findings are important because these studies performed seasonally relevant, retrospective monitoring for THMs one year after birth for the critical exposure period, providing a more individualized exposure metric than Chisholm *et al.* (2008). Despite this contrast, Chisholm *et al.* (2008) did not discuss their only significant finding for a specific birth defect in relation to the contradictory findings from Dodds and King (2001), Dodds *et al.* (1999).

Chisholm *et al.* (2008) only mention the foregoing studies in a broader context in that they "*show significant effects for several adverse birth outcomes at levels of exposure to THMs that have been observed in Australian metropolitan areas, such as Perth.*" They also noted that Dodds and King (2001) found a significant association of neural tube defects with exposure to >20 μg/L of BDCM, "*a level at which BDCM is found within Perth water supplies.*" Chisholm *et al.* (2008) did not explain in making this comment that their own finding for nervous system defects was an OR = 1.08, CI: 0.41–2.85, a clearly non-significant relationship for Perth drinking water. As outlined above, the Chisholm *et al.* (2008) study appears to provide an inadequate evidentiary basis to make any confident claims about DBPs in drinking water causing birth defects in Perth.

Another study (Joyce *et al.* 2008) involving some of the same researchers as Chisholm *et al.* (2008) attempted to link DBPs with adverse pregnancy outcomes for Perth drinking water and came under criticism (Heitz and Kristiana, 2008). Despite stating in their summary "*these preliminary results suggest a relation between the development of term PROM [prelabor rupture of membranes] and exposure to water disinfection by-products*" the only direct measure of DBPs they reported was for THMs, in which case they found no association. Rather, Joyce *et al.* (2008) found a modest association of PROM with nitrate and argued that nitrate was a surrogate for nitrogenous DBPs, stating: "*The results from independent sampling (not shown) indicated that there was an extremely high correlation between nitrate and nitrogen in Perth waters, which lends support to the theory that nitrate may be a potential surrogate indicator of nitrogen-containing disinfection by-products.*" Heitz and Kristiana (2008) noted that nitrate was the main component of total nitrogen in Perth water, comprising up to 90% of total nitrogen in the higher concentration samples and Joyce *et al.* (2008) presented no evidence to support their hypothesis that nitrates were a viable surrogate for nitrogenous DBPs.

Joyce *et al.* (2008) declined to exercise their opportunity to reply to the scientific criticism in the letter from Heitz and Kristiana (2008) choosing only to advise the Editor that: "*Heitz and Kristiana are financially affiliated with and are commercially obligated to the Western Australian Water Corporation, a company directly affected by the contents of the letter*." This comment does nothing to address the lack of scientific merit to the Joyce *et al.* (2008) claim that nitrate is a viable surrogate for nitrogenous DBPs. The unusual wording the journal attributed to the paper's authors is an oddly pejorative description of the normal research funding provided to Curtin University by Water Corporation, a water utility owned by the State government to serve its population. The apparent implication that readers of this journal should disbelieve the scientific merits of the letter because it challenged the validity of Joyce *et al.* (2008) for strictly "commercial" motives fails to explain why the paper's authors provided no scientific answer to the valid scientific criticism contained in the letter.

11.5.4 Short term exposure and degree of caution

For public health professionals, it is important to recognize no matter which evidence or interpretation may be preferred, the level of precaution for THM4 based on toxicology evidence is very large. In particular, public health practitioners should take note of the recommendations of Keegan *et al.* (1998) for drinking water concentrations as a basis to trigger a 1 day health advisory for chloroform and BDCM. These values are intended to provide advice to relevant public health and water professionals about levels of drinking water exposure that might cause harm through a short term (1 day) exposure. The recommendations were based on a NOAEL (chloroform: 30,000 μg/kg-d; BDCM: 41,000 μg/kg-d) and a 100 fold uncertainty factor (Keegan *et al.* 1998). They proposed a one day exposure health advisory level of 3000 μg/L for a 10 kg child and 10,000 μg/L for a 70 kg adult for chloroform and 4,000 μg/L for a 10 kg child and 14,000 μg/L for a 70 kg adult for BDCM.

Even if Keegan *et al.* (1998) are wrong by a factor of 10 or more, it should be clear that exceeding current guideline/regulatory values by even 100% (i.e. a factor of 2), for short periods, would not be expected to cause any adverse health effect in humans. Exceeding THM limits (80 μg/L as a running annual average, USA to 250 μg/L as a peak value, Australia) by 20 or 30% for short periods would certainly not call for emergency actions based on any expectation of adverse health outcomes

However, the nature of the risk communication challenge facing frontline public health personnel under circumstances when a guideline/regulatory value is exceeded should not be underestimated. The public will clearly expect that guideline/regulatory values should not be exceeded and, by definition, exceeding a guideline/regulatory limit is a problem (Chapter 3). An appropriate communication position must take this as the starting point, but responsible recommendations about what should be done on an immediate basis must be made with the understanding of how imminent (or not) is the health threat. Frontline public health personnel will not be serving the public interest by reacting as if they believe that there is an immediate danger to individuals where the toxicity evidence does not support that position.

11.5.5 Comparing predictions with experience

Overall, the epidemiology related to cancer outcomes has proven to be an interesting story over the past 30 years. Early studies were launched with an expectation that cancer outcomes should be found because chloroform/THMs had been found to be carcinogenic in rodent bioassays. Only in the past 15 years has our improved understanding of the importance of the mode of action observed in rodent bioassay studies tempered expectations of cancer outcomes from exposure to chlorinated drinking water. Meanwhile, epidemiologic evidence has largely proven to be equivocal for the possibility of colon and rectal cancer. The evident disagreement between the two major studies has left these possibilities somewhat in limbo: Hildeshiem *et al.* 1998 which supports rectal cancer, but does not support colon cancer vs. King *et al.* (2000a) which supports colon cancer in males, not females, but does not support rectal cancer. Part of the explanation for evident discrepancies in results may be caused by the inevitable differences in the mix of chlorination DBPs that exist among different water supplies. Differing composition of the NOM that forms the precursors for chlorination DBPs and the presence of bromide will influence the proportion of brominated DBPs that are formed which may alter risk. Evidence for other cancer sites is weak to non-existent, leaving urinary bladder cancer as the most plausible cancer risk associated with chlorination DBPs.

At present, a causal link between bladder cancer and some component of chlorine disinfected drinking water remains a viable working hypothesis with various elements of support primarily from the number of significant epidemiologic findings. Overall, the consistency of findings on urinary bladder cancer is noteworthy, but the specificity and plausibility, as to causal agent, are weak to negative and the strength of association is generally low enough to be susceptible to even minor bias or confounding.

As noted in Section 11.3.3 the upper bound cancer risk estimates made by using CSF for DBPs were only 92 cases of any type of cancer per year for a surface and groundwater served population of 263 million (USEPA, 2005) as compared to between approximately 1400 and 12,000 cases of bladder cancer per year in the U.S.A. that would be attributed to DBPs in chlorinated drinking water if the epidemiologic PAR estimates were accurate (Odom *et al.* 1999; USEPA, 2006). Bull *et al.* (2001) note with reference to the current regulatory levels for THMs and HAAs in the U.S.: "*there is no evidence that decreasing THM and HAA concentrations in drinking water will reduce the risk from bladder cancer. There are no data to indicate that any of these compounds can contribute to bladder cancer by any mechanism. More focused attention on identifying the cause of bladder cancer would directly resolve the question of whether drinking water disinfection inevitably leads to unacceptable risk or whether those risks can be rationally mitigated*."

The bottom line is that the recent regulatory focus on THMs has been rationalized, in large part, as providing a means to reduce the occurrence of bladder cancer. However, the toxicological evidence clearly indicates that there is no support for a causal connection between chloroform (the dominant THM) and bladder cancer which means that reducing THMs alone cannot be assured to achieve any reduction in population bladder cancer incidence. If there are other chlorination DBPs that are responsible for causing bladder cancer, reduction of THMs may or may not reduce these other chlorination DBPs. Only mitigative measures such as reduction of chlorination DBP precursors (i.e. raw water natural organic matter) are likely to assure concurrent reduction of THMs and unknown chlorination DBPs. Other measures specifically targeting reduced THM formation, such as aeration or chloramination, may not achieve any reduction of the unknown chlorination DBPs and, in the case of chloramination, may yield an increase in other more toxic chlorination DBPs, such as nitrosamines.

More focused attention on causes of bladder cancer is necessary because a large proportion of the comparisons of high chlorination DBP exposures with lower chlorination DBP exposures involve comparing exposure to disinfected surface water vs. lightly or non-disinfected groundwater. Such comparisons in North America will inevitably carry some, up to a very large, element of urban vs. rural residence. Such comparisons involve differences in a wide variety socioeconomic and cultural factors between populations, some of which may bear on health outcomes. Any attempt to mathematically adjust for such non-specific factors will be imperfect because the models of such factors will not be precise nor perfectly accurate. Particularly where the strength of association (magnitude of the OR) is low (being generally less than 2, with lower confidence intervals often approaching 1.0), anything less than perfect adjustment for confounding by factors other than drinking water quality could certainly allow for weak spurious associations to be observed.

Given the notable consistency of findings at a number of locations by different investigators in several epidemiology studies, the chance of spurious associations should be reduced. However, a substantial concern remains that these studies, with only one small and substantially qualified exception (Chevrier *et al.* 2004), have all relied upon similar means to achieve the low-end exposures to chlorination DBPs. Until epidemiology studies are completed with substantial numbers of participants residing in larger urban areas who have had low to negligible chlorination DBP exposure because of alternate disinfection and water treatment practices (ozonation, UV or no disinfection), the possibility will remain of a small systematic bias sufficient to explain the consistent, but comparatively weak association (generally OR < 2) of urinary bladder cancer with chlorination DBPs.

Richardson *et al.* (2007) have thoroughly reviewed evidence on the genotoxicity and carcinogenicity of DBPs. This massive compilation of literature illustrates that many questions remain to be resolved and certainly reinforces the complexity of the issue. For example, there is evidence that chlorination produces more total mutagenicity in the overall complex mixture of produced DBPs than alternate disinfection processes. The implications of this for causing human cancer remain unclear because of the difficulty of accurately translating various measures of mutagenicity into human cancer risk (e.g. MX, Chapter 9). More thorough direct experimentation with complex DBP mixtures, including multiple exposure routes, may be necessary to provide a better basis for assessing human cancer risk from toxicological evidence.

The foregoing needs are being addressed in a major research effort by the USEPA (Miltner *et al.* 2008; Simmons *et al.* 2008; Richardson *et al.* 2008) to explore the toxicology associated with a complex mixture of DBPs. Some early results are discussed in Chapter 2 (Section 2.1.5).

11.6 CONCLUSIONS

Despite causing some distractions, overall the chlorination DBP debate for the past 35 years has made major and valuable contributions to understanding water chemistry leading to substantial improvements in water treatment practices and finished water quality. The resulting attention focused on improving water quality has been clearly beneficial for consumers. Considerable challenges for improving and maintaining the aesthetic quality of public water supplies remain to avoid any further erosion of market share to bottled water.

Continuing improvements in water quality need to be pursued, with full recognition and acknowledgement of the substantial level of precaution that is evident in the current regulatory control over chlorination DBPs. Public health risk management is well justified to proceed with substantial precaution on the DBP issue; water consumers have every right to expect such precaution. However, given a continuing commitment to an open and fully precautionary approach, there is no justification for artificial over-statement of otherwise limited health risk evidence regarding chlorination DBPs.

In the meantime, drinking water professionals (both suppliers and regulators) must be careful to avoid blindly pursuing alternative disinfection practices that have received no comprehensive risk assessment for possible adverse health effects. Likewise, it is important that health professionals be careful about misleading the public into spending its limited discretionary income on reducing exposure to DBPs, either by using bottled water or installing home water treatment devices. Such choices should be made on the basis of preferences for aesthetic quality where such alternatives may offer a verifiable, tangible improvement over community water supplies.

Major take home points for frontline drinking water professionals are:

- Epidemiology is an unavoidably insensitive and potentially misleading tool for assessing small and very complex human health risks such as may be posed by chlorination DBPs. Epidemiological evidence needs to be interpreted with caution. Rules of epidemiologic practice are not like laws of physics; they require considerable judgement in application. However, healthy scepticism is needed because the basic premise of epidemiology, even though it may be complicated by sophisticated mathematical methods, is statistical assessment of correlation between an exposure and an outcome. Inherent weakness in the evidence regarding either exposure or outcome cannot be replaced by mathematical manipulation.

- The chlorination DBP public health risk issue is inherently complex and is, unfortunately, not amenable to the application of the best epidemiologic tools available, i.e. randomized, double blind, experimental studies. Other than practical and cost implications, ethical constraints governing intentional experimental exposures are a major factor preventing the use of the most powerful epidemiologic methods on the DBP issue.
- The main cancer outcome which has been possibly associated with chlorinated DBPs is bladder cancer. The supporting case for this is based primarily on the consistency of a comparatively small number (6 to 8) of case-control studies with comparatively small measures of strength of association (generally with observed OR less than 2) without the benefit of having identified a toxicologically plausible agent upon which to base exposure. Notwithstanding all these limitations, a causal association between bladder cancer and drinking water exposure to chlorination DBPs remains a viable hypothesis.
- If chlorination DBPs are a factor in causing urinary bladder cancer, it must be acknowledged that they are not a very large one in relation to overall population incidence of bladder cancer.
- Numerous studies exploring severe reproductive effects (spontaneous abortion, stillbirth, congenital defects) and chlorination DBPs have shown weak to no evidence supporting a causal relationship. Some measures of fetal growth (low birth weight, intrauterine growth retardation and small for gestational age) have shown limited evidence in support of a causal relationship with chlorination DBPs.
- The pervasive nature of drinking water DBP exposure compels us to remain precautionary to assume that there is a causal connection with some adverse health outcomes and therefore to limit DBP exposures to the extent reasonably possible.
- The foundations of epidemiology were established by John Snow through his remarkable demonstration of the role of contaminated drinking water in the spread of cholera in London of the mid-1800s. Snow was successful because of his unwavering commitment to challenging the evidence either supporting or contradicting his meticulous, hypothesis-driven research (Vinten-Johansen *et al.* 2003). Current epidemiologic research efforts addressing the human health effects of DBPs in drinking water must demand no less.

REFERENCES

Aggazzotti G., Righi E., Fantuzzi G., Biasotti B., Ravera G., Kanitz S., Barbone F., Sansebastiano G., Battaglia M. A., Leoni V., Fabiani L., Triassi M., Sciacca S. and Collaborative Group for the Study of Chlorinated Drinking Waters (2004). Chlorination by-products (CBPs) in drinking water and adverse pregnancy outcomes in Italy. *Journal of Water & Health*, **2**(4), 233–247.

Alavanja M., Goldstein I. and Susser M. (1978). A case control study of gastrointestinal and urinary tract cancer mortality and drinking water chlorination. In: Water Chlorination, Environmental Impact and Health Effects, Vol 2, Ann Arbor Science Publishers Inc, Ann Arbor, Michigan, pp. 395–409.

American Cancer Society. (2011). Cancer Facts & Figures 2011. American Cancer Society, Atlanta. http://www.cancer.org/acs/groups/content/@epidemiologysurveilance/documents/document/acspc-029771.pdf (accessed 13 February 2012).

Amy G., Graziano N., Craun G., Krasner S., Cantor K., Weyer P. and King W. (2006). Improved Exposure Assessment on Existing Cancer Studies, Report on Project 2582, American Water Association Research Foundation, Denver, CO.

Angell M. (1997). Science on Trial – The Clash of Medical Evidence and the Law in the Breat Implant Case. W.W. Norton., London, 256.

Anon. (2005). Court backs Virginia city immunity to DBP claims. *American Water Works Association Journal*, **97**(2), 18.

Arbuckle T. E., Hrudey S. E., Krasner S. W., Nuckols J. R., Richardson S. D., Singer P., Mendola P., Dodds L., Weisel C., Ashley D. L., Froese K. L., Pegram R. A., Schultz I. R., Reif J., Bachand A. M., Benoit F. M., Lynberg M., Poole C. and Waller K. (2002). Assessing exposure in epidemiologic studies to disinfection by-products in drinking water: report from an international workshop. *Environmental Health Perspectives*, **110**(Suppl 1), 53–60.

Aschengrau A., Zierler S. and Cohen A. (1989). Quality of community drinking water and the occurrence of spontaneous abortion. *Archives of Environmental Health*, **44**(5), 283–290.

Aschengrau A., Zierler S. and Cohen A. (1993). Quality of community drinking water and the occurrence of late adverse pregnancy outcomes. *Archives of Environmental Health*, **48**(2), 105–113.

Backer L. C., Ashley D. L., Bonin M. A., Cardinali F. L., Kieszak S. M. and Wooten J. V. (2000). Household exposures to drinking water disinfection by-products: whole blood trihalomethane levels. *Journal of Exposure Analysis and Environmental Epidemiology*, **10**(4), 321–326.

Bader E. L., Hrudey S. E. and Froese K. L. (2004). Urinary excretion half life of trichloroacetic acid as a biomarker of exposure to chlorinated drinking water disinfection by-products. *Occupational & Environmental Medicine*, **61**, 715–716.

Bean J. A., Isacson P., Hausler W. J. and Kohler J. (1982). Drinking water and cancer incidence in Iowa. *American Journal of Epidemiology*, **116**(6), 912–923.

Bonita R., Beaglehole R. and Kjellstrom T. (2006). Basic Epidemiology. 2nd edn, World Health Organization, Geneva.

Bove F. J., Fulcomer M. C., Klotz J. B., Esmart J., Dufficy E. M. and Savrin J. E. (1995). Public drinking water contamination and birth outcomes. *American Journal of Epidemiology*, **141**(9), 850–862.

Bove G. E., Jr., Rogerson P. A. and Vena J. E. (2007a). Case control study of the geographic variability of exposure to disinfectant byproducts and risk for rectal cancer. *International Journal of Health Geographics [Electronic Resource]*, **6**, 18.

Bove G. E., Rogerson P. A. and Vena J. E. (2007b). Case-control study of the effects of trihalomethanes on urinary bladder cancer risk. *Archives of Environmental & Occupational Health*, **62**(1), 39–47.

Bove F., Shim Y. and Zeitz P. (2002). Drinking water contaminants and adverse pregnancy outcomes: a review. *Environmental Health Perspectives*, **110**(Suppl 1), 61–74.

Bradford-Hill S. A. (1965). The environment and disease: association or causation? *Proceedings of the Royal Society of Medicine*, **58**, 295–300.

Brenniman G. R., Vasilomanolakis-Lagos J., Amsel J., Namekata T. and Wolff A. H. (1980). Case-control study of cancer deaths in Illinois communities served by chlorinated or nonchlorinated water. Water Chlorination – Environmental Impact and Health Effects. Jolley, R. L. (ed.) Ann Arbor Science, Ann Arbor, MI, **3**, 1043–1057.

Bull R. J., Krasner S. W., Daniel P. A. and Bull R. J. (2001). Health Effects and Occurrence of Disinfection By-Products. AWWA Research Foundation, Denver, 397.

Butterworth B. E., Conolly R. B. and Morgan K. T. (1995). A strategy for establishing mode of action of chemical carcinogens as a guide for approaches to risk assessments. *Cancer Letters*, **93**, 129–146.

Cantor K. P., Villanueva C. M., Silverman D. T., Figueroa J. D., Real F. X., Garcia-Closas M., Malats N., Chanock S., Yeager M., Tardon A., Garcia-Closas R., Serra C., Carrato A., Castano-Vinyals C., Samanic C., Rothman N. and Kogevinas M. (2010). Polymorphisms in GSTT1, GSTZ1, and CYP2E1, disinfection by-products, and risk of bladder cancer in spain. *Environmental Health Perspectives*, **118**(11), 1545–1550.

Cantor K. P., Hoover R., Hartge P., Mason T. J., Silverman D., Altman R., Austin D. F., Child M. A., Key C. R. Marrett L. D., Myers M. H., Narayana A. S., Levin L. I., Sullivan J. W., Swanson G. M., Thomas D. B. and West D. W. (1987). Bladder cancer, drinking water source, and tap water consumption: a case-control study. *Journal of the National Cancer Institute*, **79**(6), 1269–1279.

Cantor K. P., Hoover R., Mason T. J. and McCabe L. J. (1978). Associations of cancer mortality with halo methanes in drinking water. *Journal of the National Cancer Institute*, **61**(4), 979–986.

Cantor K. P., Lynch C. F., Hildesheim M. E., Dosemeci M., Lubin J., Alavanja M. and Craun G. (1998). Drinking water source and chlorination byproducts. I. Risk of bladder cancer. *Epidemiology*, **9**(1), 21–28.

Carlo G. L. and Mettlin C. J. (1980). Cancer incidence and trihalomethane concentrations in a public drinking water system. *American Journal of Public Health*, **70**(5), 523–525.

Cedergren M. I., Selbing A. J., Loefman O. and Kaellen B. A. (2002). Chlorination byproducts and nitrate in drinking water and risk for congenital cardiac defects. *Environmental Research*, **89**(2), 124–130.

Chang C. C., Ho S. C., Wane L. Y. and Yang C. Y. (2007). Bladder cancer in Taiwan: relationship to trihalomethane concentrations present in drinking-water supplies. *Journal of Toxicology and Environmental Health, Part A*, **70**(19–20), 1752–1757.

Chevrier C., Junod B. and Cordier S. (2004). Does ozonation of drinking water reduce the risk of bladder cancer? *Epidemiology*, **15**(5), 605–614.

Chisholm C., Cook A., Bower C. and Weinstein P. (2008). Risk of birth defects in Australian communities with high brominated disinfection by-product levels. *Environmental Health Perspectives*, **116**(9), 1267–1273.

Colford J. M., Roy S. M., Beach M. J., Hightower A., Shaw S. and Wade T. J. (2006). A review of household drinking water intervention trials and an approach to the estimation of waterborne gastroenteritis in the United States. *Journal of Water & Health*, **4**(Suppl 2), 71–88.

Colford J. M., Wade T. J., Sandhu S. K., Wright C. C., Lee S., Shaw S., Fox K., Burns S., Benker A., Brookhart M. A., van der Laan M. and Levy D. A. (2005). A randomized, control trial of in-home drinking water intervention to reduce gastrointestinal disease. *American Journal of Epidemiology*, **161**(5), 472–482.

Cordier S., Clavel J., Limasset J. C., Boccon-Gibod L. V., Le Moual N., Mandereau L. and Hemon D. (1993). Occupational risks of bladder cancer in France: a multicentre case-control study. *International Journal of Epidemiology*, **22**, 403–411.

Costet N., Villanueva C. M., Jaakokola J. J., Kogevinas M., Cantor K. P., King W. D., Lynch C. F., Nieuwenhuijsen M. J. and Cordier S. (2011). Water disinfection by-products adn bladder cancer: is there a European specificity? A pooled and meta-analysis of European case-control studies. *Occupational & Environmental Medicine*, **68**, 379–385.

Cragle D. L., Shy C. M., Struba R. J. and Siff E. J. (1985). A case-control study of colon cancer and water chlorination in North Carolina. In: *Water Chlorination – Chemistry, Environmental Impact and Health Effects*, Jolley, R. L. (ed), Lewis Publishers, Chelsea, MI, Vol 5, pp. 153–159.

DeAngelo A. B., Geter D. R., Rosenberg D. W., Crary C. K. and George M. H. (2002). The induction of aberrant crypt foci (ACF) in the colons of rats by trihalomethanes administered in the drinking water. *Cancer Letters*, **187**(1–2), 25–31.

Dodds L. and King W. D. (2001). Relation between trihalomethane compounds and birth defects. *Occupational & Environmental Medicine*, **58**(7), 443–446.

Dodds L., King W. D., Allen A. C., Armson B. A., Fell D. B. and Nimrod C. (2004). Trihalomethanes in public water supplies and risk of stillbirth. *Epidemiology*, **15**(2), 179–186.

Dodds L., King W. D., Woolcoot C. and Pole J. (1999). Trihalomethanes in public water supplies and adverse birth outcomes. *Epidemiology*, **10**(3), 233–237.

Doyle T. J., Zheng W., Cerhan J. R., Hong C. P., Sellers T. A., Kushi L. H. and Folsom A. R. (1997). The association of drinking water source and chlorination by-products with cancer incidence among postmenopausal women in Iowa: a prospective cohort study. *American Journal of Public Health*, **87**(7), 1168–1176.

Driedger S. M. and Eyles J. (2003). Different frames, different fears: communicating about chlorinated drinking water and cancer in the Canadian media. *Social Science & Medicine*, **56**, 1279–1293.

Egger M., Smith G. D. and Phillips A. N. (1997). Meta-analysis: principles and procedures. *BMJ*, **315**, 1533–1537.

Flaten T. P. (1992). Chlorination of drinking water and cancer incidence in Norway. *International Journal of Epidemiology*, **21**(1), 6–15.

Freedman D. M., Cantor K. P., Lee N. L., Chen L.-S., Lei H.-H., Ruhl C. E. and Wang S. S. (1997). Bladder cancer and drinking water: a population-based case-control study in Washington County, Maryland (United States). *Cancer Causes & Control*, **8**(5), 738–744.

Froese K., Sinclair M. and Hrudey S. E. (2002). Trichloroacetic acid as a biomarker of exposure to DBPs in drinking water: the Adelaide Water Exposure Trial (AWET). *Environmental Health Perspectives*, **110** (7), 679–687.

Gallagher M. D., Nuckols J. R., Stallones L. and Savitz D. A. (1998). Exposure to trihalomethanes and adverse pregnancy outcomes. *Epidemiology*, **9**(5), 484–489.

George M. H., Olson G. R., Doerfler D., Moore T., Kilburn S. and DeAngelo A. B. (2002). Carcinogenicity of bromodichloromethane administered in drinking water to male F344. *International Journal of Toxicology*, **21**(3), 219–230.

Geter D. R., George M. H., Moore T. M., Kilburn S., Huggins-Clark G. and DeAngelo A. B. (2004a). Vehicle and mode of administration effects on the induction of aberrant crypt foci in the colons of male F344. *Journal of Toxicology and Environmental Health, Part A*, **67**(1), 23–29.

Geter D. R., George M. H., Moore T. M., Kilburn S. R., Huggins-Clark G. and DeAngelo A. B. (2004b). The effects of a high animal fat diet on the induction of aberrant crypt foci in the colons of male F344. *Toxicology Letters*, **147**(3), 245–252.

Geter D. R., Moore T. M., George M. H., Kilburn S. R., Allen J. W., Nelson G. M., Winkfield E. and DeAngelo A. B. (2005). Tribromomethane exposure and dietary folate deficiency in the formation of aberrant crypt foci in the colons of F344/N rats. *Food and Chemical Toxicology: An International Journal Published for the British Industrial Biological Research Association, 2005 Sep, **43**(9), 1405–1412.*

Gottlieb M. S. and Carr J. K. (1982). Case-control cancer mortality study and chlorination of drinking water in Louisiana. *Environmental Health Perspectives*, **46**, 169–177.

Gottlieb M. S., Carr J. K. and Clarkson J. R. (1982). Drinking water and cancer in Louisiana – a retrospective mortality study. *American Journal of Epidemiology*, **116**(4), 652–657.

Gottlieb M. S., Carr J. K. and Morris D. T. (1981). Cancer and drinking water in Louisiana: colon and rectum. *International Journal of Epidemiology*, **10**, 117–125.

Graves C. G., Matanoski G. M. and Tardiff R. G. (2001). Weight of evidence for an association between adverse reproductive and developmental effects and exposure to disinfection by-products: a critical review. *Regulatory Toxicology & Pharmacology*, **34**(2), 103–124.

Grazuleviciene R., Nieuwenhuijsen M. J., Vencloviene J., Kostopoulu-Karadanelli M., Krasner S. W., Danilevicute A., Balcius G. and Kapustinskiene V. (2011). Individual exposures to drinking water trihalomethanes, low birth weight and small for gestational age risk: a prospective Kaunas cohort study. *Environmental Health*, **10**, 32.

Grellier J., Bennett J., Patelarou E., Smith R. B., Toledano M. B., Rushton L., Briggs D. J. and Nieuwenhuijsen M. J. (2010). Exposure to disinfection by-products, fetal growth, and prematurity. *Epidemiology*, **21**(3), 200–313.

Greenland S. (1987). Quantitative methods in the review of epidemiologic literature. *Epidemiologic Reviews*, **9**, 1–30.

Greenland S. (1994). Invited commentary: a critical look at some popular metaanalytic methods. *American Journal of Epidemiology*, **140**, 290–296.

Guest D. (2008). Tap water 'increases risk of birth defects'. *The West Australian,* Perth, WA, Australia, pages 1 and 4, 22 April, 2008.

Health and Welfare Canada. (1978). Guidelines for Canadian Drinking Water Quality. Minister of Supply and Servces Canada, Canadian Government Publishing Centre, Hull, Quebec.

Heitz A. and Kristiana I. (2008). Letter to the Editor re: water disinfection by-products and prelabor rupture of membranes. *American Journal of Epidemiology*, **169**, 122–123.

Hellard M. E., Sinclair M. I., Forbes A. B. and Fairley C. K. (2001). A randomized, blinded, controlled trial investigating the gastrointestinal health effects of drinking water quality. *Environmental Health Perspectives*, **109**(8), 773–778.

Hildesheim M. E., Cantor K. P., Lynch C. F., Dosemeci M., Lubin J. Alavanja M. and Craun G. (1998). Drinking water source and chlorination byproducts. II. Risk of colon and rectal cancers. *Epidemiology*, **9**(1), 29–35.

Hinckley A. F., Bachand A. M., Nuckols J. R. and Reif J. S. (2005). Identifying public water facilities with low spatial variability of disinfection by-products for epidemiological investigations. *Occupational & Environmental Medicine*, **62**(7), 494–499.

Hoffman C. S., Mendola P., Savitz D. A., Herring A. H., Loomis D., Hartmann K. E., Singer P. C., Weinberg H. S. and Olshan A. F. (2008a). Drinking water disinfection by-product exposure and duration of gestation. *Epidemiology*, **19**(5), 738–746.

Hoffman C. S., Mendola P., Savitz D. A., Herring A. H., Loomis D., Hartmann K. E., Singer P. C., Weinberg H. S. and Olshan A. F. (2008b). Drinking water disinfection by-product exposure and fetal growth. *Epidemiology*, **19**(5), 729–737.

Howards P. P. and Hertz-Picciotto I. (2006). Invited commentary: disinfection by-products and pregnancy loss – lessons. *American Journal of Epidemiology*, **164**(11), 1052–1055.

Hrudey S. E. (2008). Chlorination Disinfection By-Products (DBPs) in Drinking Water and Public Health in Canada – A Primer for Public Health Practitioners Reviewing Evidence from over 30 Years of Research. National Collaborating Centre on Environmental Health, Vancouver, BC. http://ncceh.ca/en/practice_policy/ncceh_reviews/chlorination_disinfection_by-products (accessed 14 March 2012).

Hrudey S. E. and Hrudey E. J. (2004). Safe Drinking Water – Lessons from Recent Outbreaks in Affluent Nations. IWA Publishing, London.

Hrudey S. E. and Leiss W. (2003). Risk management and precaution – insights on the cautious use of evidence. *Environmental Health Perspectives*, **111**(13), 1577–1581.

Hung R. J., Boffetta P., Brennan L. M., Malaveille C., Hautefeuille A., Donato F., Gelatti U., Spaliviero M., Pacidi D., Carta A., Scotto di Carlo A. and Porru S. (2005). GST, NAT, SULT1A1, CYP1B1 genetic polymorphisms, interactions with environmental exposures and bladder cancer risk in a high-risk population. *International Journal of Cancer*, **110**, 598–604.

Huslin A. (2002). In a Glass of Water, A Mother's Worst Fear. *Washington Post*. Washington, DC, C01.

Hwang B. F. and Jaakkola J. J. (2003). Water chlorination and birth defects: a systematic review and meta-analysis. *Archives of Environmental Health*, **58**(2), 83–91.

Hwang B.-F., Jaakkola J. and Guo H.-R. (2008). Water disinfection by-products and the risk of specific birth defects: a population-based cross-sectional study in Taiwan. *Environmental Health*, **7**, 23.

Hwang B. F., Magnus P. and Jaakkola J. J. (2002). Risk of specific birth defects in relation to chlorination and the amount of natural organic matter in the water supply. *American Journal of Epidemiology*, **156**(4), 374–382.

IARC (1991). Chlorinated drinking-water; chlorination by-products; some other halogenated compounds; cobalt and cobalt compounds. IARC Monographs on the Evaluation of Carcinogenic Risks to Humans, International Agency for Research on Cancer, Lyon, France, Vol 52, pp. 1–544.

IARC (2004). Some drinking-water disinfectants and contaminants, including Arsenic. IARC Monographs on the Evaluation of Carcinogenic Risks to Humans, International Agency for Research on Cancer, Lyon, France, Vol 84, 512 pp.

IARC (2011). Globocan Cancer Statistics. Interational Agency for Research on Cancer, Lyon, France. http://globocan.iarc.fr/ (accessed 5 February 2012).

Infante-Rivard C. (2004). Drinking water contaminants, gene polymorphisms, and fetal growth. *Environmental Health Perspectives*, **112**(11), 1213–1216.

Isacson P., Bean J. A. and Lynch C. (1983). Relationships of cancer incidence rates in Iowa municipalities to chlorination status of drinking water. Water Chlorination: Environmental Impact and Health Effects, Vol 4, Book 2, Environment, Health Ann Arbor Science Publishers, Ann Arbor, Michigan.

Jaakkola J. J., Magnus P., Skrondal A., Hwang B. F., Becher G. and Dybing E. (2001). Foetal growth and duration of gestation relative to water chlorination. *Occupational & Environmental Medicine*, **58**(7), 437–442.

Jorgenson T. A., Meierhenry J. R., Rushbook C. J., Bull R. J. and Robinson M. (1985). Carcinogenicity of chloroform in drinking water to male Osborne-Mendel rats and female B6C3F1 mice. *Fundamental & Applied Toxicology*, **5**, 760–769.

Joyce S. J., Cook A., Newnham J., Brenters M., Ferguson C. and Weinstein P. (2008). Water disinfection by-products and prelabor rupture of membranes. *American Journal of Epidemiology*, **168**, 514–521.

Kallen B. A. J. and Robert E. (2000). Drinking water chlorination and delivery outcome: a registry-based study in Sweden. *Reproductive Toxicology*, **14**(4), 303–309.

Kanarek M. S. and Young T. B. (1982). Drinking water treatment and risk of cancer death in Wisconsin. *Environmental Health Perspectives*, **46**, 179–186.

Kanitz S., Franco Y., Patrone V., Caltabellotta M., Raffo E., Riggi C., Timitilli D. and Vavera G. (1996). Association between drinking water disinfection and somatic parameters at birth. *Environmental Health Perspectives*, **104**(5), 516–520.

Keegan T. E., Simmons J. E. and Pegram R. A. (1998). NOAEL and LOAEL determinations of acute hepatoxicity for chloroform and bromodichloromethane delivered in an aqueous vehicle to F344 rats. *Journal of Toxicology & Environmental Health, Part A*, **55**, 65–75.

Kim H., Haltmeier P., Klotz J. B. and Weisel C. P. (1999). Evaluation of biomarkers of environmental exposures: urinary haloacetic acids associated with ingestion of chlorinated drinking water. *Environmental Research*, **80**, 187–195.

King W. D., Marrett L. D. and Woolcott C. G. (2000a). Case-control study of colon and rectal cancers and chlorination by-products in treated water. *Cancer Epidemiology, Biomarkers & Prevention*, **9**(8), 813–818.

King W. D., Dodds L. and Allen A. C. (2000b). Relation between stillbirth and specific chlorination by-products in public water supplies. *Environmental Health Perspectives*, **108**(9), 883–886.

King W. D., Dodds L., Allen A. C., Armson B. A., Fell D. and Nimrod C. (2005). Haloacetic acids in drinking water and risk for stillbirth. *Occupational & Environmental Medicine*, **62**(2), 124–127.

King W. D. and Marrett L. D. (1996). Case-control study of bladder cancer and chlorination by-products in treated water (Ontario, Canada). *Cancer Causes & Control*, **7**(6), 596–604.

Klotz J. B. and Pyrch L. (1999). Neural tube defects and drinking water disinfection by-products. *Epidemiology*, **10**, 383–390.

Koivusalo M., Hakulinen T., Vartiainen T., Pukkala E., Jaakkola J. J. and Tuomisto J. (1998). Drinking water mutagenicity and urinary tract cancers: a population-based case-control study in Finland. *American Journal of Epidemiology*, **148**(7), 704–712.

Koivusalo M., Jaakkola J. J., Vartiainen T., Hakulinen T., Karjalainen S., Pukkala E. and Tuomisto J. (1994). Drinking water mutagenicity and gastrointestinal and urinary tract cancers: an ecological study in Finland. *American Journal of Public Health*, **84**(8), 1223–1300.

Koivusalo M., Pukkala E., Vartiainen T., Jaakkola J. J. and Hakulinen T. (1997). Drinking water chlorination and cancer – a historical cohort study in Finland. *Cancer Causes & Control*, **8**(2), 192–200.

Kramer M. D., Lynch C. F., Isacson P. and Hanson J. W. (1992). The association of waterborne chloroform with intrauterine growth retardation. *Epidemiology*, **3**(5), 407–413.

Krasner S. W., Weinberg H. S., Richardson S. D., Pastor S. J., Chinn R., Sclimenti M. J., Onstad G. D. and Thruston A. D. Jr. (2006). Occurrence of a new generation of disinfection byproducts. *Environmental Science & Technology*, **40**(23), 7175–7185.

Kuzma R. J., Kuzma C. M. and Buncher R. (1977). Ohio drinking water source and cancer rates. *American Journal of Public Health*, **67**(8), 725–729.

Lewis C. Suffet I. H., Hoggatt K. and Ritz B. (2007). Estimated effects of disinfection by-products on preterm birth in a population served by a single water utility. *Environmental Health Perspectives*, **115**(2), 290–295.

Lewis C. Suffet I. H. and Ritz B. (2006). Estimated effects of disinfection by-products on birth weight in a population served by a single water utility. *American Journal of Epidemiology*, **163**(1), 38–47.

Lynch C., Woolson R., O'Gorman T. and Cantor K. (1989). Chlorinated drinking water and bladder cancer: effect of misclassification on risk estimates. *Archives of Environmental Health*, **44**(4), 252–259.

Magnus P., Jaakkola J. J., Skrondal A., Alexander J., Becher G., Krogh T. and Dybing E. (1999). Water chlorination and birth defects. *Epidemiology*, **10**(5), 513–517.

McGeehin M. A., Reif J. S., Becher J. C. and Mangione E. J. (1993). Case-control study of bladder cancer and water disinfection methods in Colorado. *American Journal of Epidemiology*, **138**(7), 492–501.

Messite J. and Stellman S. D. (1996). Accuracy of death certificate completion. *JAMA. The Journal of the American Medical Association*, **275**(10), 794–796.

Michaud D. S., Kogevinas M., Cantor K. P., Villanueva C. M., Garcia-Closas M., Rothman N., Malats N., Real F. X., Serra C., Garcia-Closas R., Tardon A., Carrato A., Dosemeci M. and Silverman D. T. (2007). Total fluid and water consumption and the joint effect of exposure to disinfection by-products on risk of bladder cancer. *Environmental Health Perspectives*, **115**(11), 1569–1572.

Michaud D. S., Spiegelman D., Clinton S. K., Rimm E. B., Curhan G. C., Willett W. C. and Giovannucci E. L. (1999). Fluid intake and the risk of bladder cancer in men. *New England Journal of Medicine*, **140**(18), 1390–1397.

Mills C. J., Bull R., Cantor K., Reif J., Hrudey S. E., Huston P. and an Expert Working Group (1998). Health risks of drinking water chlorination byproducts: report of an expert working group. *Chronic Diseases in Canada*, **19**(3), 91–102.

Miltner R. J., Speth T. F., Richardson S. D., Krasner S. W., Weinberg H. S. and Simmons J. E. (2008). Integrated disinfection by-products research: disinfection of drinking waters by chlorination and ozonation/postchlorination treatment scenarios. *Journal of Toxicology and Environmental Health – Part A*, **71**(17), 1133–1148.

Morris R. D., Audet A. M., Angelillo I. F., Chalmers T. C. and Mosteller F. (1992). Chlorination, chlorination by-products, and cancer: a meta-analysis. *American Journal of Public Health*, **82**(7), 955–963.

Murta-Nascimento C., Schmitz-Drager B. J., Zeegers M. P., Steineck G., Kogevinas M., Real F. X. and Malats N. (2007). Epidemiology of urinary bladder cancer: from tumor development to patient's death. *World Journal of Urology*, **25**, 285–295.

Nieuwenhuijsen M. J. (2010). HiWATE Publishable Final Report. Health Impacts of Long-Term Exposure to Disinfection By-Products in Drinking Water. FP6 Food Quality and Safety. EU Sixth Framework Programme for Research and Technological Development. http://www.hiwate.eu/news/hiwate-publishable-report-255.htm (accessed 21 January 2012).

Nieuwenhuijsen M. J., Grellier J., Smith R., Iszatt N., Bennett J., Best N. and Toledano M. (2009a). The epidemiology and possible mechanisms of disinfection by-products in drinking water. *Philosophical Transactions of The Royal Society - A*, **367**, 4043–4076.

Nieuwenhuijsen M. J., Martinez D. Grellier J., Bennett J., Best N. Iszatt N., Vrijheid M. and Toledano M. B. (2009c). Chlorination disinfection by-products in drinking water and congenital anomalies: review and meta-analyses. *Enviromental Health Perspectives*, **117**(10), 1486–1493.

Nieuwenhuijsen M. J., Smith R., Golfinopoulus S., Best N., Bennett J., Aggazotti G., Righi E., Fantuzzi G., Bucchini L., Cordier S., Villanueva C., Moreno V., LaVecchia C., Bosetti C., Vartiainen T., Rautiu R., Toledano M., Iszatt N., Grazuleviciene R. and Kogevinas M. (2009b). Health impacts of long term exposure to disinfection by-products in dirnking water in Europe: HIWATE. *Journal of Water & Health*, **7**(2), 185–207.

Nieuwenhuijsen M. J., Toledano M., Bennett J., Best N., Hambly P., de Hoogh C., Wellesley D., Boyd P., Abramsky L., Dattani N., Fawell J., Briggs D., Jarup L. and Elliott P. (2008). Chlorination disinfection by-products and risk of congenital anomalies in England and Wales. *Enviromental Health Perspectives*, **116**(2), 216–222.

Nieuwenhuijsen M. J., Toledano M., Eaton N., Fawell J. and Elliott P. (2000a). Chlorination disinfection by-products in water and their association with adverse reproductive outcomes: a review. *Occupational and Environmental Medicine*, **57**, 73–85.

Nieuwenhuijsen M. J., Toledano M. B., Eaton N. E., Fawell J. and Elliott P. (2000b). Chlorination disinfection byproducts in water and their association with adverse reproductive outcomes: a review. *Occupational & Environmental Medicine*, **57**(2), 73–85.

Nieuwenhuijsen M. J., Toledano M. B. and Elliott P. (2000c). Uptake of chlorination disinfection by-products; a review and a discussion of its implications for exposure assessment in epidemiological studies. *Journal of Exposure Analysis & Environmental Epidemiology*, **10**(6 Pt 1), 586–599.

NTP (1987). Toxicology and Carcinogenesis Studies of Bromodichloromethane in F344/N Rats and B6C3F1 Mice. National Toxicology Program, Research Triangle Park, NC.

NTP (2006). Toxicology and Carcinogenesis Studies of Bromodichloromethane (CAS No. 75–27-4) in Male F344/N Rats and Female B6C3F1 Mice (Drinking Water Studies). National Toxicology Program, Research Triangle Park, NC.

Odom R., Regli S., Messner M., Cromwell J. and Javdan M. (1999). Benefit-cost analysis of the Stage 1 D/DBP Rule. *Journal American Water Works Association*, **91**(4), 137–147.

Ottenbaacher K. J. (1998). Quantitative evaluation of multiplicity in epidemiology adn public health research. *American Journal of Epidemiology*, **147**(7), 615–619.

Patelarou E., Kargaki S., Stephanou E. G., Nieuwenhuijsen M., Sourtzi P., Gracia E., Chatzi L., Koutis A. and Kogevinas M. (2011). Exposure to brominated trihalomethanes in drinking water and reproductive outcomes. *Occupational & Environmental Medicine*, **68**, 438–445.

Ploeg M., Aben K. K. H. and Kiemeney L. A. (2009). The present and future burden of urinary bladder cancer in the world. *World Journal of Urology*, **27**, 289–293.

Pontius F. W. (2000). Chloroform: science, policy and politics. *Journal American Water Works Association*, **92**(5), 12, 14, 16, 18.

Poole C. and Greenland S. (1999). Random-effects meta-analyses are not always conservative. *American Journal of Epidemiology*, **150**(5), 469–475.

Porter C. K., Putnam S. D., Hunting K. L. and Riddle M. R. (2005). The effect of trihalomethane and haloacetic acid exposure on fetal growth in a Maryland county. *American Journal of Epidemiology*, **162**(4), 334–344.

Ranmuthugala G., Pilotto L., Smith W., Vimalasiri T., Dear K. and Douglas R. (2003). Chlorinated drinking water and micronuclei in urinary bladder epithelial cells. *Epidemiology*, **14**(5), 617–622.

Rahman M. B., Driscoll T., Cowie C. and Armstrong B. K. (2010). Disinfection by-products in drinking water and colorectal cancer: a meta-analysis. *International Journal of Epidemiology*, **39**, 733–745.

Reif J. S., Hatch M. C., Bracken M., Holmes L. B., Schwetz B. A. and Singer P. C. (1996). Reproductive and developmental effects of disinfection by-products in drinking water. *Environmental Health Perspectives*, **104**(10), 1056–1061.

Richardson S. D., Plewa M. J., Wagner E. D., Schoeny R. and DeMarini D. M. (2007). Occurrence, genotoxicity, and carcinogenicity of regulated and emerging disinfection by-products in drinking water: a review and roadmap for research. *Mutation Research – Reviews in Mutation Research*, **636**(1–3), 178–242.

Richardson S. D., Thruston A. D., Krasner S. W., Weinberg H. S., Miltner R. J., Schenck K. M., Narotsky M. G., McKague A. B. and Simmons J. E. (2008). Integrated disinfection by-products research: comprehensive characterization of water concentrates prepared by chlorinated and ozonated/post chlorinated drinking water. *Journal of Toxicology and Environmental Health – Part A*, **71**(17), 1165–1186.

Savitz D. A., Andrews K. W. and Pastore L. M. (1995). Drinking water and pregnancy outcome in central North Carolina: source, amount, and trihalomethane levels. *Environmental Health Perspectives*, **103**(6), 592–596.

Savitz D. A., Singer P., Hartmann K. E., Herring A., Weinberg H., Makarushka C., Hoffman C. S., Chan R. and Maclehose R. (2005). Drinking Water Disinfection By-Products and Pregnancy Outcome. AWWA Research Foundation, U.S. Environmental Protection Agency, Denver, 209 pp.

Savitz D. A., Singer P. C., Herring A. H., Hartmann K. E., Weinberg H. S. and Makarushka C. (2006). Exposure to drinking water disinfection by-products and pregnancy loss. *American Journal of Epidemiology*, **164**(11), 1043–1051.

Shaw G. M., Ranatunga D., Quach T., Neri E., Correa A. and Neutra R. R. (2003). Trihalomethane exposures from municipal water supplies and selected congenital malformations. *Epidemiology*, **14**(2), 191–199.

Simmons J. E., Richardson S. D., Teuschler L. K., Miltner R. J., Speth T. F., Schenk K. M., Hunter E. S. and Rice G. (2008). Research issues underlying the four-lab study: Integrated disinfection by-products research. *Journal of Toxicology and Environmental Health – Part A*, **71**(17), 1125–1132.

Suarez-Varela M. M. M., Llopis Gonzalez A., Tejerizo Perez M. L. and Ferrer Caraco E. (1994). Chlorination of drinking water and cancer incidence. *Journal of Environmental Pathology Toxicology and Oncology*, **13**(1), 39–41.

Swan S. H. and Waller K. (1998). Disinfection by-products and adverse pregnancy outcomes: what is the agent and how should it be measured?[comment]. *Epidemiology*, **9**(5), 479–481.

Tardiff R. G., Carson M. L. and Ginevan M. E. (2006). Updated weight of evidence for an association between adverse reproductive and developmental effects and exposure to disinfection by-products. *Regulatory Toxicology & Pharmacology*, **45**(2), 185–205.

Taubes G. (1995). Epidemiology faces its limits. *Science*, **269**, 164–169.

Thomas S. and Hrudey S. E. (1997). Risk of Death in Canada – What We Know and How We Know It. University of Alberta Press, Edmonton, 292 pp.

Toledano M. B., Nieuwenhuijsen M. J., Best N., Whitaker H., Hambly P., de Hoogh C., Fawell J., Jarup L. and Elliot P. (2005). Relation of trihalomethane concentrations in public water supplies to stillbirth and birth weight in three water regions in England. *Environmental Health Perspectives*, **113**(2), 225–232.

USPHS (1964). Smoking and health: Report of the advisory committee to the Surgeon General of the Public Health Service. PHS Publication, United States Public Health Service, Washington, DC,.

USEPA (1979). National Interim Primary Drinking Water Regulations; Control of Trihalomethanes in Drinking Water. *Federal Register*, **44**(231), 68624, (November 29).

USEPA (1998). 40 CFR Parts 9, 141, and 142 – disinfectants and disinfection byproducts; final rule. National Primary Drinking Water Regulations. *Federal Register*, **63**, 69389–69476.

USEPA (2004). An Examination of EPA Risk Assessment Principles and Practices. Office of the Science Advisor Staff Paper. U.S. Environmental Protection Agency, Washington, DC, 182.

USEPA (2005). Economic Analysis for the Final Stage 2 Disinfectants and Disinfection Byproducts Rule. Office Water, U.S. Environmental Protection Agency. EPA 815-R-05-010.

USEPA (2006). 40 CFR Parts 9, 141, and 142: national primary drinking water regulations: stage 2 disinfectants and disinfection byproducts rule. environmental protection agency. January 4, 2006, *Federal Register*, **71**(2), 388–494.

Vena J. E., Graham S., Freudenheim J., Marshall J., Zielenzny M., Swanson M. and Sufrin G. (1993). Drinking water, fluid intake, and bladder cancer in western New York. *Archives of Environmental Health*, **48**(3), 191–198.

Villanueva C. M., Cantor K. P., Cordier S., Jaakkola J. J. K., King W. D., Lynch C. F., Porru S. and Kogevinas M. (2004). Disinfection byproducts and bladder cancer – a pooled analysis. *Epidemiology*, **15**(3), 357–367.

Villanueva C. M., Cantor K. P., Grimalt J. O., Malats N., Silverman D., Tardon A., Garcia-Closas R., Serra C., Carrato A., Castano-Vinyals G., Marcos R., Rothman N., Real F. X., Dosemeci M. and Kogevinas M. (2007). Bladder cancer and exposure to water disinfection by-products through ingestion, bathing, showering, and swimming in pools. *American Journal of Epidemiology*, **165**(2), 148–156.

Villanueva C. M., Cantor K. P., King W. D., Jaakkola J. J. K., Cordier S., Lynch C. F., Porru S. and Kogevinas M. (2006). Total and specific fluid consumption as determinants of bladder cancer risk. *International Journal of Cancer*, **118**, 2040–2047.

Villanueva C. M., Fernandez F., Malats N., Grimalt J. O. and Kogevinas M. (2003). Meta-analysis of studies on individual consumption of chlorinated drinking water and bladder cancer. *Journal of Epidemiology & Community Health*, **57**(3), 166–173.

Villanueva C. M., Gracia-Lavedan E., Ibarluzea J., Santa Marina L., Ballester F., Llop S., Tardon A., Fernandez M. F., Freire C., Goni F., Basagana X., Kogevinas M., Grimalt J. O. and Sunyer J. (2011). Exposure to trihalomethanes through different waer uses adn birth weight, small for gestational age and preterm delivery in Spain. *Environmental Health Perspectives*, **119**(12), 1824–1830.

Vinceti M., Fantuzzi G., Monici L., Cassinadri M., Predieri G. and Aggazzotti G. (2004). A retrospective cohort study of trihalomethane exposure through drinking water and cancer mortality in northern Italy. *Science of the Total Environment*, **330**(1–3), 47–53.

Vinten-Johansen P., Brody H., Paneth N., Rachman S. and Rip M. (2003). Cholera, Chloroform and the Science of Medicine – A Life of John Snow. Oxford University Press, Oxford.

von Elm E., Egger M., Altman D. G., Pocock S. J., Gozsche P. C. and Vandenbroucke J. P. (2007). Strengthening the reporting of observational studies in epidemiology (STROBE) statement: guidelines for reporting observational studies. *BMJ*, **335**, 806–808.

Waller K., Swan S. H., DeLorenze G. and Hopkins B. (1998). Trihalomethanes in drinking water and spontaneous abortion. *Epidemiology*, **9**(2), 134–140.

Waller K., Swan S. H., Windham G. and Fenster L. (2001a). Influence of exposure assessment method in an epidemiologic study of trihalomethane exposure and spontaneous abortion. *American Journal of Epidemiology*, **153**(11), 540.

Waller K., Swan S. H., Windham G. and Fenster L. (2001b). Influence of exposure assessment methods on risk estimates in an epidemiologic study of total trihalomethane exposure and spontaneous abortion. *Journal of Exposure Analysis and Environmental Epidemiology*, **11**(6), 522–531.

WHO (1993). Guidelines for Drinking-Water Quality. 2nd edn, World Health Organization, Geneva.

WHO (1984). Guidelines for Drinking-Water Quality. World Health Organization, Geneva.

Wigle D. (1998). Safe drinking water: a public health challenge. *Chronic Diseases in Canada*, **19**(3), 103–107.

Wilkins J. R., 3rd and Comstock G. W. (1981). Source of drinking water at home and site-specific cancer incidence in Washington County, Maryland. *American Journal of Epidemiology*, **114**(2), 178–190.

Willett W., Greenland S., MacMahon B., Trichopoulos D., Rothman K., Thomas D., Thun M., Weiss N., Rapp N., Miller R. W., Saah A. J. and Gori G. B. (1995). The discipline of epidemiology. Letters in reply to Taubes (1995). *Science*, **269**, 1325–1328.

Wright J. M., Schwartz J. and Dockery D. W. (2003). Effect of trihalomethane exposure on fetal development. *Occupational & Environmental Medicine*, **60**(3), 173–180.

Wright J. M., Schwartz J. and Dockery D. W. (2004). The effect of disinfection by-products and mutagenic activity on birth weight and gestational duration. *Environmental Health Perspectives*, **112**(8), 920–925.

Wu X., Ros M., Gu J. and Kiemeney L. (2008). Epidemiology and genetic susceptibility to bladder cancer. *BJUI – British Journal of Urology International*, **102**, 1207–1215.

Yang C.-Y., Cheng B.-H., Tsai S.-S., Wu T.-N., Lin M.-C. and Lin K.-C. (2000). Association between chlorination of drinking water and adverse pregnancy outcome in Taiwan. *Environmental Health Perspectives*, **108**(8), 765–768.

Yang C., Chiu H., Cheng M. and Tsai S. (1998). Chlorination of drinking water and cancer mortality in Taiwan. *Environmental Research, Section A*, **78**(1), 1–6.

Yang C. Y., Xiao Z. P., Ho S. C., Wu T. N. and Tsai S. S. (2007). Association between trihalomethane concentrations in drinking water and adverse pregnancy outcome in Taiwan. *Environmental Research*, **104**(3), 390–395.

Young T. B. and Kanarek M. S. (1983). Matched pair case control study of drinking water chlorination and cancer mortality. Water Chlorination: Environmental Impact and Health Effects, Volume 4, Book 2, Environment, Health Ann Arbor Science Publishers, Ann Arbor, Michigan.

Young T. B., Kanarek M. S. and Tsiatis A. A. (1981). Epidemiologic study of drinking water chlorination and Wisconsin female cancer mortality. *Journal of the National Cancer Institute*, **67**(6), 1191–1198.

Young T. B., Wolf D. A. and Kanarek M. S. (1987). Case-control study of colon cancer and drinking water trihalomethanes in Wisconsin. *International Journal of Epidemiology*, **16**(2), 190–197.

Zierler S., Danley R. A. and Feingold L. (1986). Type of disinfectant in drinking water and patterns of mortality in Massachusetts. *Environmental Health Perspectives*, **69**, 275–279.

Zierler S., Feingold L., Danley R. A. and Craun G. (1988). Bladder cancer in Massachusetts related to chlorinated and chloraminated drinking water – a case-control study. *Archives of Environmental Health*, **43**(2), 195–200.

Zhang W., Gabos S., Schopflocher D., Li X. F., Gati W. P. and Hrudey S. E. (2009a). Reliability in using urinary and blood trichloroacetic acid as a biomarker of exposure to chlorinated drinking water disinfection by-products. *Biomarkers*, **14**(6), 355–36.

Zhang W., Gabos S., Schopflocher D., Li X. F., Gati W. P. and Hrudey S. E. (2009b). Validation of urinary trichloroacetic acid as a biomarker of exposure to drinking water disinfection by-products. *Journal of Water & Health*, **7**(3), 359–371.

Chapter 12

Concluding thoughts on DBPs, water quality and public health risks

S. E. Hrudey and J. W. A. Charrois

12.1 PROGRESS UNDERSTANDING DBPs SINCE 1974

Chapter 1 introduces this book by reviewing the context for the emergence of DBPs in drinking water as a public health issue. Progress in understanding microbial pathogens over the past 35 to 40 years has been substantial and has led to very tangible improvements in our understanding of previously unrecognized microbial pathogens, which has in turn demonstrably improved our ability to protect consumers from waterborne disease. We know the absolute and tangible negative consequences of not getting disinfection right. Readers will have to judge for themselves whether equivalent progress has been achieved in understanding human health risks from DBPs after reading this book.

There have clearly been many important advances in the scientific evidence relevant to the question of DBPs and human health risks for drinking water consumers, perhaps none more substantial than having gained an understanding that chloroform, the original DBP in drinking water, does not pose a cancer risk to consumers at current regulatory concentrations. This understanding has been largely achieved through toxicology studies that have established the mode of action of chloroform (Chapter 10) in causing tumours. This knowledge has allowed regulatory agencies worldwide to revise their rationale for regulating chloroform, ultimately incorporating a threshold approach and abandoning the use of a cancer slope factor (CSF). The levels of precaution applied to estimating the chloroform threshold are substantial. For example Health Canada (2006) applied a composite uncertainty factor of 2100 to the lowest animal dose (lowest observed adverse effect level, LOAEL) for chloroform based on long term experiments with beagles (Heywood *et al.* 1979) to develop Canada's limit for total THMs. This is certainly a large margin of caution (even for application to a LOAEL) considering how extensively chloroform toxicology has been studied. The health research evidence indicates that the regulatory limits for total THMs have been and continue to be precautionary.

Unfortunately, the knowledge that THM or chloroform limits in drinking water are not based on cancer risk is not as widely known or understood as it should be. A tangible example of that is a recent paper (Chowdhury and Hall, 2010), which purported to estimate the annual number of cases of cancer caused by drinking water exposure to THMs in Canada's 20 largest cities. There were several problems with this paper. Not surprisingly, the THM monitoring data that were relied upon for these calculations was dominated by chloroform, making the calculation of cancer cases largely dependent upon assuming that chloroform causes cancer from these drinking water exposures. The paper claimed to have found a cancer slope factor (CSF) for chloroform in the Integrated Risk Information System (IRIS) data system (USEPA, 2009), but there was then and is no longer a CSF for chloroform ingestion in that data base. Chowdhury and Hall (2010) used a value for CSF that was numerically equal to the threshold, Reference Dose (RfD), which was found in IRIS, although the units of a RfD are the inverse of those for a CSF.

Because the corresponding author had been made aware of this fundamental error, but had not informed the journal Editor, this paper was retracted by the journal. Publication of a second paper (Chowdhury *et al.* 2011) verified that the corresponding author knew there was no CSF for chloroform in IRIS by acknowledging its absence and using a "*previously reported*" CSF. No citation was given for the "*previously reported*" CSF that was used, but its use was justified in order to be "*protective against cancer risks*". IRIS explains that the threshold RfD it provides for chloroform was chosen to be protective for avoiding the mode of action of chloroform that might ultimately lead to cancer.

This 2011 paper estimated that THMs in drinking water were causing 703 cases of cancer per year in Canada's provinces and territories. A commentary critical of the findings of this paper (Bull *et al.* in press) noted that in addition to ignoring the substantial evidence that chloroform does not cause cancer at such drinking water exposure levels, this paper (as Chowdhury and Hall, 2010 had done) presented the cancer cases calculated with the invalid CSF as being cases of cancer per year, rather than an upper bound estimate of cancer cases per 70 year lifetime that the CSF method does yield. Consequently, when corrected, the original 703 cancer cases per year claimed for Canada would drop to ~3 cases when the chloroform CSF is eliminated and the remaining cancer case predictions are properly expressed per year, relying on CSFs for the other THMs (Chowdhury, in press). As noted in Chapter 10, even the validity of the CSFs for other THMs is questionable. Regardless, an upper bound estimate of ~3 cases of cancer per year (the lower bound estimate being zero) in Canada being attributable to THMs is essentially negligible compared with approximately 7200 actual new cases of bladder cancer and 177,800 new cases of all cancers expected to be diagnosed in 2011 for Canada (CCS, 2011).

The reality that these papers claiming to estimate cases of cancer caused by chloroform (as the dominant THM in the data set used) could be peer reviewed and published in major international journals in 2010 and 2011 suggests that the advances in knowledge about the mode of action of chloroform in drinking water have not been widely understood or clearly communicated, even among drinking water researchers who publish or peer review such papers. What chance of understanding such esoteric, yet critically important issues is there for front-line drinking water professionals who have a water treatment plant to run and paying customers to serve?

We will return to our opening question about the overall progress in understanding human health risks for DBPs at the end of this Chapter after discussing what this book contains.

12.2 IDENTIFYING AND REGULATING DBPs

Chapter 2 provides a general overview of the vital role of analytical chemistry progress on the DBP issue and a summary of the regulatory history and current status of DBP regulations and guidelines worldwide. Before Rook (1974), water scientists were blind to chloroform in drinking water because the previous analytical method for trace organics in drinking water relied upon using chloroform as the extraction solvent. While this may appear amusing to young water professionals today, this anecdote accurately illustrates the continuing challenge of analytical chemistry that you must have some idea of what you are looking for to be able to expect any trace analytical method to find it. Yet without finding specific DBPs in drinking water, how can we know to look for them? Likewise, without knowing that specific DBPs are present in drinking water, how can we prioritize toxicology research to identify whether any particular DBP is potent enough to represent a potentially relevant human health risk in drinking water?

Against this background of uncertainty, developing regulatory criteria for DBPs requires knowledge that specific DBPs actually occur in drinking water at substantive concentrations and that their toxicity profile and potency warrants controlling them in drinking water. When the DBP issue is viewed in this context, it is not entirely surprising that after the original relatively rapid response to the discovery of THMs, regulatory programs for DBPs have lagged somewhat behind the generation of research knowledge about DBPs. The clarity and meaning of that new knowledge has not always been obvious.

With advances in instrumentation, detection limits are being lowered (now into parts per trillion and parts per quadrillion ranges) and more DBPs are inevitably detected. Today we are aware of >600 DBPs (Richardson, 2011) but monitoring even the handful that are regulated is expensive, making comprehensive and meaningful monitoring programs difficult. Incorporating some of these emerging

DBPs into an epidemiology study would be difficult if not impossible. DBP identification and occurrence studies are important for advancing our understanding of the largely unanswered questions about DBPs and human health. However, as many have warned, we must be careful not to draw operational attention away from the inevitable health threats posed by pathogens. It should also be clear that simply generating more and more criteria for individual DBPs is not likely to improve overall confidence in the protection of public health. That approach is only certain to improve employment prospects for laboratory personnel dedicated to compliance monitoring.

The challenge for the encouraging global trend towards a risk management, water safety plan approach for insuring safe drinking water (Breach, 2012; NHMRC, 2011; WHO, 2011) is to generate knowledge about enough of the emerging, potential DBP problems that ill-informed process changes can be avoided. Ultimately, a water safety plan approach will be effective when it increases the knowledge that operational personnel have about their own systems, how they function, what they are vulnerable to and what challenges they are likely to face.

12.3 A WATER UTILITY COPING WITH DBPs

Chapter 3 provides the perspective of a utility water quality manager who must meet DBP regulatory requirements for over 200 water systems statewide while facing challenging source water conditions across 1/3 of the dry continent of Australia. Given a current regulatory focus on THMs, managing them becomes "the DBP issue" because a water utility is obliged to meet its regulatory limits. Failing to meet prescribed regulatory limits is viewed as a water quality failure, whether or not any tangible human health risk is involved. Even if the regulatory agency understands that THMs do not pose an imminent human health risk in drinking water, consumer expectations will reasonably be that limits are set for a reason and they must be satisfied. Regulatory agencies should not believe otherwise; once they set the limits, the water provider will be obliged to meet them, even if they are called "guidelines". In today's Internet-connected world, even limits developed by other jurisdictions for different circumstances may be cited by concerned consumers.

Given these realities, regulators need to recognize the current state of uncertainty about DBPs and the quandary faced by water providers. If utilities are unreasonably pushed towards chasing regulated DBPs (THMs or HAAs) by any means possible, they may ultimately create other more serious water quality problems.

Unintended consequences with implementation of process changes to satisfy regulatory imperatives are always a possibility if compliance with new DBP regulations and guidelines is pursued in isolation. For example, experience has demonstrated increased lead corrosion in distribution and plumbing systems when free chlorine was changed to chloramines largely to improve THM compliance (Edwards and Dudi, 2004). Subsequently, this impact was implicated as the cause of increased blood lead levels in exposed children (Edwards *et al.* 2009), an outcome that no one could find beneficial. Other examples covered in this book include the potential, depending on site-specific conditions, for increased formation of more toxic iodinated THMs (Chapter 4) and authentic animal carcinogens, the *N*-nitrosamines (Chapters 5, 6 & 7) if chloramines are substituted for free chlorine.

12.4 COMPLEXITY OF HALOGENATED DBPs

Chapter 4 provides an overview of the complexity of DBP formation, notably with regard to the halogens bromine and iodine, which can participate in DBP formation in addition to chlorine. Essentially, the more we look, the more we find out about formation of new DBPs by various mechanisms and interactions of water treatment processes and source water quality. This is an analytical chemist's delight and a water treatment engineer's nightmare.

The examples provided illustrate that any given water utility seeking to manage DBPs needs to know considerably more about its source water quality and behaviour of its treatment processes than many of us might have imagined even a decade ago. To put a positive interpretation on what might otherwise be depressing news for the water treatment profession, these examples illustrate the needs and the benefits of knowing your own system, which is the essence of the water safety plan approach to assuring drinking water safety (Breach, 2012). These cases illustrate that unintended consequences abound for anyone looking for a quick fix to DBP problems. They certainly reinforce the knowledge, which has long been appreciated by experienced water professionals, that process changes must not be implemented without performing meaningful, site-specific trials at bench and pilot scale. Local source water conditions and process realities, including their variability, cannot be adequately predicted from generalizations adapted from published research or technology vendors. Perhaps one of the more notable examples of the dangers of ignoring this sage advice is the view that changing the coagulant from alum to polyaluminum chloride at full scale was an important contributing factor to the massive 1993 Milwaukee cryptosporidiosis outbreak to the extent that this change contributed to inconsistent turbidity removal performance during Milwaukee's first experience with the new coagulant facing challenging spring raw water quality conditions (Fox and Lytle, 1996).

12.5 COMPLEXITY OF NITROGENOUS DBPs

Chapter 5 introduces and provides an overall perspective on the issues of nitrogenous DBPs. The original fixation on halogenated DBPs was driven mainly by the initial focus on the DBPs, like THMs and HAAs that are formed primarily by reaction with humic materials commonly referred to as natural organic matter (NOM), which generally contain little nitrogen. Disinfectants, which are, after all, effective against microorganisms because they are reactive oxidants, can and will react with other organic matter commonly found in source water. Studying reactions with DBP precursors found in other sources such as algal organic matter and effluent organic matter revealed the issue of nitrogenous DBPs.

This overview suggested four major themes: all disinfection techniques may form DBPs; combining disinfectants may be helpful, but DBP tradeoffs are likely; gaining understanding of how DBPs form may lead to cost-effective solutions; and the DBPs of current interest may not be those of greatest importance for health risk. In essence, these findings are another perspective on the benefits of knowing your own system.

12.6 NITROGENOUS DBPs – A PRIMER

Chapter 6 explores the study of nitrogenous DBPs in more detail. This line of inquiry has revealed a storehouse of knowledge about trace water chemistry, which has greatly expanded our understanding of water quality in general. Likewise, finding of notably carcinogenic DBPs like the *N*-nitrosamines has shown that prevalent beliefs that halogenated compounds are inherently more toxic or dangerous than non-halogenated compounds are simply inaccurate and seriously misleading.

In addition to *N*-nitrosamines, Chapter 6 addresses nitrogenous DBPs including halonitriles, cyanogens halides, halonitroalkanes, haloamines and non-halogenated nitriles. For each, their occurrence, analysis, treatment and control and current regulatory status are reviewed.

The diversity and complexity of nitrogenous DBPs now recognized, combined with the range of precursors and reaction mechanisms involved, make a strong case for knowing site-specific water chemistry. Research into nitrogenous DBPs, *N*-nitrosamines in particular, has provided other revelations by showing how important understanding contaminants/precursors in source water can be as well as highlighting an important role for water treatment chemicals as DBP precursors in some cases. Again, the more you know about your system, the better able you are to judge potential consequences from unexpected (or expected) changes.

12.7 NITROSAMINES – AN OBVIOUS CANCER RISK?

Chapter 7 explores the relevance of finding *N*-nitrosamines, primarily *N*-nitrosodimethylamine (NDMA), as DBPs. On the face of it, NDMA should be a serious health risk issue, given its documented genotoxic carcinogenicity and high potency in experimental animals. Unlike the THMs and HAAs, there is no controversy about *N*-nitrosamines being authentic, genotoxic animal carcinogens.

However, population health risk is strongly influenced by the prevalence of NDMA exposure. NDMA does not occur universally in disinfected drinking water; only in about 10% of samples, but 27% of studied water systems have shown an NDMA detection among the 18,000 samples collected and analyzed as part of the second unregulated contaminant monitoring rule (UCMR2) program (USEPA, 2011). Where detected, NDMA concentrations are low, generally less than 10 ng/L. Other *N*-nitrosamines are even more infrequent and found at much lower concentrations.

NDMA provides a good illustration of how the 10^{-6} (one-in-a-million) lifetime cancer risk, a long-time foundation of US EPA environmental risk policy, despite not having any published, documented rationale (Hrudey, 1998; Kelly, 1991), may prove unhelpful regarding public health risk. The issue here is the distinction between individual risk and population risk, the latter being the basis for most public health issues beyond environmental health.

From one perspective, an NDMA concentration of 0.7 ng/L consumed for a lifetime will yield a lifetime upper bound cancer risk of 10^{-6}, which is typically the lower threshold for receiving regulatory attention from the US EPA. From another perspective, if the entire U.S. population consumed drinking water at 10 ng/L of NDMA for a lifetime, only 60 cases of cancer per year would be predicted for the entire U.S. population, compared with about 70,000 new cases of bladder cancer diagnosed per year in the U.S. (American Cancer Society, 2011). However, only a fraction (likely less than 20%) of the U.S. population might be exposed to 10 ng/L of NDMA in drinking water for a lifetime, making the upper bound estimate for cancer cases (i.e. only ~12 per year) that might realistically be attributable to NDMA in drinking water shrink to an insignificant fraction (0.02%) of new annual bladder cancer cases per year. These data show that despite the remarkably high potency (i.e. a large CSF) for NDMA, the *N*-nitrosamines in drinking water cannot explain any meaningful fraction of population bladder cancer incidence.

Additionally, if the small relative source contribution of drinking water to total NDMA exposure is considered, which has been estimated to be less than 1% (Fristachi and Rice, 2007), the population health risk case for regulating NDMA in drinking water to a stringent level (i.e. <10 ng/L) is not compelling even though US EPA policy specifies that relative source contribution should not be considered for setting maximum contaminant levels in drinking water for genotoxic carcinogens like NDMA.

The differing perspectives, which are revealed between considering individual risk versus population risk are not unique to NDMA, but this example provides a good data base for illustrating the differences. Much of that difference can be explained with an understanding that a 1 in a million (10^{-6}) lifetime cancer risk is a remarkably small risk target for a disease that is estimated to pose an individual overall lifetime risk in North America of more than 2 in 5, i.e. >40% (American Cancer Society, 2011; Canadian Cancer Society, 2011).

The original meaning of the 10^{-6} lifetime cancer risk level (Hrudey, 1998) was intended to function as a level below which cancer risks could reasonably be considered to be negligible (i.e. *de minimis* risk), but over time, some have interpreted any risk estimate above 10^{-6} lifetime cancer risk to be unacceptable.

Anyone who understands the nature of the evidence that must be used and the process by which it is used would not seriously claim this criterion should be a sharp dividing line between "acceptable" and "unacceptable". In fact, Dr. Joe Rodricks, who was one of the pioneers in establishing the cancer risk estimation procedures for regulatory purposes, stated about the modeling process that normally underlies estimation of the lifetime cancer risk from experimental animal data (Rodricks, 2007): "*The linearized multistage model was selected because it seemed to have some basis in the leading mechanistic hypotheses regarding the carcinogenic process, and also because it seemed highly likely that the model – because of its 'linearization' at low dose – would not underestimate low dose risk, that it would, in fact, place an upper bound on low-dose risk. Actual risk might be as large as the upper bound, but could be lower and could even be zero. It is not the case that risk assessors, at least those who truly understood the problem of low-dose extrapolation, have ever claimed that risks predicted in this fashion are known to be accurate, even ignoring the uncertainties introduced by the fact that most risk assessments are based on animal, not human data.*" [emphasis added]

12.8 CHALLENGE OF CHARACTERIZING NEW DBPs

Chapter 8 describes the research effort involved in validating an analytical method to document the occurrence of a class of DBPs that were predicted to occur in drinking water, but which had not been demonstrated before. Previously Bull *et al.* (2006), reviewed toxicological data on related compounds, supplemented by use of a quantitative structure toxicity relationship (QSTR) program to identify chemicals either found in drinking water or predicted to be formed as DBPs which also had a high probability of being chronically toxic and/or carcinogenic. These included the halogenated benzoquinones (Bull *et al.* 2011), forming the basis for the research described.

Chapter 8 documents how challenging even a focused, targeted search can be, demanding the latest advances in analytical technology. The confirmation that some halogenated benzoquinones are found in disinfected drinking water is a very important finding that justifies more investigation to characterize the occurrence and potential human exposure via drinking water. However, like *N*-nitrosamines, even if these compounds are found to have the high toxicity predicted, the low concentrations observed so far suggest that population exposure will not be high enough to explain the epidemiological predictions for bladder cancer.

12.9 CANCER RISK OF A POTENT MUTAGEN – MX?

Chapter 9 addresses MX and related halogenated furanones. MX became notorious when it was isolated as the most mutagenic component of chlorinated pulp and paper effluent. Given the anticipated connection between mutagenicity measured *in vitro* and genotoxic carcinogenicity, *in vivo*, MX would have appeared, at the time, to have had at least some of the elements to qualify as the missing magic bullet DBP needed to explain epidemiologic estimates for cancer. Chapter 9 reviews how that evidence has developed, noting that the anticipated connection has not proven to be as direct or clear as some observers might have expected. As noted for other DBPs, understanding, in some detail, what the mode of action is for a carcinogenic response is vital to developing an informed risk assessment and resulting risk management approach.

Low dose cancer risk assessments have been completed by various agencies and some researchers yielding widely varying results considering that all have been based upon the same animal experimental data. A 280-fold difference from lowest to highest risk prediction included a 10-fold difference in adopted risk criterion (i.e. 10^{-5} vs. 10^{-6} upper bound lifetime cancer risk), but this still leaves almost a 30-fold difference based on policy assumptions for dealing with the experimental data. Such differences among jurisdictions in applying policy assumptions to risk assessment are not unusual – the same situation occurred with the development of an NDMA water quality objective in Ontario where differing analyses of the same toxicology evidence yielded a 5100-fold difference in the recommended criteria (ACES, 1992).

At least, such realities should highlight for regulatory and public health personnel that short term numerical values exceeding a limit by less than say 100% (i.e. 2-fold) should demand attention and explanation, but should not normally require urgent action with regard to lifetime risk. Likewise, criteria expressed as two significant figures are not accurately representing the limited underlying confidence that is warranted by the methods and the evidence.

In any case, the most restrictive proposal for MX of "7.8 ng/L" for a 10^{-6} upper bound lifetime cancer risk is 10-fold higher than the similar criterion for NDMA. Occurrence data for MX (and other halogenated furanones) is nowhere near as comprehensive as that now available for NDMA and other nitrosamines (USEPA, 2011) but MX does not appear to be a pervasive occurrence at concentrations posing a concern in drinking water. These data make it clear that if NDMA (Section 12.7) cannot explain the gap between toxicological cancer risk predictions and the epidemiological cancer risk predictions for bladder cancer that might be attributable to chlorinated drinking water DBPs, MX will not provide the missing explanation either.

12.10 CANCER RISK – EXPERIMENTAL EVIDENCE

Chapter 10 explores the often confusing and somewhat overwhelming body of experimental toxicology evidence about DBPs and what insights can be gained on the magnitude of cancer risk that is suggested by that evidence in combination with the epidemiological evidence on chlorinated drinking water. While showing that it is very unlikely that THMs contribute to bladder cancer risk or that reducing THM exposures via drinking water will result in any lowering of bladder cancer incidence, this chapter does not call into question whether chlorinated drinking water contributes to causing bladder cancer. The range of population attributable risk (PAR) estimates for the proportion of bladder cancer that might be attributable to exposure to chlorinated drinking water (ranging from 2% to 17%) prepared for the US EPA DBP regulatory rationale (Odom *et al.* 1999) was used to derive a central estimate of at least a 1 in a 1000 lifetime cancer risk that may be attributable to chlorinated drinking water. This value, derived using an estimated proportion (PAR) of observed cancer cases, is at least 10-fold higher than the 1 in 10,000 upper bound, lifetime cancer risk that USEPA normally uses as a threshold for determining that regulatory action on a given contaminant or contaminant group is necessary. Of course, a key problem in this case is that it is not yet clear which DBP, or group of DBPs, needs to be regulated because none of those identified to date can explain a lifetime cancer rate as high as 1 in a 1000. The confidence to be placed in these epidemiologic estimates, which give rise to the PAR values is clearly important. Chapter 11 addresses those issues in some detail.

Chapter 10 also provides a rigorous analysis of why the important recent epidemiology study considering possible susceptibility based on genetic polymorphisms (Cantor *et al.* 2010) did not demonstrate, as some readers may infer, that THMs were responsible for the strikingly high odds ratios that were observed for those who were judged to have higher susceptibility to drinking water chemical exposures. This analysis does not downplay the importance of the Cantor *et al.* study, it only challenges the misinterpretation that the study demonstrates evidence in support of THMs causing bladder cancer. Finally, this Chapter with its assessment of an enormous body of evidence provides a basis to advocate for a more rigorous, rational, hypothesis-driven research agenda to resolve the questions of DBPs and cancer risk after more than 35 years of health research on the topic.

12.11 HEALTH RISK – EPIDEMIOLOGIC EVIDENCE

Chapter 11 provides a primer on epidemiologic methods and the inferences derived from epidemiology studies regarding the critical issue of whether a hypothesized cause may be an authentic cause of any disease outcome. This background is provided because most drinking water professionals, a key component of our target audience for this book, are not exposed to epidemiology to any extent in their science education. Yet the importance of understanding the role of human evidence regarding this question is clearly pivotal and drinking water professionals are often those who are left to explain any of this to their consumers.

A taxonomy of evidence about health risks (Thomas and Hrudey, 1997) is explained because we often do not ask how it is we come to know what we claim to know about human health. Epidemiology, no matter how complicated it may be in the details of its application, derives its quantitative conclusions about the strength of an association between a hypothesized cause and a health outcome as an assessment of the correlation between them (Angell, 1997; Taubes, 1995). Causal inference, the aspect that particularly matters in this book, relies upon the rigorous and self-critical assessment of a number of relevant factors, which are generally not absolute; most are inherently "expert" judgments. The overview provided in Chapter 11 demonstrates that, at least for long term chronic diseases like cancer, the tools available to study a pervasive, but individually variable exposure like contaminants in drinking water are inherently limited.

In reality, only observational epidemiology studies are possible and for a disease like bladder cancer, which only occurs at a substantial rate in late middle age[1] and likely requires more than 30 years of exposure to a hypothesized cause, experimental studies are neither feasible nor likely to be ethically possible. Observational studies are largely limited to the case-control design. Those restrictions are major limitations, because case-control studies are subject to serious problems with bias that cannot be eliminated with only improved data collection and mathematical analysis techniques.

The findings reviewed in Chapter 11 affirm the general consensus that it is mainly bladder cancer that has shown any consistency of results among epidemiological studies addressing chlorinated drinking water or DBPs as a cause of cancer. The confidence that can or should be placed upon the strength of causal inference for these epidemiology studies is more open to debate. Chapter 11 provides a number of questions about the bladder cancer epidemiology studies, which bear directly on how confident we can be in considering that these studies have demonstrated that chlorinated drinking water is a cause of bladder cancer. Published reviews of these studies, often, as might be expected, written by those who have performed one or more of these studies, have not fully addressed or resolved many relevant questions that should be considered. Exhaustive self-criticism is not easy, but as rigorously demonstrated by the pioneer of epidemiology, John Snow, is one key to ultimate credibility (Vinten-Johansen *et al.* 2003).

Confidence in the PAR estimates as a cause of bladder cancer, which have been relied upon to make the public health case for regulating DBPs, depends entirely on the merits of the causal inference derived from the epidemiology studies. These estimates can likely be taken as an upper bound on what the bladder cancer risk attributable to chlorinated drinking water could be. However, the USEPA (2006) acknowledged in citing the PAR for estimating the benefits of regulating DBPs, (adopting a PAR of 16%) that the true number of attributable bladder cancer

[1]Probability of developing bladder cancer is 19-fold higher for males 40–59 year old compared with males from birth-39 and increases another 2.4-fold for males aged 60–69 and another 3.95-fold for males over 70, i.e. males over 70 are 180 times more likely to experience bladder cancer than males under 40 (American Cancer Society, 2011).

cases defining the benefits of regulating DBPs "*may be as low as zero since causality has not yet been established between exposure to chlorinated water and bladder cancer*". The epidemiology evidence up to 2011, reviewed in Chapter 11, provides no substantive grounds to conclude any differently about the lower boundary being zero and the key question of causation.

On the issue of adverse reproductive effects being attributable to DBPs, the epidemiology literature is somewhat more clear. The evidence for the outcomes of stillbirth, spontaneous abortion and congenital anomalies, all of which can be devastating for parents, has been inconsistent and unconvincing despite the numerous studies that have been performed. Less devastating adverse outcomes of low birth weight, preterm delivery and small for gestational age have shown only limited evidence in support for the latter being "associated with" DBP exposures. Because of the general lack of consistency over the wide range of adverse reproductive outcome studies, the case for DBPs being causal of any of these outcomes is not evident.

Despite the lack of incontrovertible evidence of adverse human health outcomes resulting from exposure to DBPs in drinking water, the case for limiting human exposures to DBPs in drinking water remains entirely valid on precautionary grounds because drinking water is such a pervasive exposure across society. The challenge, as always with environmental health issues, is to decide how precautionary we need to be (Hrudey and Leiss, 2003).

12.12 DBPs AND HEALTH – WHERE ARE WE NOW?

With so much complexity and uncertainty in the health relevant evidence, there is more than ample scope for well-informed scientists to disagree on how to interpret and judge the evidence. However, there should be reasonable unanimity about some thoroughly untenable and misguided interpretations related to DBPs and human health. A few examples will be provided, although many readers will probably have their own collection of negative experience on this topic.

A journal titled *Medical Hypotheses* published a letter (Katona, 2009) titled: "*Step aside tobacco, chlorine could be man's next great carcinogen*." This letter argues the premise regarding chlorinated drinking water cancer risk that: "*It took decades to establish and publicise the link between tobacco and cancer. Let's not make the same mistake twice*." While acknowledging that this journal is intended for publishing hypotheses, not evidence, and that the title uses "*could be*" rather than "*is*", the comparison with tobacco is spurious. There is simply no validity to comparing DBPs in drinking water with lung cancer being caused by tobacco smoking, which is responsible for over 90% (Doll and Peto, 1976) of an estimated 220,000 new cases and about 157,000 lung cancer deaths in the U.S. for 2011 alone (American Cancer Society, 2011). If the overwhelming epidemiological evidence was not enough, looking back over the past 75 years shows that lung cancer death rates in males peaked in the early 1990s and are substantially declining while similar rates in females appear to have peaked in 2005 and may finally be declining, with both related to the comparative timelines of smoking behaviour (population rates of starting and quitting) in men and women. We should not expect any such compelling evidence to be found for bladder cancer arising from exposure to DBPs.

On November 25, 2011 an editorial was published in the Tennant and District Times referring to a boil water alert called by the water utility in Tennant Creek, Northern Territory, Australia, noting that finding "*16 (yes, just 16) microscopic E. coli organisms*" had caused the water utility to over-react by assuring that the water was "*doused with chlorine*" and a "*scare-mongering notice*" was sent to local agencies. This commentary went on to explain that a 2004 referendum showed that 80% of residents "*were against chlorination which produces carcinogenic by-products*." Finally, the editorial observed that the water utility found it easier to "*throw some poison into the tanks*" than to find a green solution to the health of the town's water supply. This media perspective must be considered in light of the experience in Walkerton, Ontario, Canada in May 2000 (elaborated in Section 1.2.1, Chapter 1), where the failure to maintain and monitor chlorination, in part because the Operator Foreman objected to chlorinating the town's water supply, contributed to a fatal disaster (Walker and Hrudey, 2005).

Those selling home water treatment devices have been known to exploit the health literature on DBPs provided by Health Canada, which serves as the scientific secretariat to Canada's Federal-Provincial-Territorial Committee on Drinking Water which establishes national drinking water guidelines. In one case (CBC, 2011), a company selling such devices provided its sales personnel with a laminated sheet showing the Health Canada masthead at the top and providing a number of quotes about the harmful effects of chlorine and DBPs. This was provided for the purpose of convincing prospective customers of their need to purchase expensive reverse osmosis water treatment equipment to avoid the health risks of chlorination and DBPs described by Health Canada. Although the quotes from Health Canada were taken out of context, these quotes were accurately transcribed because the original language sounded frightening enough for the sales purposes.

This final chapter opened by referring to a question about how much progress has been made towards better understanding of human health risks from DBPs. Clearly, we have learned much about new DBPs and their toxic properties, while learning not nearly as much about the big question concerning DBPs causing cancer. Perhaps the only thing we can be very certain about is that uncertainty will not be resolved any time soon.

Our state of knowledge on human health impacts of DBPs in drinking water is not amenable to many claims of certainty, with a few important exceptions. For example, there is clearly as much certainty as there can be about the ability of microbial pathogens in drinking water to cause human illness, hence the absolute imperative for disinfection of public water supplies. In contrast, no evidence for any DBP is remotely close to that kind of certainty regarding causation of disease through drinking water exposure (Hrudey *et al.* 2011). There is also reasonable consensus based on solid toxicological and mode of action evidence to conclude that chloroform is not a cancer risk at the level of regulated drinking water concentrations in place (Chapter 2) throughout the developed world.

Faced with currently intractable uncertainty on many other questions of DBPs and human health, there is no viable alternative but to maintain a sensible precautionary approach regarding DBPs. In these circumstances, health regulators surely have an obligation to be more forthcoming with drinking water providers and the public about the limited strength and uncertainty of the health evidence. Likewise, health researchers must be fully committed to the scientific method that obliges investigators to entertain the possibility that there might be no adverse health effects caused by any particular DBP at realistic drinking water exposures. Different researchers can and will hold different views about how likely the no-effect scenario may be, but all scientists need to be fully committed to conducting their research in a manner that includes the null response as a conceivable outcome.

We should all work in accordance with the sage advice of Carl Sagan about the scientific method that we quoted to close our introductory comments in Chapter 1.

REFERENCES

ACES (1992). A Standard for *N*-Nitrosodimethylamine (NDMA) – A Recommendation to the Minister of the Environment. Advisory Committee on Environmental Standards. ACES Report 92-01, January 1992. ISBN 0-7729-9367-X.

American Cancer Society (2011). Cancer Facts & Figures 2011. American Cancer Society, Atlanta. http://www.cancer.org/acs/groups/content/@epidemiologysurveilance/documents/document/acspc-029771.pdf (accessed 3 February 2012).

Angell M. (1997). Science on Trial – The Clash of Medical Evidence and the Law in the Breast Implant Case. W.W. Norton, London, 256.

Breach B. (ed.) (2012). Drinking Water Quality from Catchment to Consumer – A Practical Guide for Utilities Based on Water Safety Plans. IWA Publishing, London, UK.

Bull R. J., Cotruvo J. A., Fawell J. and Hrudey S. E. (in press). *Journal of Hazardous Materials.* published online, doi: 10.1016/j.jhazmat.2011.07.040. Re: S. Chowdhury et al. (2011). Disinfection byproducts in Canadian provinces: associated cancer risks and associated medical expenses. *Journal of Hazardous Materials*, **187**, 574–584.

Bull R. J., Reckhow D. A., Li X.-F., Humpage A. R., Joll C. and Hrudey S. E. (2011). Potential carcinogenic hazards of non-regulated disinfection by-products: haloquinones, halo-cyclopentene and cyclohexene derivatives, *N*-halamines, halonitriles and heterocyclic amines. *Toxicology*, **286**, 1–19.

Bull R. J., Reckhow D. A., Rotello V., Bull O. M. and Kim J. (2006). Use of Toxicological and Chemical Models to Prioritize DBP Research. Project 2867. Report to the AwwaResearch Foundation, Denver, CO.

Canadian Cancer Society (2011). Canadian Cancer Statistics 2011. Canadian Cancer Society's Steering Committee on Cancer Statistics. Canadian Cancer Society,

Toronto. http://www.cancer.ca/Canada-wide/Publications/Alphabetical%20list%20of%20publications/Canadian%20Cancer%20Statistics.aspx?sc_lang=en (accessed 3 February 2012).

Cantor K. P., Villanueva C. M., Silverman D. T., Figueroa J. D., Real F. X., Garcia-Closas M., Malats N., Chanock S., Yeager M., Tardon A., Garcia-Closas R., Serra C., Carrato A., Castano-Vinyals C., Samanic C., Rothman N. and Kogevinas M. (2010). Polymorphisms in GSTT1, GSTZ1, and CYP2E1, disinfection by-products, and risk of bladder cancer in Spain. *Environmental Health Perspectives*, **118**(11), 1545–1550.

CBC (2011). *Marketplace*. Canadian Broadcasting Corporation. '*Clean Water – Dirty Tricks*' telecast 27 January 2011. Available on-line: http://www.cbc.ca/marketplace/2011/cleanwaterdirtytricks/#readcomment (accessed 10 February 2012).

Chowdhury S. (in press). Response to HAZMAT-D-11-01781. *Journal of Hazardous Materials*. In press, published on-line. doi: 10.1016/j.jhazmat.2011.09.047

Chowdhury S. et al. (2011). Disinfection byproducts in Canadian provinces: associated cancer risks and associated medical expenses. *Journal of Hazardous Materials*, **187**, 574–584.

Chowdhury S. and Hall K. (2010). Human health risk assessment from exposure to trihalomethanes in Canadian cities. *Environment International*, **36**, 453–460.

Doll R. and Peto R. (1976). Mortality in relation to smoking: twenty years observations on male British doctors. *British Medical Journal*, **2**(6051), 1525–1536.

Edwards M. and Dudi A. (2004). Role of chlorine and chloramines in corrosion of lead-bearing plumbing materials. *Journal American Water Works Association*, **96**(10), 69–81.

Edwards M., Triantafyllidou S. and Best D. (2009). Elevated blood lead in your children due to lead-contaminated drinking water: Washington, D.C. 2001–2004. *Environmental Science & Technology*, **43**, 1618–1623.

Fox K. R. and Lytle D. A. (1996). Milwaukee's crypto outbreak: investigation and recommendations. *Journal American Water Works Association*, **88**(9), 87–94.

Fristachi A. and Rice G. (2007). Estimation of the total daily oral intake of NDMA attributable to drinking water. *Journal Water & Health*, **5**(3), 341–355.

Health Canada (2006). Guidelines for Canadian Drinking Water Quality: Guideline Technical Document – Trihalomethanes. Water Quality and Health Bureau, Healthy Environments and Consumer Safety Branch, Health Canada, Ottawa, Ontario. www.hc-sc.gc.ca/ewh-semt/alt_formats/hecs-sesc/pdf/pubs/water-eau/trihalomethanes/trihalomethanes-eng.pdf accessed 10 February 2012

Heywood R., Sortwell R. J., Noel P. R., Street A. E., Prentice D. E., Roe F. J., Wadsworth P. F. and Worden A. N. (1979). Safety evaluation of toothpaste containing chloroform III. Long-term studies in beagle dogs. *Journal of Environmental Pathology and Toxicology*, **2**, 835–851.

Hrudey S. E. (1998). Quantitative Cancer Risk Assessment – Pitfalls and Progress. *Issues in Environmental Science and Technology*, R. E. Hester and R. M. Harrison (eds.), Vol. **9**, Royal Society of Chemistry Publishing, Cambridge, U.K., pp. 57–90.

Hrudey S. E., Conant B., Douglas I. P., Fawell J., Gillespie T., Hill D., Leiss W., Rose J. B. and Sinclair M. (2011). Managing uncertainty in the provision of safe drinking water. *Water Science & Technology: Water Supply*, **11**(6), 675–681.

Hrudey S. E. and Leiss W. (2003). Risk management and precaution – insights on the cautious use of evidence. *Environmental Health Perspectives*, **111**, 1577–1581.

Hrudey S. E. and Walker R. (2005). Walkerton – 5 years later. Tragedy could have been prevented. *Opflow*, **31**(6), **1**, 4–7.

Katona S. (2009). Step aside tobacco, chlorine could be man's next great carcinogen. *Medical Hypotheses*, **72**, 753.

Kelly K. E. (1991). The myth of 10^{-6} as a definition of acceptable risk. Air and Waste Management Association, 84th Annual Meeting, 16–21 June 1991, Vancouver, BC.

NHMRC/NRMMC (2011) Australian Drinking Water Guidelines. http://www.nhmrc.gov.au/guidelines/publications/eh52 (accessed 07 February 2012).

Odom R., Regli S., Messner M., Cromwell J. and Javdan M. (1999). Benefit-cost analysis of the Stage 1 D/DBP Rule. *Journal American Water Works Association*, **91**(4), 137–147.

Richardson S. D. (2011). Disinfection by-products: Formation and occurrence in drinking water. In: *The Encyclopedia of Environmental Health*, J. O. Nriagu (ed.), Vol. **2**, Elsevier, Burlington, pp. 110–136.

Rodricks J. V. (2007). The 2006 American College of Toxicology Distinguished Service Award Lecture: has quantitative risk assessment been of benefit to the science of toxicology. *International Journal of Toxicology*, **26**, 3–12.

Rook J. J. (1974). Formation of haloforms during chlorination of natural waters. *Water Treatment Examination*, **23**, 234–243.

Taubes G. (1995). Epidemiology faces its limits. *Science*, **269**, 164–169.

Thomas S. and Hrudey S. E. (1997). Risk of Death in Canada – What We Know and How We Know It. University of Alberta Press, Edmonton, 292.

USEPA (2009). Integrated Risk Information System (IRIS) Online database. US Environmental Protection Agency. http://www.epa.gov/IRIS/ (accessed 15 December 2009).

USEPA (2011). Unregulated Contaminant Monitoring Rule 2 (UCMR2) http://water.epa.gov/lawsregs/rulesregs/sdwa/ucmr/ucmr2/ (accessed 3 February 2012).

Vinten-Johansen P., Brody H., Paneth N., Rachman S. and Rip M. (2003). Cholera, Chloroform and the Science of Medicine – A Life of John Snow. Oxford University Press, Oxford.

WHO (2011). Guidelines for Drinking Water Quality. 4th edn, World Health Organization, Geneva. http://www.who.int/water_sanitation_health/publications/2011/dwq_guidelines/en/index.html (accessed 07 February 2012).

Index

D

E

Lightning Source UK Ltd.
Milton Keynes UK
UKOW051209271112

202812UK00003B/56/P

9 781843 395195